The Guenons: Diversity and Adaptation in African Monkeys

DEVELOPMENTS IN PRIMATOLOGY: PROGRESS AND PROSPECTS

Series Editor:

Russell H. Tuttle
University of Chicago, Chicago, Illinois

This peer-reviewed book series will meld the facts of organic diversity with the continuity of the evolutionary process. The volumes in this series will exemplify the diversity of theoretical perspectives and methodological approaches currently employed by primatologists and physical anthropologists. Specific coverage includes: primate behavior in natural habitats and captive settings; primate ecology and conservation; functional morphology and developmental biology of primates; primate systematics; genetic and phenotypic differences among living primates; and paleoprimatology.

ALL APES GREAT AND SMALL
VOLUME I: AFRICAN APES
Edited by Birute M. F. Galdikas, Nancy Erickson Briggs, Lori K. Sheeran. Gary L. Shapiro and Jane Goodall

THE GUENONS: DIVERSITY AND ADAPTATION IN AFRICAN MONKEYS
Edited by Mary E. Glenn and Marina Cords

The Guenons: Diversity and Adaptation in African Monkeys

Edited by

MARY E. GLENN
Humboldt State University
Arcata, CA

and

MARINA CORDS
Columbia University
New York, NY

Kluwer Academic / Plenum Publishers
New York, Boston, Dordrecht, London, Moscow

Library of Congress Cataloging-in-Publication Data

The Guenons: Diversity and Adaptation in African Monkeys
              p.   Edited by Mary E. Glenn and Marina Cords
     Includes bibliographical references and index.
     ISBN 0-306-47346-1
      1.   2.

ISBN 0-306-47346-1

©2002 Kluwer Academic/Plenum Publishers
233 Spring Street, New York, NY 10013

http://www.wkap.nl./

10 9 8 7 6 5 4 3 2 1

Printed in the United States of America

Contributors

Pascal Baguma
  Makerere University Biological Field
  Station, P.O. Box 409, Fort Portal,
  Uganda

Keith J. Bensen
  Windward Islands Research and
  Education Foundation, 11 East Main
  Street, Bayshore, NY 11706, USA

Thomas M. Butynski
  Zoo Atlanta's Africa Biodiversity
  Conservation Program, National
  Museums of Kenya, P.O. Box 24434,
  Nairobi, Kenya

Paul J. Buzzard
  Department of Anthropology, Columbia
  University, New York, NY 10027, USA

Ann A. Carlson
  Large Animal Research Group,
  Department of Zoology, University of
  Cambridge, Downing Street, Cambridge
  CB2 3EJ, UK

Colin A. Chapman
  Department of Zoology, University of
  Florida, Gainesville, FL 32611, USA

Lauren J. Chapman
  Department of Zoology, University of
  Florida, Gainesville, FL 32611, USA

Dorothy L. Cheney
  Department of Biology, University of
  Pennsylvania, Philadelphia, PA 19104,
  USA

Janice Chism
  Department of Biology, Winthrop
  University, Rock Hill, SC 29733, USA

Marc Colyn
  UMR 6552, CNRS – Université de
  Rennes I, Station Biologique, 35380
  Paimpont, France

Marina Cords
  Department of Ecology, Evolution
  and Environmental Biology,
  Columbia University, New York, NY
  10027, USA

Sheila H. Curtin
  942 Shevlin Drive, El Cerrito, CA 94530,
  USA

Pierre Deleporte
  UMR 6552, CNRS – Université de
  Rennes I, Station Biologique, 35380
  Paimpont, France

Kate M. Detwiler
  Department of Anthropology, New York
  University, 25 Waverly Place, New York,
  NY 10003, USA

Todd R. Disotell
  Department of Anthropology, New York
  University, 25 Waverly Place, New York,
  NY 10003, USA

Steffen Förster
  Department of Zoology, Technische
  Universität Braunschweig,
  Braunschweig, Germany

Joel Mwangi Gathua
Formerly, Mammalogy Department, National Museums of Kenya, Nairobi, Kenya

Jean-Pierre Gautier
UMR 6552, CNRS – Université de Rennes I, Station Biologique, 35380 Paimpont, France

Annie Gautier-Hion
UMR 6552, CNRS – Université de Rennes I, Station Biologique, 35380 Paimpont, France

Mary E. Glenn
Department of Anthropology, Humboldt State University, Arcata, CA 95521, USA

K. Ann Horsburgh
Department of Anthropology, 701 East Kirkwood Avenue, Indiana University, Bloomington, IN 47405, USA

Lynne A. Isbell
Department of Anthropology, University of California, Davis, CA 95616, USA

Beth A. Kaplin
Department of Environmental Studies, Antioch New England Graduate School, 40 Avon Street, Keene, NH 03431, USA

Joanna E. Lambert
Department of Anthropology, University of Oregon, Eugene, OR 97403, USA

Michael J. Lawes
School of Botany and Zoology, Forest Biodiversity Programme, University of Natal, Pietermaritzburg, South Africa

Jeremiah S. Lwanga
Makerere University Biological Field Station, P.O. Box 409, Fort Portal, Uganda

Mairi C. Macleod
School of Life Sciences, University of Surrey Roehampton, West Hill, London SW15 3SN, UK

Elizabeth Matisoo-Smith
Department of Anthropology, The University of Auckland, Private Bag 92019, Auckland, New Zealand

Reiko Matsuda
Department of Anthropology, Graduate School and University Center of the City University of New York, New York, NY 10016, USA

W. Scott McGraw
Department of Anthropology, The Ohio State University, 1680 University Drive, Mansfield, OH 44906, USA

Don J. Melnick
Center for Environmental Research and Conservation, Columbia University, New York, NY 10027, USA

Juan Carlos Morales
Center for Environmental Research and Conservation, Columbia University, New York, NY 10027, USA

Karen Pazol
Yerkes Regional Primate Research Center, Emory University, Lawrenceville, GA 30043, USA

Ryan L. Raaum
Department of Anthropology, New York University, 25 Waverly Place, New York, NY 10003, USA

Karyn Rode
Department of Zoology, University of Florida, Gainesville, FL 32611, USA

William Rogers
Department of Biology, Winthrop University, Rock Hill, SC 29733, USA

Caroline Ross
School of Life Sciences, University of Surrey Roehampton, West Hill, London SW15 3SN, UK

Robert M. Seyfarth
Department of Psychology, University of Pennsylvania, Philadelphia, PA 19104, USA

Christophe Thébaud
UMR 5552 CNRS/UPS, 13 Avenue du
Colonel Roche, BP 4072, F-31029,
Toulouse Cedex 4, France

Anthony J. Tosi
Department of Anthropology,
Columbia University, New York, NY
10027, USA

Adrian Treves
Center for Applied Biodiversity Science,
Conservation International, 65
Greenway Terrace, Princeton, NJ 08540,
USA

Caroline E.G. Tutin
Centre International de Recherches
Médicales de Franceville, Gabon

Tharcisse Ukizintambara
School of Biological Sciences,
University of East Anglia,
Norwich NR4 7TJ, UK

Régine Vercauteren Drubbel
Université Libre de Bruxelles –
Anthropologie, CP 192 – B 1050
Bruxelles, Belgium

Lee J.T. White
Wildlife Conservation Society, 185th
Street and Southern Boulevard, Bronx,
NY 10460, USA

Tammy L. Windfelder
Department of Anthropology, University
of Notre Dame, Notre Dame, IN 46556,
USA

Eric A. Worch
Education Department, University of
Michigan, Flint, MI 48502, USA

Toni E. Ziegler
Department of Psychology and
Wisconsin Regional Primate Research
Center, University of Wisconsin, 1223
Capitol Court, Madison, WI 53715, USA

Klaus Zuberbühler
School of Psychology, University of St.
Andrews, St. Andrews, Fife KY16 9JU,
UK

Contents

Part III Ecology

Part IV Conservation

Preface

The African monkeys commonly referred to as the guenons make up the largest and most diverse primate group on the African continent. Their pelage variations are arguably the greatest of any group of mammals in the world, ranging from drab browns to multicolored experiments in body design. Some species live mainly in savanna-woodlands while others live almost entirely in the upper canopy of rain forests. Social organization varies from monogamy to unimale/multifemale groups to multimale/multifemale groups. The fascinating diversity of these monkeys has not been lost on scientists interested in evolutionary biology. Field and laboratory researchers have begun to examine many aspects of the biology of guenons, from molecular variation within and among species, to their roles in the ecosystems of which they are a part.

Interestingly, and most likely because many of the species live in dense forests, the vast majority of research on guenons has taken place during the past few decades. For example, the first and until recently last major scientific symposium devoted to the study of wild guenons occurred at the XIX International Ethological Conference in 1985. It was appropriately entitled the "Biology, Phylogeny, and Speciation of Forest Cercopithecines." Out of this symposium came the groundbreaking publication, *A Primate Radiation: Evolutionary Biology of the African Guenons*. It became, like its symposium predecessor, the first and last book dedicated to guenon field research. Almost all the major conceptual frameworks for our understanding of guenon biology were outlined within it.

Since that publication, however, new guenon species and subspecies have been discovered, previously unstudied species have become the subject of long-term research projects, knowledge of the better-known species has greatly increased, and original hypotheses have been challenged. It was for these reasons that we (Mary Glenn and Marina Cords) and Annie Gautier-Hion decided to organize another scientific symposium focusing on guenon field research at the XVIII Congress of the International Primatological Society in Adelaide, Australia in January 2001. We decided to invite as many guenon biologists as possible, including those who had been studying guenons for decades, and most importantly, those who were just starting out. So many of our colleagues agreed to present their work that we extended the symposium to two days. The excitement generated by all the new research inevitably led to the proposition of producing a new guenon book. Another call for contributions went out and we received an even greater response than before, a total of forty-nine authors representing ten different countries. The results of all the authors' tremendous efforts are presented in this volume.

The intensive research on a number of the member taxa has provided substantial grist for debate on the process of speciation as well as the definition of species. This debate continues and is evidenced by the disagreement about how the taxonomic organization of these monkeys should be arranged. Depending on which version you accept, the guenons include between twenty-three and thirty-six species. There is an even greater variety of controversial subspecies within *Cercopithecus*, *Miopithecus*, *Allenopithecus*, *Erythrocebus*, and *Chlorocebus*, the viability of the last four genera themselves being open to debate. Inevitably the question of taxonomy arose while we were organizing this book. After much debate and consultation with taxonomic experts (some of whom are also authors in this volume), we decided to use the consensus taxonomy developed for guenons by Peter Grubb, Tom Butynski, John Oates, Simon Bearder, Todd Disotell, Colin Groves, and Tom Struhsaker at the IUCN/SSC Primate Specialist Group meeting held in Orlando, Florida in February 2000. Any authors who have deviated from this taxonomy have explained their reasons in their respective chapters and generation.

We decided to follow the precedent of *A Primate Radiation*, as well as more recent field guides, to include within the guenon group all monkeys in the genera *Cercopithecus*, *Allenopithecus*, *Chlorocebus*, *Miopithecus*, and *Erythrocebus*. Over the years, the term guenon has been used to describe both this wider set of primate taxa as well as only those within *Cercopithecus*. There is no clear precedent or distinct historical root for the usage of "guenon." Our usage of the term is meant to provide a simple manner in which to refer to this group of closely related species and genera.

We organized the chapters into four general categories. Befitting any treatise on a diverse radiation of organisms, the first section deals with evolutionary biology and biogeography. The chapters in this section readily demonstrate the continuing challenge associated with understanding the complicated evolutionary relationships of guenons. The second section is devoted to behavioral studies, and, as in *A Primate Radiation*, it is the largest. Our ever-increasing appreciation of guenon behavioral diversity has led researchers to question many original ideas about their behavior, including those related to male relationships and reproductive strategies, group structure and social organization, and vocalizations. Investigations of ecological relationships make up the third section. As with the previous section, many original concepts have been reassessed with the increase in single species studies conducted over many years in multiple locations, as well as information about previously unstudied guenons. For example, dietary flexibility in some species appears to be more common than previously thought. Finally, and most notably different from *A Primate Radiation* is the fourth section devoted to conservation. The rate at which African primates, including the guenons, are becoming threatened with extinction has increased dramatically over the past decade and a half. Contributions reflect both an increasingly sophisticated understanding of extinction processes as well as the extreme urgency for immediate conservation action with regard to many of the guenon taxa.

The juxtaposition of many new discoveries with the realization that some species or subspecies of guenons may not exist by the time another book dedicated to guenon biology is published is sobering. A perusal of the chapters, however, is cause for optimism. Some of the contributors to *A Primate Radiation* are included here. Their unflagging devotion to the taxa is inspiring. In addition, some authors began their careers after the original guenon book was published. Some of them are so new to the field that this is their first guenon publication. Most encouraging are the chapters by African authors. The continued interest of so many people representing the past, present, and future of guenon biology can only be a cause for hope and reaffirms the importance of studying these fascinating beings.

There are a number of people who deserve our gratitude. We wish to thank the contributing authors for their hard work, diligence and excellent manuscripts. Many of the authors also provided comments and reviews for other chapters in the book, for which we are grateful. We are indebted to the series editor, Russell Tuttle, for his meticulous copyediting and reviews, hard work, support and patience. We also thank Thelma Rowell, Keith Bensen and several anonymous reviewers for their helpful criticism and insightful comments on the manuscripts; our volume profited greatly from their suggestions, and we appreciate the amount of time it took to review so many manuscripts. Special thanks must go to Alisha Clompus for allowing us to publish her beautiful sketches, and to Tammy Smart for her assistance. Finally, this publication would not have been possible without help and encouragement from Andrea Macaluso, Life Sciences Editor, and the many other staff members at Kluwer Academic Plenum Publishers.

We close with a dedication of this volume to three biologists who have contributed greatly to our knowledge of guenon biology. Two are senior researchers who did not prepare chapters for this volume. Still, their academic and research progeny are well represented in our list of authors. Thelma Rowell was one of the first scientists to study guenons in the wild and in captivity more than 30 years ago: her work inspired curiosity and questions, and her enthusiasm inspired a next-generation of students and their research. Tom Struhsaker was similarly a pioneer in field studies of wild guenons, working in multiple sites across the African continent. He has been a dedicated advocate for conservation of guenon habitat, indeed for African forests in general. This book owes much, both directly and indirectly, to the inspiration of both Thelma and Tom.

We also dedicate this volume to Joel Mwangi Gathua, who was just starting his research career. He received his first field training as an undergraduate from Thelma Rowell, and went on to complete his Ph.D. studying the ecology of red-tailed monkeys in a forest in his native Kenya. As a biologist from a guenon-habitat country, with a deep interest in understanding and conserving forest organisms and ecosystems, he stands as an example of where future research must go. Mwangi died tragically young, but we hope his example of dedication, care and energy will inspire other young African scholars especially for many years to come.

Mary E. Glenn
Marina Cords

Arcata, California and New York, 2002

Part I

Evolutionary Biology and Biogeography

The Guenons: An Overview 1
of Diversity and Taxonomy

THOMAS M. BUTYNSKI

Diversity among the Guenons

The guenons are a large, diverse group of monkeys endemic to sub-Saharan Africa (Groves, 2001; Grubb *et al.*, 2002). They occupy a wide geographic range that extends from Mauritania (green monkey, *Cercopithecus aethiops sabaeus*; patas monkey, *Erythrocebus patas*) to the Cape of Good Hope (vervet monkey, *C. aethiops pygerythrus*; samango monkey, *C. mitis labiatus*) (Dandelot, 1971; Wolfheim, 1983; Kingdon, 1997).

Guenons live in a wide variety of habitats. These range from woodlands (patas monkey, vervet monkey), to mangrove forests at sea level (northern talapoin monkey, *Miopithecus ogouensis*), to swamp forests (Allen's swamp monkey, *Allenopithecus nigroviridis*), through lowland, mid-altitude and montane forests, to bamboo forests at >3300 m (golden monkey, *Cercopithecus mitis kandti*; owl-faced monkey, *C. hamlyni*), to alpine moorland at 4500 m (Bale Mountains grivet monkey, *C. aethiops djamdjamensis*). The guenons also occur along the arid edges of the Kalahari and Sahara Deserts (green monkey, vervet monkey, patas monkey) as well as in the wettest places in Africa, such as Mt. Cameroon and Bioko Island, where mean annual rainfall is *ca.* 1000 cm (Preuss's monkey, *Cercopithecus preussi*; red-eared monkey, *C. erythrotis*)

THOMAS M. BUTYNSKI • Zoo Atlanta's Africa Biodiversity Conservation Program, National Museums of Kenya, P.O. Box 24434, Nairobi, Kenya.
The Guenons: Diversity and Adaptation in African Monkeys, edited by Glenn and Cords. Kluwer Academic/Plenum Publishers, New York, 2002.

(Wolfheim, 1983; Butynski, 1988; Lee *et al.*, 1988; Rowe, 1996; Kingdon, 1997; Gautier-Hion *et al.*, 1999).

Many guenons rank among the most graceful and colorful of the world's primates (Diana monkey, *Cercopithecus diana*; crowned monkey, *C. pogonias pogonias*), while others are heavy-set and drab (Allen's swamp monkey, owl-faced monkey) (Kingdon, 1997). The group includes monkeys that range from 800 g (female northern talapoin monkey) to 13 kg (male patas monkey). Although the majority of the guenons are highly arboreal, a few can dive and swim (northern talapoin monkey), and some are semiterrestrial (patas monkey, Allen's swamp monkey, l'Hoest's monkey, *Cercopithecus lhoesti*) (Kingdon, 1988a,b, 1997; Rowe, 1996; Gautier-Hion *et al.*, 1999; McGraw, 2002). The patas monkey can attain speeds on the ground of at least 55 km per hour, making it the fastest primate in the world (Kingdon, 1971).

Although most guenons have diets consisting largely of fruit and seeds (typically 40–70% of the diet), arthropods (typically 10–25% of the diet), and leaves, they also eat pith, grass, gum, flowers, roots, mushrooms, fish, crabs, shrimp, and worms, and sometimes eggs and small vertebrates (amphibians, reptiles, birds, mammals) (Butynski, 1982, 1988; Cords, 1987; Harvey *et al.*, 1987; Gautier, 1988; Gautier-Hion *et al.*, 1999; Chapman *et al.*, 2002). The patas monkey probably deviates most from the typical guenon diet in that in some places it eats large amounts of gum and galls from acacias (Isbell, 1998).

Most species of guenons live in groups comprising a single adult male and several adult females and their offspring, although additional adult males may join groups temporarily during the breeding season. A few species live regularly in multimale–multifemale groups (Allen's swamp monkey, northern talapoin monkey, vervet monkey), and one species, at least sometimes, lives in family groups of one adult male and one adult female (DeBrazza's monkey, *Cercopithecus neglectus*) (Cords, 1987, 1988; Kingdon, 1997; Gautier-Hion *et al.*, 1999). The sizes of home ranges vary greatly, from 3 ha (Campbell's monkey, *Cercopithecus campbelli*) to >52 km^2 (patas monkey); similarly, the mean distance moved per day varies considerably, from about 500 m per day (DeBrazza's monkey) to more than 4300 m per day (patas monkey) (Wolfheim, 1983; Harvey *et al.*, 1987; Lee *et al.*, 1988; Butynski, 1990; Rowe, 1996).

Guenon Taxonomy

Although the first fossil guenons are from the Pliocene, *ca.* 3 mya, most of the evolutionary radiation of guenons probably occurred within the last 1 million years (Leakey, 1988), and they are likely still in an active stage of speciation (Gautier-Hion *et al.*, 1988). As such, research on the taxonomy and systematics of the guenons is both interesting and challenging.

One of the prerequisites to understanding the broad patterns linking ecology, behavior, morphology and social organization among closely related species and the adaptations these patterns represent is a taxonomic classification that is more or less agreed upon. A working taxonomy is also a prerequisite to establishing conservation priorities, and to making informed decisions for conservation actions *in situ* and *ex situ* (Butynski, 1996, 2002; Oates, 1996; Hilton-Taylor, 2000).

Attaining a widely agreed upon taxonomic classification for a group of species as large and as diverse as the guenons is a formidable task. It has been >230 years since Linnaeus described and named the first species of guenons, yet we continue to find new taxa of guenons, and to debate and revise the taxonomy.

The most recent taxonomy for guenons, and the one adopted for use in this volume, was compiled in February 2000 during the IUCN/SSC Primate Specialist Group's workshop, *Primate Taxonomy for the New Millennium*, held in Orlando, Florida (Grubb *et al.*, 2002). Two objectives of the workshop were to assess the diversity of the Order Primates, and to reach some level of agreement on the most appropriate current taxonomy to apply to the Order. Members of the workshop reviewed available information on the biogeography, morphology, physiology, ecology, behavior, and genetics for each taxon, held discussions, and then made a decision on whether to recognize the taxon. In many cases there was a consensus decision, but this was not always possible. The taxonomy for the guenons, as compiled during this workshop, is in Table I.

When we speak of guenons, to what taxa are we referring? There are many opinions. According to the narrowest definition, a guenon is a monkey within *Cercopithecus*. By this definition, other taxa within the Cercopithecini, namely *Miopithecus ogouensis, M. talapoin, Allenopithecus nigroviridis* and *Erythrocebus patas*, are not guenons. This definition is complicated by the fact that there is disagreement both as to what species should be placed within *Cercopithecus*, and whether *Miopithecus, Allenopithecus* and *Erythrocebus* are valid genera (see below).

According to the broadest, and currently most widely adopted definition, a guenon is a monkey within the Cercopithecini. This is the definition used in this volume.

Taxonomy is based not only upon differences, but also upon similarities and commonalties. It is important, therefore, to know what morphological characteristics all species in the Cercopithecini (the guenons) share. The Cercopithecini belong to the Cercopithecinae (cheek-pouched monkeys) together with the Papionini, which comprise *Macaca, Cercocebus, Mandrillus, Lophocebus, Papio*, and *Theropithecus*. Members of these two tribes have the following morphological characteristics in common: cheek pouches present; no lingual enamel on the lower incisors; elongated nasal bones; low cranial vault; elongated face; and mesaxonic feet, i.e., the third ray is the longest (Groves, 2001).

Table I.   Summary of the Taxonomy for the Guenons (Tribe Cercopithecini) as Compiled during the IUCN/SSC Primate Specialist Group's Workshop, *Primate Taxonomy for the New Millennium* (February, 2000) (Grubb *et al.*, 2002). Family Cercopithecidae Gray, 1821: Old World Monkeys; Subfamily Cercopithecinae Gray, 1821: Cheek-pouched Monkeys; Tribe Cercopithecini Gray, 1821: Guenons.

Genus and species of guenon	Vernacular name
Allenopithecus Lang, 1923	Swamp Monkey
Allenopithecus nigroviridis (Pocock, 1907)	Allen's Swamp Monkey
Miopithecus I. Geoffroy Saint-Hilaire, 1842	Talapoin Monkeys
Miopithecus talapoin (Schreber, 1774)	Southern Talapoin Monkey
Miopithecus ogouensis Kingdon, 1997	Northern Talapoin Monkey
Erythrocebus Trouessart, 1897	Patas Monkey
Erythrocebus patas (Schreber, 1774)	Patas Monkey
Cercopithecus Linnaeus, 1758	Cercopithecus Monkeys or Guenons
Aethiops-Group	
Cercopithecus aethiops (Linnaeus, 1758)	Vervet, Grivet, Tantalus, Green, and Malbrouck Monkeys
C. a. aethiops (Linnaeus, 1758)	Grivet Monkey
C. a. djamdjamensis Neumann, 1902	Bale Mountains Grivet
C. a. tantalus Ogilby, 1841	Tantalus Monkey
C. a. sabaeus Linnaeus, 1766	Green Monkey
C. a. cynosurus (Scopoli, 1786)	Malbrouck Monkey
C. a. pygerythrus (F. Cuvier, 1821)	Vervet Monkey
Lhoesti-Group	
Cercopithecus lhoesti Sclater, 1899	l'Hoest's Monkey
Cercopithecus preussi Matschie, 1898	Preuss's Monkeys
C. p. preussi Matschie, 1898	Cameroon Preuss's Monkey
C. p. insularis Thomas, 1910	Bioko Preuss's Monkey
Cercopithecus solatus Harrison, 1988	Sun-tailed Monkey
Diana-Group	
Cercopithecus diana (Linnaeus, 1758)	Diana and Roloway Monkeys
C. d. diana (Linnaeus, 1758)	Diana Monkey
C. d. roloway (Schreber, 1774)	Roloway Monkey
Dryas-Group	
Cercopithecus dryas Schwarz, 1932	Dryad Monkey
Neglectus-Group	
Cercopithecus neglectus Schlegel, 1876	De Brazza's Monkey
Hamlyni-Group	
Cercopithecus hamlyni Pocock, 1907	Owl-faced Monkeys
C. h. hamlyni Pocock, 1907	Northern Owl-faced Monkey
C. h. kahuziensis Colyn & Rahm, 1987	Mt. Kahuzi Owl-faced Monkey
Cephus-Group	
Cercopithecus petaurista (Schreber, 1774)	Lesser Spot-nosed Monkeys
C. p. petaurista (Schreber, 1774)	Eastern Lesser Spot-nosed Monkey
C. p. buettikoferi Jentink, 1886	Western Lesser Spot-nosed Monkey
Cercopithecus erythrogaster Gray, 1866	White-throated Monkeys
C. e. erythrogaster Gray, 1866	Red-bellied Monkey
C. e. pococki Grubb, Lernould & Oates, 2000	Nigeria White-throated Monkey

(Cont.)

Genus and species of guenon	Vernacular name
Cercopithecus sclateri Pocock, 1904	Sclater's Monkey
Cercopithecus erythrotis Waterhouse, 1838	Red-eared Monkeys
C. e. erythrotis Waterhouse, 1838	Bioko Red-eared Monkey
C. e. camerunensis Hayman in Sanderson, 1940	Cameroon Red-eared Monkey
Cercopithecus cephus (Linnaeus, 1758)	Moustached Monkeys
C. c. cephus (Linnaeus, 1758)	Red-tailed Moustached Monkey
C. c. cephodes Pocock, 1907	Grey-tailed Moustached Monkey
C. c. ngottoensis Colyn, 1999	White-nosed Moustached Monkey
Cercopithecus ascanius (Audebert, 1799)	Red-tailed Monkeys
C. a. ascanius (Audebert, 1799)	Black-cheeked Red-tailed Monkey
C. a. schmidti Matschie, 1892	Schmidt's Red-tailed Monkey
C. a. whitesidei Thomas, 1909	Yellow-nosed Red-tailed Monkey
C. a. katangae Lönnberg, 1919	Katanga Red-tailed Monkey
C. a. atrinasus Machado, 1965	Black-nosed Red-tailed Monkey
***Mona*-Group**	
Cercopithecus campbelli Waterhouse, 1838	Campbell's and Lowe's Monkeys
C. c. campbelli Waterhouse, 1838	Campbell's Monkey
C. c. lowei Thomas, 1923	Lowe's Monkey
Cercopithecus mona (Schreber, 1775)	Mona Monkey
Cercopithecus pogonias Bennett, 1833	Crowned Monkeys
C. p. pogonias Bennett, 1833	Golden-bellied Crowned Monkey
C. p. nigripes Du Chaillu, 1860	Black-footed Crowned Monkey
C. p. grayi Fraser, 1850	Gray's Crowned Monkey
C. p. denti Thomas, 1907	Dent's Monkey
C. p. wolfi Meyer, 1891	Congo Basin Wolf's Monkey
C. p. elegans Dubois & Matschie, 1912	Lomami River Wolf's Monkey
C. p. pyrogaster Lönnberg, 1919	Fire-bellied Wolf's Monkey
***Nictitans*-Group (*Nictitans*-Subgroup)**	
Cercopithecus nictitans (Linnaeus, 1766)	Putty-nosed Monkeys
C. n. nictitans (Linnaeus, 1766)	Eastern Putty-nosed Monkey
C. n. stampflii Jentink, 1888	Stampfli's Putty-nosed Monkey
C. n. ludio Gray, 1849	Ludio Putty-nosed Monkey
C. n. martini Waterhouse, 1838	Bioko Putty-nosed Monkey
***Nictitans*-Group (*Mitis/Albogularis*-Subgroup)**	
Cercopithecus mitis Wolf, 1822	Blue, Sykes's and Golden Monkeys, Samango
***Albogularis* Section**	Sykes's Monkeys, Samango
C. m. albotorquatus de Pousargues, 1896	Pousargues's Sykes's Monkey
C. m. kolbi Neumann, 1902	Kolb's White-collared Monkey
C. m. albogularis (Sykes, 1831)	Zanzibar Sykes's Monkey
C. m. monoides I. Geoffroy Saint-Hilaire, 1841	Tanzania Sykes's Monkey
C. m. francescae Thomas, 1902	Red-eared Sykes's Monkey
C. m. moloneyi Sclater, 1893	Moloney's White-collared Monkey
C. m. erythrarchus Peters, 1852	Stairs's White-collared Monkey
C. m. labiatus I. Geoffroy Saint-Hilaire, 1842	Samango
***Heymansi* Section**	Blue Monkey
C. m. heymansi Colyn & Verheyen, 1987	Lomami River Blue Monkey

(Cont.)

Genus and species of guenon	Vernacular name
Mitis **Section**	Blue Monkeys
C. m. opisthostictus Sclater, 1894	Rump-spotted Blue Monkey
C. m. mitis Wolf, 1822	Pluto Blue Monkey
Boutourlinii **Section**	Blue Monkey
C. m. boutourlinii Giglioli, 1887	Boutourlini's Blue Monkey
Stuhlmanni **Section**	Blue Monkeys, Golden Monkey
C. m. stuhlmanni Matschie, 1893	Stuhlmann's Blue Monkey
C. m. schoutedeni (Schwarz, 1928)	Schouteden's Blue Monkey
C. m. doggetti Pocock, 1907	Doggett's Blue Monkey
C. m. kandti Matschie, 1905	Golden Monkey

There are two primary characters that distinguish members of the Cercopithecini from members of the Papionini. First, the third lower molar in the Cercopithecini is four-cusped and lacks a hypoconulid, whereas in the Papionini the third lower molar nearly always has a hypoconulid. Second, in the Cercopithecini, the diploid chromosome number varies, but is always more than 42, whereas in the Papionini the diploid chromosome number is always 42 (Groves, 2001).

The taxonomic classifications of the past 35 years recognize one to five genera in the Cercopithecini:

- One genus (or supergenus): *Cercopithecus* (Dandelot, 1971; Gautier, 1988; Gautier-Hion, 1988; Kingdon, 1988a,b; Lernould, 1988);
- Two genera: *Cercopithecus, Erythrocebus* (Ruvolo, 1988);
- Four genera: *Cercopithecus, Miopithecus, Allenopithecus, Erythrocebus* (Hill, 1966; Napier, 1981; Dutrillaux *et al.*, 1988; Martin and MacLarnon, 1988; Oates, 1996; Kingdon, 1997; Gautier-Hion *et al.*, 1999; Grubb *et al.*, 2002);
- Five genera: *Cercopithecus, Miopithecus, Allenopithecus, Erythrocebus, Chlorocebus* (Groves, 1989, 1993, 2001; Morales *et al.*, 1999).

As indicated by the large number of recent taxonomic classifications and synonyms (Hill, 1966; Napier and Napier, 1967; Dandelot, 1968, 1971; Napier, 1981; Lernould, 1988; Groves, 1993, 2001; Kingdon, 1997; Grubb *et al.*, 2002), the classification of the guenons has been a major challenge for primate taxonomists, not only at the generic level but also at the specific level.

Table II is a comparison of the six most recent taxonomic classifications for the guenons. They differ considerably both in which species are distinguished as well as the total number of species. The number of species of guenons varies from 23 (Grubb *et al.*, 2002) to 36 (Groves, 2001). Much of the difference depends on how the authors define a species. There is particular uncertainty, conflicting evidence, disagreement, and need for research concerning the taxonomy of the *aethiops, mona,* and *nictitans* species-groups (superspecies).

Table II.   Summary of the Six most Recent Classifications of Guenon Species. The Xs indicate the species recognized by each author. These six classifications also differ in the genera that are recognized. Therefore, the classification of Grubb *et al.* (2002) is adopted here for the genera.

Species	Lernould (1988)	Groves (1993)	Oates (1996)	Kingdon (1997)	Groves (2001)	Grubb *et al.* (2002)
1. *Allenopithecus nigroviridis*	X	X	X	X	X	X
2. *Erythrocebus patas*	X	X	X	X	X	X
Miopithecus (talapoin) superspecies	X					
3. *Miopithecus talapoin*	X	X	X	X	X	X
4. *Miopithecus ogouensis*		X	X	X	X	X
Cercopithecus (aethiops) superspecies/group	X			X		X
5. *Cercopithecus (a.) aethiops*	X	X	X	X	X	X
6. *Cercopithecus (a.) djamdjamensis*				X	X	
7. *Cercopithecus (a.) pygerythrus*	X			X	X	
8. *Cercopithecus (a.) tantalus*	X			X	X	
9. *Cercopithecus (a.) sabaeus*	X			X	X	
10. *Cercopithecus (a.) cynosuros*					X	
Cercopithecus (cephus) superspecies/group	X			X	X	X
11. *Cercopithecus (c.) cephus*	X	X	X	X	X	X
12. *Cercopithecus (c.) sclateri*	X	X	X	X	X	X
13. *Cercopithecus (c.) erythrotis*	X	X	X	X	X	X
14. *Cercopithecus (c.) ascanius*	X	X	X	X	X	X
15. *Cercopithecus (c.) petaurista*	X	X	X	X	X	X
16. *Cercopithecus (c.) erythrogaster*	X	X	X	X	X	X
17. *Cercopithecus (c.) signatus*					X	
Cercopithecus (diana) superspecies/group	X				X	X
18. *Cercopithecus (d.) diana*	X	X	X	X	X	X
19. *Cercopithecus (d.) roloway*					X	
Cercopithecus (dryas) group					X	X
20. *Cercopithecus dryas*	X	X	X	X	X	X
Cercopithecus (lhoesti) superspecies/group	X			X	X	X
21. *Cercopithecus (l.) lhoesti*	X	X	X	X	X	X
22. *Cercopithecus (l.) preussi*	X	X	X	X	X	X
23. *Cercopithecus (l.) solatus*	X	X	X	X	X	X
Cercopithecus (mona) superspecies/group	X			X	X	X
24. *Cercopithecus (m.) mona*	X	X	X	X	X	X
25. *Cercopithecus (m.) campbelli*	X	X	X	X	X	X
26. *Cercopithecus (m.) lowei*				X	X	
27. *Cercopithecus (m.) pogonias*	X	X	X	X	X	X
28. *Cercopithecus (m.) denti*				X	X	
29. *Cercopithecus (m.) wolfi*	X	X	X	X	X	

(Cont.)

Species	Lernould (1988)	Groves (1993)	Oates (1996)	Kingdon (1997)	Groves (2001)	Grubb et al. (2002)
Cercopithecus (nictitans) **superspecies/group**	X			X	X	X
30. *Cercopithecus (n.) nictitans*	X	X	X	X	X	X
31. *Cercopithecus (n.) mitis*	X	X	X	X	X	X
32. *Cercopithecus (n.) dogetti*					X	
33. *Cercopithecus (n.) kandti*					X	
34. *Cercopithecus (n.) albogularis*	X		X		X	
Cercopithecus (neglectus) **group**					X	X
35. *Cercopithecus neglectus*	X	X	X	X	X	X
Cercopithecus (hamlyni) **group**					X	X
36. *Cercopithecus hamlyni*	X	X	X	X	X	X
Total species	27	24	25	30	36	23

Groves (2001) and Grubb *et al.* (2002) provide the two most recent discussions, reviews and summaries of the taxonomy of the guenons. They also present comprehensive reviews of subspecies, synonyms, authorities, references, and English names.

Since the publication of Allen's (1939) *Catalogue of African Mammals* more than 60 years ago, seven taxa of guenons thought to be invalid and treated at that time as synonyms have come to be generally recognized as valid. They include one species (Sclater's monkey, *Cercopithecus sclateri*) and six subspecies (Grubb *et al.*, 2002). In addition, 12 new names for guenons have been proposed during 1939–2001. Of these, one species (sun-tailed monkey, *Cercopithecus solatus*) and four subspecies are considered by Grubb *et al.* (2002) to be valid taxa.

Summary

The guenon monkeys (Tribe Cercopithecini) represent a relatively recently evolved, widespread, and taxon-rich group of species with a diverse morphology, ecology, behavior, and social organization. Not surprisingly, the guenons also have a complicated and controversial taxonomy. The most recent taxonomy lists four genera, 23 species, and 55 subspecies (Grubb *et al.*, 2002). Given the large number of species and the considerable variability within and among the species, the guenons represent an excellent taxonomic group within the Order Primates for the study of patterns linking morphology, ecology, behavior, and social organization. Much more research on the systematics and taxonomy of the guenons, both at the species and generic levels, must be undertaken if the current biodiversity represented by the guenons is to be appreciated and conserved.

ACKNOWLEDGMENTS

I thank John Oates, Mary Glenn, Marina Cords and two anonymous reviewers for their valuable comments on the draft manuscript, and Zoo Atlanta, Conservation International, The National Museums of Kenya, and the IUCN Eastern Africa Regional Office for support during the writing of this chapter.

References

Allen, G. M. 1939. A checklist of African mammals. *Bull. Mus. Comp. Zool. Harv.* **83**:1–763.

Butynski, T. M. 1982. Vertebrate predation by primates: A review of hunting patterns and prey. *J. Human Evol.* **11**:421–430.

Butynski, T. M. 1988. Guenon birth seasons and correlates with rainfall and food. In: A. Gautier-Hion, F. Bourlière, J.-P. Gautier, and J. Kingdon (eds.), *A Primate Radiation: Evolutionary Biology of the African Guenons*, pp. 284–322, Cambridge University Press, Cambridge.

Butynski, T. M. 1990. Comparative ecology of blue monkeys (*Cercopithecus mitis*) in high- and low-density subpopulations. *Ecol. Monogr.* **60**:1–26.

Butynski, T. M. 1996. International trade in CITES Appendix II African primates. *African Primates* **2**:5–9.

Butynski, T. M. 2002. Conservation of the guenons: An overview of status, threats, and recommendations. In: M. E. Glenn, and M. Cords (eds.), *The Guenons: Diversity and Adaptation in African Monkeys*, pp. 411–424. Kluwer Academic Publishers, New York.

Chapman, C. A., Chapman, L. J., Cords, M., Gathua, J. M., Gautier-Hion, A., Lambert, J. E., Rode, K., Tutin, C. E. G., and White, L. J. T. 2002. Variation in the diets of *Cercopithecus* species: Differences within forests, among forests and across species. In: M. E. Glenn, and M. Cords (eds.), *The Guenons: Diversity and Adaptation in African Monkeys*, pp. 325–350. Kluwer Academic Publishers, New York.

Cords, M. 1987. Forest guenons and patas monkeys: Male-male competition in one-male groups. In: B. B. Smuts, D. L. Cheney, R. M. Seyfarth, R. W. Wrangham, and T. T. Struhsaker (eds.), *Primate Societies*, pp. 98–111. The University of Chicago Press, Chicago.

Cords, M. 1988. Mating systems of forest guenons: A Preliminary review. In: A. Gautier-Hion, F. Bourlière, J.-P. Gautier and J. Kingdon (eds.), *A Primate Radiation: Evolutionary Biology of the African Guenons*, pp. 323–339. Cambridge University Press, Cambridge.

Dandelot, P. 1968. Primates: Anthropoidea. In: J. Meester (ed.). *Smithsonian Institution Preliminary Identification Manual for African Mammals*, pp. 1–80. Smithsonian Institution Press, Washington, DC, part 24.

Dandelot, P. 1971. Order Primates. In: J. Meester and H. W. Setzer (eds.), *The Mammals of Africa: An Identification Manual*, pp. 1–45. Smithsonian Institution Press, Washington, DC, part 3.

Dutrillaux, B., Muleris, M., and Couturier, J. 1988. Chromosomal evolution of Cercopithecinae. In: A. Gautier-Hion, F. Bourlière, J.-P. Gautier and J. Kingdon (eds.), *A Primate Radiation:*

Evolutionary Biology of the African Guenons, pp. 150–159. Cambridge University Press, Cambridge.

Gautier, J.-P. 1988. Interspecific affinities among guenons as deduced from vocalizations. In: A. Gautier-Hion, F. Bourlière, J.-P. Gautier and J. Kingdon (eds.), *A Primate Radiation: Evolutionary Biology of the African Guenons*, pp. 194–226. Cambridge University Press, Cambridge.

Gautier-Hion, A. 1988. The diet and dietary habits of forest guenon. In: A. Gautier-Hion, F. Bourlière, J.-P. Gautier and J. Kingdon (eds.), *A Primate Radiation: Evolutionary Biology of the African Guenons*, pp. 257–283. Cambridge University Press, Cambridge.

Gautier-Hion, A., Bourlière, F., Gautier, J.-P., and Kingdon, J. (eds.). 1988. Introduction. *A Primate Radiation: Evolutionary Biology of the African Guenons*, pp. 1–3. Cambridge University Press, Cambridge.

Gautier-Hion, A., Colyn, M., and Gautier, J.-P. 1999. *Histoire Naturelle des Primates d'Afrique Centrale*, ECOFAC, Libreville.

Groves, C. P. 1989. *A Theory of Human and Primate Evolution*, Clarendon Press, Oxford.

Groves, C. P. 1993. Order Primates. In: D. E. Wilson and D. M. Reeder (eds.), *Mammalian Species of the World: A Taxonomic and Geographic Reference* (2nd edition), pp. 243–277. Smithsonian Institution Press, Washington DC

Groves, C. P. 2001. *Primate Taxonomy*, Smithsonian Institution Press, Washington, DC.

Grubb, P., Butynski, T. M., Oates, J. F., Bearder, S. K., Disotell, T. R., Groves, C. P., and Struhsaker, T. T. 2002. An assessment of the diversity of African primates. Unpublished report. IUCN/SSC Primate Specialist Group, Washington, DC.

Harvey, P. H., Martin, R. D., and Clutton-Brock, T. H. 1987. Life histories in comparative perspective. In: B. B. Smuts, D. L. Cheney, R. M. Seyfarth, R. W. Wrangham and T. T. Struhsaker (eds.), *Primate Societies*, pp. 181–196. The University of Chicago Press, Chicago.

Hill, W. C. O. 1966. *Primates. Comparative Anatomy and Taxonomy. 6. Catarrhini, Cercopithecoidea, Cercopithecinae*, Edinburgh University Press, Edinburgh.

Hilton-Taylor, C. 2000. *2000 IUCN Red List of Threatened Animals*, IUCN, Gland, Switzerland.

Isbell, L. A. 1998. Diet for a small primate: Insectivory and gummivory in the (large) patas monkey (*Erythrocebus patas pyrrhonotus*). *Am. J. Primatol.* **45**: 381–398.

Kingdon, J. 1971. *East African Mammals: An Atlas of Evolution in Africa*, vol. 1. Academic Press, London.

Kingdon, J. 1988a. Comparative morphology of hands and feet in the genus *Cercopithecus*. In: A. Gautier-Hion, F. Bourlière, J.-P. Gautier and J. Kingdon (eds.), *A Primate Radiation: Evolutionary Biology of the African Guenons*, pp. 184–193. Cambridge University Press, Cambridge.

Kingdon, J. 1988b. What are face patterns and do they contribute to reproductive isolation in guenons? In: A. Gautier-Hion, F. Bourlière, J.-P. Gautier and J. Kingdon (eds.), *A Primate Radiation: Evolutionary Biology of the African Guenons*, pp. 227–245. Cambridge University Press, Cambridge.

Kingdon, J. 1997. *The Kingdon Field Guide to African Mammals*, Academic Press, London.

Leakey, M. 1988. Fossil evidence for the evolution of the guenons. In: A. Gautier-Hion, F. Bourlière, J.-P. Gautier and J. Kingdon (eds.), *A Primate Radiation: Evolutionary Biology of the African Guenons*, pp. 7–12. Cambridge University Press, Cambridge.

Lee, P. C., Thornback, J., and Bennett, E. L. 1988. *Threatened Primates of Africa: The IUCN Red Data Book*, IUCN, Gland, Switzerland.

Lernould, J.-M. 1988. Classification and geographical distribution of guenons: A review. In: A. Gautier-Hion, F. Bourlière, J.-P. Gautier and J. Kingdon (eds.), *A Primate Radiation: Evolutionary Biology of the African Guenons*, pp. 54–78. Cambridge University Press, Cambridge.

Martin, R. D., and MacLarnon, A. M. 1988. Quantitative comparisons of the skull and teeth in guenons. In: A. Gautier-Hion, F. Bourlière, J.-P. Gautier, and J. Kingdon (eds.), *A Primate*

Radiation: Evolutionary Biology of the African Guenons, pp. 160–183. Cambridge University Press, Cambridge.

McGraw, W. S. 2002. Diversity of guenon positional behavior. In: M. E. Glenn and M. Cords (eds.), *The Guenons: Diversity and Adaptation in African Monkeys*, pp. 113–131. Kluwer Academic Publishers, New York.

Morales, J. C., Disotell, T. R., and Melnick, D. J. 1999. Molecular phylogenetic studies of nonhuman primates. In: P. Dolhinow and A. Fuentes (eds.), *The Nonhuman Primates*, pp. 18–28. Mayfield Publishing Company, Mountain View, California.

Napier, J. R. and Napier, P. H. 1967. *A Handbook of Living Primates*, Academic Press, London.

Napier, P. H. 1981. *Catalogue of Primates in the British Museum (Natural History) and Elsewhere in the British Isles. Part 2: Family Cercopithecidae, Subfamily Cercopithecinae*, British Museum (Natural History), London.

Oates, J. F. 1996. *African Primates: Status Survey and Conservation Action Plan*, Revised Edition, IUCN/SSC Primate Specialist Group, IUCN, Gland, Switzerland.

Rowe, N. 1996. *The Pictorial Guide to the Living Primates*. Pogonias Press, East Hampton, New York.

Ruvolo, M. 1988. Genetic evolution in the African guenons. In: A. Gautier-Hion, F. Bourlière, J.-P. Gautier and J. Kingdon (eds.), *A Primate Radiation: Evolutionary Biology of the African Guenons*, pp. 127–139. Cambridge University Press, Cambridge.

Wolfheim, J. H. 1983. *Primates of the World: Distribution, Abundance, and Conservation*, University of Washington Press, Seattle.

Y-chromosomal Window onto the History of Terrestrial Adaptation in the Cercopithecini

2

ANTHONY J. TOSI, PAUL J. BUZZARD,
JUAN CARLOS MORALES, and DON J. MELNICK

Introduction

Although there is an extensive literature dedicated to the origins of terrestrial bipedalism in early hominids, the change from life in the trees to life on the ground among nonhuman primates has received much less attention. Consequently, we lack a sense for how readily primates in general can make the transition from an arboreal habitus to a terrestrial one. Guenon systematics provide a valuable case study in this respect. *Allenopithecus* and *Miopithecus*,

ANTHONY J. TOSI and PAUL J. BUZZARD • Department of Anthropology, Columbia University, New York, NY 10027 and New York Consortium in Evolutionary Primatology (NYCEP), USA. JUAN CARLOS MORALES • Center for Environmental Research and Conservation, Columbia University, New York, NY 10027 and New York Consortium in Evolutionary Primatology (NYCEP), USA. DON J. MELNICK • Department of Anthropology, Center for Environmental Research and Conservation, and Department of Biological Sciences, Columbia University, New York, NY 10027 and New York Consortium in Evolutionary Primatology (NYCEP), USA.

The Guenons: Diversity and Adaptation in African Monkeys, edited by Glenn and Cords. Kluwer Academic/Plenum Publishers, New York, 2002.

the two most primitive guenon taxa in terms of chromosome number (Dutrillaux, 1979; Dutrillaux *et al.*, 1980; Szalay and Delson, 1979), retention of molar flare and female sexual swellings (Strasser and Delson, 1987; Groves, 1989, 2000), and multimale social system (Gautier, 1985; Rowell, 1988), maintain a postcranial anatomy typical of arboreal guenons (Gebo and Sargis, 1994). The limb morphology of these two genera, combined with their likely basal divergence from the rest of the (mostly tree-living) Cercopithecini, lead one to reconstruct the guenon common ancestor as an arboreal taxon. In contrast, three extant species, or species groups (*Erythrocebus patas*, the *Cercopithecus aethiops* group, and the *C. lhoesti* group), spend a significant amount of time on the ground to the extent that their postcranial skeletons exhibit clear evidence of terrestrial adaptation (Gebo and Sargis, 1994). One may then ask whether these three, patas, vervets, and l'Hoest's monkeys, share a recent common ancestor, such that terrestriality arose only once among

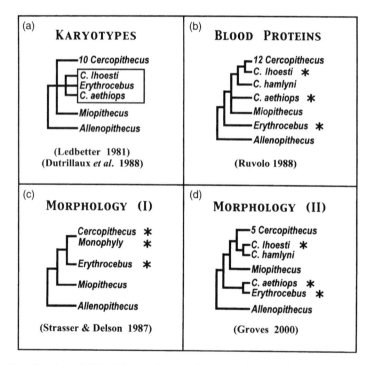

Fig. 1. Previously-published hypotheses of cercopithecin phylogeny condensed to highlight the relationships depicted among the three terrestrial taxa: *Erythrocebus patas*, *Cercopithecus aethiops*, and *C. lhoesti*. Karyotype studies (a) suggest these taxa form a monophyly (indicated by box). The other analyses (b–d), however, suggest these taxa (highlighted with asterisks) are more distantly related, implying two or three independent transitions from an arboreal habitus to a terrestrial one. [The karyotype tree (a) is a strict consensus of the studies by Ledbetter (1981) and Dutrillaux *et al.* (1988).]

the extant guenons, or whether they represent multiple, independent departures from an arboreal lifestyle.

The phylogenetic hypotheses of earlier researchers (Fig. 1) suggested different answers to this question. The karyotypic studies [Fig. 1(a)] of Ledbetter (1981) and Dutrillaux *et'al.* (1988) posited a close evolutionary relationship of *Erythrocebus patas*, *Cercopithecus aethiops*, and *C. lhoesti* exclusive to the rest of the Cercopithecini, implying only a single adaptive transition to the ground. In contrast, Ruvolo's (1988) analysis of blood proteins [Fig. 1(b)] showed three separate origins for these taxa, thus, three origins of terrestriality. Morphological phylogenies [Fig. 1(c) and (d)] differ from those based on karyotypes and blood proteins, as well as from each other. The tree [Fig. 1(c)] of Strasser and Delson (1987) joins *Erythrocebus patas* to the base of a monophyletic clade of *Cercopithecus*—a position therefore distinct from *C. aethiops* and *C. lhoesti*— resulting in a pattern consistent with (at least) two independent departures from the inferred arboreal lifestyle of the guenon common ancestor. In comparison, the morphological analysis [Fig. 1(d)] of Groves (2000) indicated three cranial synapomorphies to cluster *Erythrocebus patas* as sister taxon with *Cercopithecus aethiops*, while other features cluster *C. lhoesti* with the arboreal *Cercopithecus* spp. Although again suggesting two unique transitions to a terrestrial habitus, this topology depicts a set of relationships among the ground-living taxa different to the tree of Strasser and Delson (1987).

It is unclear which of these studies provides the strongest phylogenetic hypothesis. To investigate whether nuclear sequence data would agree with one of these hypotheses, thereby highlighting which is the most reasonable by congruence of multiple independent data sets (Miyamoto and Fitch, 1995), we subjected to phylogenetic analysis ~3.1 kb of two Y-chromosome loci (TSPY and SRY) from 16 members of the Cercopithecini and 7 outgroup taxa. The most parsimonious topologies recovered from this Y-chromosome data set depict a pattern of relationships among the terrestrial taxa that agrees with only one previously published hypothesis.

Materials and Methods

Samples

We analyzed the Y-chromosome DNA of 23 individuals (Table I); five subjects were newly surveyed, and their sequences added to those of 18 other subjects from Tosi *et al.* (2002a). The sampling includes multiple representatives of each of the four cercopithecin genera: *Allenopithecus*, *Miopithecus*, *Erythrocebus*, and *Cercopithecus*. Outgroup taxa include four papionins (*Macaca mulatta*, *Mandrillus sphinx*, *Theropithecus gelada*, *Papio hamadryas*), two colobines (*Presbytis melalophos*, *Trachypithecus cristatus*) and one hominoid (*Homo sapiens*).

Table I. Genetic Samples List

No.	Taxon	Origin	Designation	Source	GENBANK# (TSPY, SRY)
1	*Allenopithecus nigroviridis*	Congo, Africa	R146/97	Nat'l. Museum, Scotland	AF284280, AF284331
2	*Allenopithecus nigroviridis*[a]	Congo, Africa	94042	Dept. Anthropology, NYU	AY048053
3	*Cercopithecus aethiops*	East Africa	*C. aethiops*	Dept. Anthropology, NYU	AY048060, AY048070
4	*Cercopithecus aethiops*	East Africa	97.015	Dept. Anthropology, NYU	AY048061, AY048071
5	*Cercopithecus ascanius*[a,b]	Kibale, Uganda	DM3376	Columbia University, CERC	AY048054
6	*Cercopithecus lhoesti*[b]	Central Africa	Antwerp	Antwerp Zoo, Belgium	AY048055, AY048067
7	*Cercopithecus lhoesti*[b]	Central Africa	94019	Oregon Zoo	AY048056, AY048068
8	*Cercopithecus mitis*	East Africa	5311.77140	Henry Doorly Zoo	AY048057, AY048069
9	*Cercopithecus mona*	Grenada	SA.BF#3	M. Glenn / K. Bensen	AF284281, AF284332
10	*Cercopithecus neglectus*[a,b]	Kenya	M43	Dept. Anthropology, NYU	AY048058
11	*Cercopithecus pogonias*[a,b]	Gabon	GAB.17	Dept. Anthropology, NYU	AY048059
12	*Erythrocebus patas*	East Africa	151	Dept. Anthropology, CUNY	AY048062, AY048072
13	*Erythrocebus patas*	East Africa	R228	Dept. Anthropology, CUNY	AY048063, AY048073
14	*Erythrocebus patas*	East Africa	R230	Dept. Anthropology, CUNY	AY048064, AY048074
15	*Homo sapiens*	Europe (?)	GENBANK	GENBANK	M98524, X53772
16	*Macaca mulatta*	Southeast China	20156	California Primate Center	AF284259, AF284310
17	*Mandrillus sphinx*	Central Africa	Mandrill	Dept. Anthropology, NYU	AF284279, AF284330
18	*Miopithecus talapoin*	Central Africa	6354	Omaha Zoo	AY048065, AY048075
19	*Miopithecus talapoin*	Central Africa	7311	Omaha Zoo	AY048066, AY048076
20	*Papio hamadryas*	East Africa	73-347	Dept. Anthropology, NYU	AF284277, AF284328
21	*Presbytis melalophos*	Johor, Malaysia	DJ.36	Malaysian Forest Service	AF284231, AF284282
22	*Theropithecus gelada*	East Africa	891096	Dept. Anthropology, NYU	AF284278, AF284329
23	*Trachypithecus cristatus*	Kedah, Malaysia	DJ.1	Malaysian Forest Service	AF284232, AF284283

[a] Samples newly-sequenced in the present study.
[b] The SRY locus of these animals was not successfully amplified by PCR.

Phylogenetic Analysis

We amplified and sequenced two loci located in the non-recombining portion of the Y-chromosome, SRY (Sex-determining Region, Y-chromosome) and TSPY (Testis-Specific Protein, Y-chromosome), using primers and protocols described by Tosi *et al.* (2000). We combined the TSPY and SRY data sets because these loci are closely linked on the Y-chromosome, suggesting that they experience a similar set of evolutionary conditions, and because partition homogeneity tests showed no significant difference in their evolutionary signal (1000 replicates, $p = 0.79$). We subjected the combined matrix to parsimony analysis using the branch-and-bound algorithm in PAUP 4.0 (Swofford, 1999). The relative phylogenetic positions of the terrestrial guenon taxa were noted in the resultant topologies. We subsequently performed branch-and-bound searches for the shortest trees depicting alternative evolutionary hypotheses (Fig. 1) using the constraint option in PAUP. We then employed the Kishino and Hasegawa (1989) test to determine whether trees recovered from the initial search were significantly different from the constrained parsimony topologies. Such a result would show that the present Y-chromosome data set unambiguously supports only the phylogenetic pattern depicted in the initial (unconstrained) trees.

Results

Alignment of the Y-chromosome sequences shows 694 variable sites, 227 of which are phylogenetically informative. The data matrix (not shown) further reveals a 6-base insertion in intron #1 of the TSPY sequences of *Erythrocebus patas*, *Cercopithecus aethiops*, and *C. lhoesti*, providing strong support for a close evolutionary relationship among these taxa.

The initial branch-and-bound search (no constraints) recovered six trees. As highlighted by the polytomous regions of their strict consensus [Fig. 2(a)], they differ only in (1) the relationships among members of an arboreal *Cercopithecus* spp. clade, and (2) whether *C. lhoesti* forms a trichotomy with *Erythrocebus patas* and *C. aethiops*, or is basal to a sister-grouping of these two taxa. Most importantly, all six trees are consistent with the karyotypic hypotheses [Fig. 1(a)] of Ledbetter (1981) and Dutrillaux *et al.* (1988) in clustering the three terrestrial guenon groups as a clade distinct from the other members of their tribe. To test whether these trees are significantly different from the evolutionary hypotheses of other researchers [Fig. 1(b–d)], we repeatedly applied the constraint option in PAUP 4.0 (Swofford, 1999) and performed the branch-and-bound searches again. We also performed a search for the shortest tree(s) without an all terrestrial clade. Constraining for these

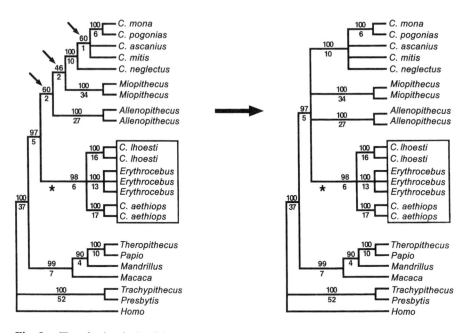

Fig. 2. Topologies derived from the present Y-chromosome data set. Bootstrap values (1000 replicates) and numbers of unambiguous changes are given above and below each branch, respectively. Asterisks indicate the origin of a 6-base insertion in intron #1 of the TSPY gene. Boxes highlight the close relationship of the three terrestrial guenon taxa. (a) The strict (and bootstrap) consensus of the six most parsimonious Y-chromosome trees. Small arrows point to weakly-supported clades. (b) The "collapsed" version of tree A.

various hypotheses resulted in topologies that were between five and 15 steps longer than the original six trees. Comparison tests in PAUP 4.0 showed each constrained topology to be significantly different from the six original trees, based on the present Y-chromosome data set (Table II).

Bootstrapping (1000 replicates) resulted in a topology congruent with the strict consensus of the most parsimonious trees [Fig. 2(a)]. Noting the weak support for the somewhat derived positions of *Allenopithecus* and *Miopithecus*, as well as considering their putatively primitive status among the guenons, we decided to investigate whether the six shortest topologies are statistically different from phylogenies depicting one or both of these taxa as basal to the other members of the Cercopithecini. Comparison tests show that none of the phylogenies placing *Allenopithecus* and/or *Miopithecus* in such basal positions is significantly different from any of the six original trees (Table II). Accordingly,

Table II. Kishino & Hasegawa Test for Phylogenetic Incongruence

Constrained topology	Treelength	Significantly different from the six shortest trees ($p < 0.05$)?[b]
(i) Six most parsimonious trees[a]	816	
(ii) Blood proteins — Fig. 1(a) (3 trees)	831	YES
(iii) Morphology I — Fig. 1(c) (33 trees)	824	YES
(iv) Morphology II — Fig. 1(d) (3 trees)	826	YES
(v) Shortest topologies WITHOUT an all terrestrial clade (9 trees)	821	YES
(vi) *Allenopithecus* basal to other Cercopithecini (3 trees)	817	NO
(vii) *Miopithecus* basal to other Cercopithecini (3 trees)	817	NO
(viii) *Allenopithecus & Miopithecus* basal to other Cercopithecini (21 trees)	818	NO

[a] In all six trees, *Erythrocebus patas, Cercopithecus aethiops, and C. lhoesti* cluster exclusive to the other seven cercopithecin species surveyed. Topologies generated under constraints "ii" through "iv" reflect the hypotheses of Ruvolo (1988), Strasser and Delson (1987), and Groves (2000), respectively. Constraint "v" recovers the shortest trees that do not include an all terrestrial clade. Comparison tests of topologies generated under constraints "vi" through "viii" — placing the putatively primitive taxa in basal positions — show that such topologies have no significant difference from the six shortest trees and therefore support the collapse of Figure 2(a) to the pattern in Figure 2(b).
[b] Answers in this column refer to all pairwise comparisons between constrained and unconstrained trees.

we believe that the present Y-chromosome data set can only reliably support a collapsed version of haplotype relationships, as in Figure 2b.

Discussion

An All Terrestrial Clade

The results of the Y-chromosome analysis (Fig. 2) agree with the karyotypic studies [Fig. 1(a)] of Ledbetter (1981) and Dutrillaux *et al.* (1988) in finding a close relationship among *Erythrocebus patas, Cercopithecus aethiops*, and *C. lhoesti*, exclusive to other members of the Cercopithecini. Such congruence among multiple, independent data sets is strong evidence that a true organismal pattern has been recovered (Miyamoto and Fitch, 1995). Two other lines of evidence also support this clade. First, in addition to six synapomorphic point mutations, the terrestrial guenons share a 6-base TSPY insertion not found in any other taxon surveyed (Fig. 2); such insertion events are evolutionarily rare and therefore considered strong indicators of shared ancestry (Rokas and Holland, 2000; Tosi *et al.*, 2002a,b). Second, not only does a branch-and-bound search confirm that the most parsimonious trees contain a patas-vervet-l'Hoest's aggregate, but subsequent comparison tests show these trees to be significantly different from the shortest topologies reflecting the alternative evolutionary scenarios of earlier researchers [Fig. 1(b–d)], as well as the shortest possible topologies (no specific hypothesis) without this three-taxon clade. Thus, the Y-chromosome data unambiguously support a close relationship among the terrestrial guenons.

Species Trees vs. Gene Trees

Shared ancestry is not the only possible explanation for the close Y-chromosome relationship of *Erythrocebus patas*, *Cercopithecus aethiops*, and *C. lhoesti*. In theory, differential sorting of genetic lineages through successive speciation events may yield a gene tree incongruent with true species phylogeny (Pamilo and Nei, 1988; Disotell and Raaum, 2002). However, this problem is unlikely to be present in cercopithecin Y-chromosome topologies because the small effective population size of paternally-inherited molecular markers leads to a rapid fixation time of genetic lineages and, consequently, a high likelihood for Y-chromosome trees to match the species phylogeny (Tosi *et al.*, 2002b).

Genetic exchange through hybridization could also have led to a potentially misleading cluster of taxa (Struhsaker *et al.*, 1988; Detwiler, 2002; Disotell and Raaum, 2002). If such interspecific mating has occurred, the long branch lengths of each terrestrial species (Fig. 2) suggest it happened deep in the past, on the order of a few million years ago. However, at least one line of reasoning suggests true phylogeny, rather than ancestral hybridization, is the more likely explanation of Y-chromosome relationships. While it may be relatively easy for a single Y-chromosome to cross a species boundary and spread to fixation in a neighboring population/species, it is far less probable for several karyotypic characters to do so. If, for example, *Cercopithecus lhoesti* and/or *C. aethiops* indeed diverged from the arboreal *Cercopithecus* clade (as independent transitions to a terrestrial lifestyle) and initially carried the several chromosomal changes unique to the common ancestor of this group (Dutrillaux *et al.*, 1980; Dutrillaux and Muleris, pers. com.), it is unlikely that the backcrossing of a few hybrid individuals (produced with early *Erythrocebus patas*) would introduce exotic chromosomal characters at high enough frequency to replace several of those native to parental populations/species (Hartl and Clark, 1997). A hybrid scenario, therefore, cannot easily account for the lack of several "arboreal *Cercopithecus*" karyotypic characters in all of the terrestrial guenons. Phylogeny then becomes the best explanation for the close karyotypic relationships among *Erythrocebus patas*, *Cercopithecus aethiops*, and *C. lhoesti*, and, by extension, for the close Y-chromosomal relationships among them (Fig. 2).

History of Terrestriality vs. Arboreality in the Cercopithecini

As mentioned above, several lines of evidence attest to the primitive nature of *Allenopithecus* and *Miopithecus*. Further, tree comparison tests (Table II: hypotheses vi–viii) show that the present Y-chromosome data set cannot refute evolutionary scenarios that posit one or both of these taxa as basal to the rest of the Cercopithecini, hence, the collapse of the consensus tree, Figure 2(a), to that in Figure 2(b). Basal positions of these taxa would lead one to most parsimoniously reconstruct the guenon common ancestor as an arboreal species, since both outgroups and (most) ingroups would share a limb morphology devoid of joint modifications associated with habitual use of the terrestrial environment. Terrestriality would then be a derived condition in the Cercopithecini, arising only once in the common ancestor to patas, vervets, and l'Hoest's monkeys.

The TSPY/SRY data, however, yield slightly stronger support for an interesting alternative hypothesis. Though not statistically shorter than topologies placing *Allenopithecus* and/or *Miopithecus* basal to the rest of the guenons, all six most parsimonious Y-chromosome trees recover an arboreal/terrestrial dichotomy in which swamp monkeys and talapoins consistently cluster with the arboreal *Cercopithecus* spp. and the terrestrial taxa form a separate clade [Fig. 2(a)]. If this pattern is a true reflection of evolutionary history, terrestriality would then be the inferred ancestral condition of the guenons since it would be shared in common between one major guenon clade and the immediate outgroup of all cercopithecins, the terrestrial papionins. However, given the weak support for clustering *Allenopithecus* and *Miopithecus* with the arboreal *Cercopithecus* spp. [Fig. 2(a)], and because such an arboreal/terrestrial dichotomy would require convergent evolution in the derived loss of molar flare and female sexual swellings in both cercopithecin clades, we presently favor the more traditional hypothesis in which swamp monkeys and talapoins are basal to the rest of the guenons and terrestriality is the derived condition. Clearly, more data are needed to definitively decide between these two competing hypotheses.

Taxonomic Status of the Terrestrial Taxa

If future studies support the Y-chromosomal relationships revealed here, and if taxonomy is to reflect phylogeny, then the generic allocations of the terrestrial guenons will require revision. The genus *Cercopithecus*, as presently defined, would be paraphyletic: *C. aethiops* and *C. lhoesti* cluster more closely with *Erythrocebus patas* than with their congeners. There are three possible ways to adjust the taxonomy. First, patas monkeys could be sunk into *Cercopithecus*. Second, patas, vervets, and l'Hoest's monkeys could be placed together in the resurrected genus *Chlorocebus* (Gray, 1870), a nomen previously used to describe *Cercopithecus aethiops* and one that holds priority over *Erythrocebus* (Trouessart, 1897) by 27 years. Third, each of the terrestrial taxa could be granted genus-level status: restore *Chlorocebus* (Gray, 1870) and *Allochrocebus* (Elliot, 1913) to vervets and l'Hoest's monkeys, respectively, while retaining *Erythrocebus* for patas monkeys.

We do not advocate the first option, since keeping patas, vervets, and l'Hoest's monkeys in *Cercopithecus* fails to emphasize their unique morphological adaptations and terrestrial lifestyle. We also do not advocate the third course of action. Elevating each of the terrestrial lineages to the genus level results in a nomenclature that lacks formal recognition of the evolutionary relationship shared among these taxa. We tentatively support the second option—placing patas, vervets, and l'Hoest's monkeys in the genus *Chlorocebus* (Gray, 1870). This set of revisions not only rectifies the taxonomy, but embodies the message that the terrestrial taxa form a clade exclusive of the other guenons.

Summary

Y-chromosomal patterns agree with earlier karyotypic studies in depicting a close relationship among patas, vervets, and l'Hoest's monkeys distinct from other members of the Cercopithecini. This relationship is suggested to be one of shared ancestry, rather than one due to past hybridization. Thus, on the working assumption that the earliest guenons were arboreal, terrestriality appears to have originated only once among the extant cercopithecins. Consequently, taxonomic revision is necessary as relates to the three terrestrial taxa: *Erythrocebus patas*, *Cercopithecus aethiops*, and *C. lhoesti*.

ACKNOWLEDGMENTS

We thank Marina Cords and Mary Glenn for inviting us to contribute to this volume. We are grateful to Todd Disotell, Cliff Jolly, and Ryan Raaum of the NYU Department of Anthropology for providing "total DNA" and offering helpful discussion. We also thank Jef Dupain and Kristel De Vleeschouwer of the Royal Zoological Society of Antwerp, and Jerry Herman of the National Museum in Edinburgh for assistance with blood and tissue samples. We are indebted to the following for providing genetic samples: Centre for Research and Conservation of the Royal Zoological Society of Antwerp; Universiti Kebangsaan Malaysia; Jabatan Perhilitan Malaysia; Oregon Zoo; Omaha Zoo; Henry Doorly Zoo; California Primate Center; Department of Anthropology, New York University; Department of Anthropology, City

University of New York; National Museum, Edinburgh, Scotland; Mary Glenn and Keith Bensen. Some of the biological materials used in this study were obtained via the "Comparative Neurobiology of Aging Resource," J. Erwin, PI, supported by NIH grant AG14308 from the National Institute on Aging. We thank Bernard Dutrillaux and Martine Muleris of the Curie Institute and Hirohisa Hirai of Kyoto University for discussions of Cercopithecin chromosomal data. A final thank you to Michael Campbell for his hospitality. The present study was funded by NSF Grant SBR-97-07883 to DJM and JCM, and was additionally supported by the Center for Environmental Research and Conservation (CERC) at Columbia University.

References

Detwiler, K. M. 2002. Hybridization between red-tailed monkeys (*Cercopithecus ascanius*) and blue monkeys (*C. mitis*) in East African forests. In: M. E. Glenn, and M. Cords (eds.), *The Guenons: Diversity and Adaptation in African Monkeys*, pp. 77–97. Kluwer Academic Publishers, New York.

Disotell, T. R., and Raaum, R. L. 2002. Molecular timescale and gene tree incongruence in the guenons. In: M. E. Glenn, and M. Cords (eds.), *The Guenons: Diversity and Adaptation in African Monkeys*, pp. 27–36. Kluwer Academic Publishers, New York.

Dutrillaux, B. 1979. Chromosomal evolution in Primates: tentative phylogeny from *Microcebus murinus* (Prosimian) to man. *Hum. Genet.* **48**:251–314.

Dutrillaux, B., Couturier, J., and Chauvier, G. 1980. Chromosomal evolution of 19 species or subspecies of Cercopithecinae. *Ann. Genet.* **23** (3):133–143.

Dutrillaux, B., Muleris, M., and Couturier, J. 1988. Chromosomal evolution of Cercopithecinae. In: A. Gautier-Hion, F. Bourlière, J.-P. Gautier, and J. Kingdon (eds.), *A Primate Radiation: Evolutionary Biology of the African Guenons*, pp. 150–159. Cambridge University Press, New York.

Elliot, D. G. 1913. *A Review of the Primates*, 3 vols. American Museum of Natural History, New York.

Gautier, J.-P. 1985. Quelques caracteristiques ecologiques du singe des marais, *Allenopithecus nigroviridis* Lang 1923. *Rev.'Ecol. (Terre Vie).* **40**:331–342.

Gebo, D. L., and Sargis, E. J. 1994. Terrestrial adaptations in the postcranial skeletons of guenons. *Am. J. Phys. Anthro.* **93**:341–371.

Gray, J. E. 1870. *Catalogue of Monkeys, Lemurs, and Fruit-eating Bats in the Collection of the British Museum*, British Museum Trustees, London.

Groves, C. P. 1989. *A Theory of Human and Primate Evolution*. Oxford University Press, New York.

Groves, C. P. 2000. The Phylogeny of the Cercopithecoidea. In: P. F. Whitehead, and C. J. Jolly (eds.), *Old World Monkeys*, pp. 77–98. Cambridge University Press, New York.

Hartl, D. L., and Clark, A. G. 1997. *Principles of Population Genetics*. 3rd Ed. Sinauer Associates, Inc. Sunderland, Massachusetts.

Kishino, H., and Hasegawa, M. 1989. Evaluation of the maximum likelihood estimate of the evolutionary tree topologies from DNA sequence data, and the branching order in Hominoidea. *J. Mol. Evol.* **29**:170–179.

Ledbetter, D. H. 1981. *Chromosomal Evolution and Speciation in the Genus* Cercopithecus *(Primates, Cercopithecinae)*. Ph.D. Thesis, University of Texas, Austin.

Miyamoto, M. M., and Fitch, W. M. 1995. Testing species phylogenies and phylogenetic methods with congruence. *Syst. Biol.* **44** (1):64–76.

Pamilo, P. and Nei, M. 1988. Relationships between gene trees and species trees. *Mol. Biol. Evol.* **5** (5):568–583.

Rokas, A., and Holland, P. W. H. 2000. Rare genomic changes as a tool for phylogenetics. *Tree* **15** (11):454–459.

Rowell, T. E. 1988. The social system of guenons, compared with baboons, macaques and mangabeys. In: A. Gautier-Hion, F. Bourlière, J.-P. Gautier, and J. Kingdon (eds.), *A Primate Radiation: Evolutionary Biology of the African Guenons*, pp. 439–451. Cambridge University Press, New York.

Ruvolo, M. 1988. Genetic evolution in the African guenons. In: A., Gautier-Hion, F. Bourlière, J.-P., Gautier, and J. Kingdon (eds.), *A Primate Radiation: Evolutionary Biology of the African Guenons*, pp. 127–149. Cambridge University Press, New York.

Strasser, E. and Delson, E. 1987. Cladistic analysis of cercopithecid relationships. *J. Hum. Evol.* **16**:81–99.

Struhsaker, T. T., Butynski, T. M., and Lwanga, J. S. 1988. Hybridization between redtail (*Cercopithecus ascanius schmidti*) and blue (*C. mitis stuhlmanni*) monkeys in the Kibale Forest, Uganda. In: A. Gautier-Hion, F. Bourlière, J.-P. Gautier, and J. Kingdon (eds.), *A Primate Radiation: Evolutionary Biology of the African Guenons*, pp. 477–497. Cambridge University Press, New York.

Swofford, D. L. 1999. *PAUP: Phylogenetic Analysis Using Parsimony. Version 4.0b2*. Illinois Natural History Survey. Champaign, Illinois.

Szalay, F. S., and Delson, E. 1979. *Evolutionary History of the Primates*. Academic Press, New York.

Tosi, A. J., Morales, J. C., and Melnick, D. J. 2000. Comparison of Y-chromosome and mtDNA phylogenies leads to unique inferences of macaque evolutionary history. *Mol. Phyl. Evol.* **17** (2):133–144.

Tosi, A. J., Buzzard, P. J., Morales, J. C., and Melnick, D. J. 2002a. Y-chromosome data and tribal affiliations of *Allenopithecus* and *Miopithecus*. *Int. J. Primatol.* **23** (6).

Tosi, A. J., Morales, J. C., and Melnick, D. J. 2002b. *Macaca fascicularis* Y-chromosome and mitochondrial markers indicate introgression with Indochinese rhesus, and a biogeographic barrier in the Isthmus of Kra. *Int. J. Primatol.* **23**:161–178.

Trouessart, E. L. 1897. Catalogus Mammalium tam Viventium quam Fossilium.

Molecular Timescale and Gene Tree Incongruence in the Guenons

3

TODD R. DISOTELL and RYAN L. RAAUM

Introduction

Guenons are well known for substantial variation in chromosome number and morphology (Ledbetter, 1981; Muleris *et al.*, 1986). While extensive multi-locus DNA sequence studies examining hominoid (Ruvolo, 1997; Satta *et al.*, 2000) and papionin (Harris and Disotell, 1998) phylogenetic relationships have been carried out, only a few sequences from multiple cercopithecin taxa have been collected. These limited studies have not examined a sufficient number of genetic systems nor enough taxa to elucidate the overall patterns of guenon phylogeny (Schätzl *et al.*, 1995; Page and Goodman, 2001). Results from an ongoing study of mitochondrial sequence variation in guenons are presented here along with estimates of the date of origin of the guenon radiation. The discussion focuses on variation in these inferred dates and recurring issues in guenon phylogeny.

Few studies including substantial numbers of guenon species have been conducted on genes or gene products, with the notable exceptions of Ruvolo's protein electrophoretic study (Ruvolo, 1988, reviewed in Disotell, 1996) and

TODD R. DISOTELL and RYAN L. RAAUM • Department of Anthropology, New York University, 25 Waverly Place, New York, NY 10003, USA.

The Guenons: Diversity and Adaptation in African Monkeys, edited by Glenn and Cords. Kluwer Academic/Plenum Publishers, New York, 2002.

the work of van der Kuyl and her colleagues on mitochondrial 12s rRNA (van der Kuyl *et al.*, 1995a,b, 2000a,b). The latter work utilized a short sequence of a slowly evolving mitochondrial gene that is well suited to identify the specific and sometimes subspecific identity of an unknown sample (van der Kuyl *et al.*, 2000b). However, these sequences are not very phylogenetically informative within the guenons due to a lack of sufficient variation (Disotell, 2000).

Current estimates of the date of divergence of the extant cercopithecin and papionin lineages range from 9.5 to 10 mya (Szalay and Delson, 1979; Kingdon, 1997). Estimates of the date of divergence of the extant cercopithecins include 6 mya (Szalay and Delson, 1979), 6.5–7 mya (Kingdon, 1997), and less than 3.5 mya (Leakey, 1988). Unfortunately, fossil evidence for cercopithecin evolution is rare, and does not exist before about 3.5 mya (Leakey, 1988). No significant new fossils have been reported since the topic was reviewed by Leakey (1988).

Methods

For phylogenetic analyses, we sequenced an 897 base pair fragment of the mitochondrial genome encompassing the 3′ end of the NADH dehydrogenase subunit (ND) 4, three tRNA genes, and the 5′ end of ND 5 (hereafter referred to as the Brown *et al.* (1982) region after the first author to use this locus for phylogenetic purposes) in 11 species of cercopithecins. This region is bounded by two *Hind*III restriction endonuclease sites in most primates (Brown *et al.*,1982). We included two representatives each of the *mona, cephus,* and *nictitans* species groups as well as a single member each of the *lhoesti, neglectus,* and *aethiops* species groups. We sequenced one individual each of *Allenopithecus*, *Miopithecus*, and *Erythrocebus* as well (Table I). We also sequenced the entire 684 base pair cytochrome oxidase subunit II (COII) gene in a subset of the taxa sequenced for the Brown region (Table I) to strengthen molecular dating estimates for key nodes in the phylogeny.

We used flanking primers previously developed in our lab to amplify the Brown and COII regions (Disotell *et al.*, 1992; Wildman, 2000). We purified PCR products using exonuclease I and shrimp alkaline phosphatase (Hanke and Wink, 1994) and cycle sequenced using ABI BigDye chemistry (Applied Biosystems). We cleaned cycle sequencing products using Centri-Sep spin columns (Princeton Separations) and then electrophoresed them on either an ABI Prism 310 or 377 automated DNA analysis system (Applied Biosystems). We sequenced both strands with an array of internal sequencing primers. We assembled contigs using Sequencher 4.1.2 (Gene Codes).

We carried out phylogenetic analyses on the Brown region via bootstrap parsimony (1000 replicates) with heuristic search algorithms with uniform weighting, the stepwise addition and total branch swapping options,

Table I. Species Sequenced for the Mitochondrial Loci Brown and COII

Latin name	Brown	COII
Allenopithecus nigroviridis	✓	✓
Miopithecus talapoin	✓	✓
Erythrocebus patas	✓	✓
Cercopithecus aethiops	✓	✓
Cercopithecus cephus	✓	✓
Cercopithecus ascanius	✓	✓
Cercopithecus mona	✓	
Cercopithecus pogonias	✓	✓
Cercopithecus lhoesti	✓	✓
Cercopithecus nictitans	✓	
Cercopithecus mitis	✓	✓
Cercopithecus neglectus	✓	✓

and a transition/transversion ratio of 12:1 under PAUP* 4.0b8 (Swofford, 2001). We also performed likelihood analyses (100 replicates) with the same parameter settings and the empirical base frequency model. Varying the parameter settings of the two types of analyses did not essentially change the results. Parsimony and likelihood analyses did not have any conflicts in their resolved clades (Fig. 1).

We calculated divergence date estimates for key nodes within the inferred guenon tree as well as for the hominoid-cercopithecoid and cercopithecin-papionin splits relative to the reasonably well accepted calibration points of *Homo-Pan* at 6 mya and *Theropithecus-Papio* at 5 mya (Fig. 2). We compared sequences from the various clades of Old World monkeys and hominoids to a New World monkey sequence to look for evidence of rate heterogeneity in the SRY and TSPY loci (Tosi *et al.*, 2000), the albumin gene (Page and Goodman, 2001), the mitochondrial 12s rRNA (van der Kuyl *et al.*, 1995a), and the combined COII-Brown regions (this study). Only the 12s rRNA data showed evidence of rate heterogeneity. We applied correction factors to the 12s rRNA estimates (Fig. 2).

Since the three nuclear loci showed relatively small amounts of sequence divergence, we made direct extrapolations based on uncorrected average pairwise nucleotide differences ("*p*" values) between members of each lineage in comparison to the calibration points' pairwise distances (Fig. 2). Earlier studies of protein coding mitochondrial genes demonstrated that third position codon bases show evidence of multiple hit saturation effects (Adkins *et al.*, 1996; Yoder *et al.*, 1996). Therefore, for the combined COII and Brown data set, we made an additional calculation in which we excluded noncoding and third codon position bases from our calculation of uncorrected pairwise distances and their corresponding divergence date estimates (Fig. 2).

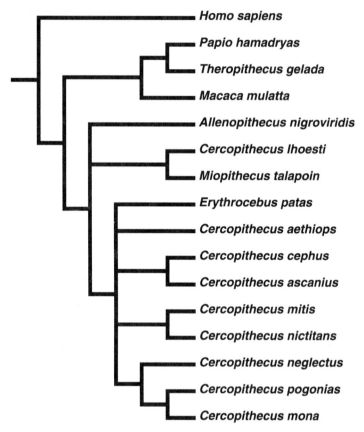

Fig. 1. Consensus tree from our study of the mitochondrial Brown and COII loci. This tree presents nodes consistent among 1000 replicate parsimony bootstrap and 100 replicate likelihood bootstrap analyses of the combined sequence dataset. All nodes have 90% or greater bootstrap support under either likelihood or parsimony analysis.

Results

The mitochondrial phylogeny shows a clade composed of the *Cercopithecus* guenons—excluding *Cercopithecus lhoesti*—and *Erythrocebus patas* (Fig. 1). Basal to this are *Allenopithecus nigroviridis* and a clade containing *Cercopithecus lhoesti* and *Miopithecus talapoin*. All resolved clades are supported by at least a 90% bootstrap measure under either parsimony or likelihood conditions.

The most internally consistent DNA sequence data for date estimation is the COII and Brown combined data excluding third codon positions. Dating analyses (Fig. 2) support a date of divergence between the papionins and cercopithecins between 10.5 and 13.5 mya and an origin for the guenon

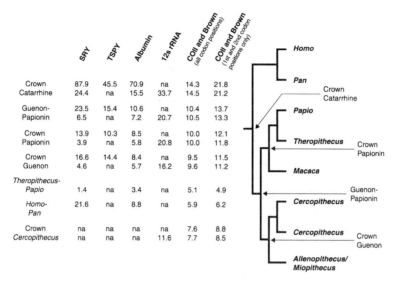

	SRY	TSPY	Albumin	12s-rRNA	COII and Brown (all codon positions)	COII and Brown (1st and 2nd codon positions only)
Crown Catarrhine	87.9	45.5	70.9	na	14.3	21.8
	24.4	na	15.5	33.7	14.5	21.2
Guenon-Papionin	23.5	15.4	10.6	na	10.4	13.7
	6.5	na	7.2	20.7	10.5	13.3
Crown Papionin	13.9	10.3	8.5	na	10.0	12.1
	3.9	na	5.8	20.8	10.0	11.8
Crown Guenon	16.6	14.4	8.4	na	9.5	11.5
	4.6	na	5.7	16.2	9.6	11.2
Theropithecus-Papio	1.4	na	3.4	na	5.1	4.9
Homo-Pan	21.6	na	8.8	na	5.9	6.2
Crown Cercopithecus	na	na	na	na	7.6	8.8
	na	na	na	11.6	7.7	8.5

Fig. 2. Date estimates of selected Old World monkey divergences. The table on the left shows estimated times of divergence for six sequence datasets at selected nodes, which are illustrated in the tree diagram to the right. Each node has been estimated from two calibration points. The top number results from calibration on the *Theropithecus-Papio* node set at 5 mya. The bottom number results from calibration on the *Homo-Pan* node set at 6 mya; na indicates that a date could not be calculated for that node based upon the sequences available.

radiation beginning at least 9.5 mya, and more likely over 11 mya. The genus *Cercopithecus* (excluding *C. lhoesti*) most likely began to diversify between 7.5 and 8.5 mya.

Discussion

The mitochondrial phylogeny (Fig. 1) conflicts with the Y chromosome phylogeny of Tosi *et al.* (2002). The most significant conflict concerns the placement of *Miopithecus talapoin, Erythrocebus patas, Cercopithecus aethiops,* and *C. lhoesti.* While the Y chromosome gene tree supports a clade containing *Erythrocebus patas, Cercopithecus aethiops,* and *C. lhoesti,* the mtDNA sequences support a contradictory placement of *Miopithecus talapoin* and *C. lhoesti* as sister taxa.

Incongruence between cytoplasmic and nuclear gene trees is not a new phenomenon. It has been seen before in the primates (*Macaca*: Tosi *et al.*, 2000) as well as many other animal and plant groups (*Drosophila*: Goto and Kimura,

2001; the beetle subgenus *Ohomopterus*: Sota and Vogler, 2001; and the legume subgenus *Glycine*: Doyle *et al.*, 1999). Tosi *et al.* (2000) attributed the incongruence between mtDNA and Y chromosome gene trees in the macaques to lineage sorting of ancestral mtDNA haplotypes with subsequent fixation of different variants in different lineages. Other studies documenting such incongruence have attributed it to hybridization (Sota and Vogler, 2001), lateral gene transfer (Doyle *et al.*, 1999), or inaccurate phylogenetic inference due to homoplasy (Goto and Kimura, 2001).

Furthermore, incongruence could be the result of inaccurate gene tree inference. In either the mtDNA or Y chromosome data sets, one or more of these species could be represented by non-orthologous sequences. Non-orthologous mtDNA sequences are usually nuclear insertions of mitochondrial sequences, or "numts" (Collura and Stewart, 1995; Bensasson *et al.*, 2001). Our laboratory methods, including mitochondrial enrichment of source DNA, the sequencing of multiple clones, and the use of very long PCR fragments, were designed to avoid such problems. None of the sequences analyzed contained premature stop codons in protein coding sequences or insertion or deletion events common in "numts". Furthermore, the substitution pattern in which third codon position substitutions are most common, followed by first position and then second position substitutions in protein coding sequences, along with the observed transition/transversion bias, all suggest these are true mtDNA sequences. Therefore, the incongruence between mtDNA and Y chromosome is probably not due to nuclear insertions of mitochondrial DNA.

The mtDNA regions sequenced have also been extensively studied in several groups of primates and other mammals (Brown *et al.*, 1982; Disotell *et al.*, 1992; Adkins and Honeycutt, 1994; Adkins *et al.*, 1996; Wildman, 2000). These new guenon sequences show no evidence of excessive homoplasy which could lead to inaccurate phylogenetic inferences (Goto and Kimura, 2001). Without evidence to the contrary, we accept that the mtDNA and Y chromosome evidence produce gene trees which accurately reflect the evolutionary history of those molecules, and we cannot explain away their incongruence by homoplasy. More sequence information from either the mitochondrial genome or the Y chromosome is unlikely to affect the phylogenetic conclusions already drawn from the loci presented here or in Tosi *et al.* (2002).

The most likely reasons for this phylogenetic incongruence are hybridization or lineage sorting. Hybridization may be viewed along a continuum ranging from full hybrid origin of a descendent species from the merger of two parental lines to the introgression of foreign genetic material through occasional cross-species mating. Both ends of this continuum are theoretically possible within the guenons. Hybridization has been observed between many guenon taxa in the wild, and in exceptional circumstances may be common (Detwiler, 2002). Whatever the extent of hybridization, the result will be that variant proportions of the descendent species' genome will derive from more than one lineage.

Lineage sorting may result when an ancestral population is polymorphic for a given locus. If the different alleles at a locus are preferentially preserved in different descendent populations, then the gene tree for the locus will differ from the true species tree. Factors conducive to lineage sorting are large ancestral effective population size and short internodes between speciation events. Both of these situations are likely to have occurred in guenon evolutionary history. In their discussion of mtDNA and Y chromosome discordance in macaques, Tosi *et al.* (2000) have argued that female philopatry and male dispersal leads to an increased effective population size for the mitochondrial genome and a decreased effective population size for the Y chromosome. Thus, they argued that lineage sorting was more likely to have occurred in mtDNA lineages than in Y chromosome lineages.

Incongruence cannot be resolved with two loci. Sampling additional individuals from throughout the species' ranges may help to identify additional alleles and to infer ancient effective population sizes. More importantly, multiple independent autosomal loci are needed. They need to be either of sufficient length or of a rapid enough rate of evolution so that they contain an adequate number of phylogenetically informative sites in order to avoid inaccurate gene tree inferences. Several laboratories, including our own, are attempting to collect such data. The conflicts between the mtDNA and Y chromosomal phylogenies do not affect the divergence date estimates inferred from the mtDNA data presented herein because none of the contested nodes are being estimated (Fig. 2).

While only the mitochondrial 12s rRNA region showed evidence of rate heterogeneity based on the relative rate test, the divergence date estimates derived from the autosomal loci (SRY, TSPY, albumin) vary widely from each other. They also vary within a locus depending on which calibration point is used. We believe this variation, both overestimating and underestimating what are realistic divergence dates, is most likely caused by the small number of substitutions observed between most lineages. For instance, there are only three, 18, and 13 substitutions observed between *Papio* and *Theropithecus* for the SRY, TSP, and albumin loci, respectively. Extrapolating from such small amounts of change is likely to yield variable results. Much longer sequences of non-coding nuclear loci will be needed to match the levels of variation found in the mitochondrial genome. For comparison, the combined COII and Brown region, which is shorter than either the albumin or TSPY regions sequenced, has 123 substitutions between *Papio* and *Theropithecus*.

The combined COII and Brown sequences underestimate the actual times for the older divergences. This underestimate is caused by the well-characterized phenomenon of multiple hit saturation in rapidly evolving genes. When the most variable third codon positions are removed from the analysis, reasonable divergence estimates emerge. These dates are older than those proposed by paleontologists. Given the extreme paucity of fossil evidence, and our inability to infer rates of morphological change, the dates for the key cercopithecine divergences based on the first and second codon positions of the COII and Brown regions may indeed be a more accurate reflection of reality.

Summary

While the phylogenetic inferences based on either the mitochondrial or Y chromosome data presented in this volume may not fully represent the species' phylogeny, the dates of key divergence points can be relatively confidently inferred. These dates are older than those suggested by extrapolation from the fossil record. Multiple independent autosomal loci from several individuals throughout a species' range will need to be sequenced in order to provide a robust phylogeny of the guenons.

ACKNOWLEDGMENTS

This work was supported by NSF grant SBR 9506892 and NIH grant R01 GM60760. We would like to thank Caro-Beth Stewart, Cliff Jolly, Derek Wildman, Paul Telfer, Colleen Noviello, Kirstin Sterner, Anthony Tosi, and Kate Detwiler for assistance, samples, discussion, and access to manuscripts.

References

Adkins, R. M., and Honeycutt, R. L. 1994. Evolution of the primate cytochrome c oxidase subunit II gene. *J. Mol. Evol.* **38**:215–231.

Adkins, R. M., Honeycutt, R. L., and Disotell, T. R. 1996. Evolution of eutherian cytochrome c oxidase subunit II: heterogeneous rates of protein evolution and altered interaction with cytochrome c. *Mol. Biol. Evol.* **13**:1393–1404.

Bensasson, D., Zhang, D.-X., Hartl, D. L., and Hewitt, G. M. 2001. Mitochondrial pseudogenes: evolution's misplaced witnesses. *Trends Ecol. Evol.* **16**:314–321.

Brown, W. M., Prager, E. M., Wang, A., and Wilson, A. C. 1982. Mitochondrial DNA sequences of primates: tempo and mode of evolution. *J. Mol. Evol.* **18**:225–239.

Collura, R. V., and Stewart, C.-B. 1995. Insertions and duplications of mtDNA in the nuclear genomes of Old World monkeys and hominoids. *Nature* **378**:485–489.

Detwiler, K. M. 2002. Hybridization between red-tailed monkeys (*Cercopithecus ascanius*) and blue monkeys (*C. mitis*) in East African forests. In: M. E. Glenn, and M. Cords (eds.), *The Guenons: Diversity and Adaptation in African Monkeys*, pp. 79–97. Kluwer Academic Publishers, New York.

Disotell, T. R. 1996. The phylogeny of Old World monkeys. *Evol. Anthropol.* **5**:18–24.

Disotell, T. R. 2000. The molecular systematics of the Cercopithecidae. In: P. F. Whitehead, and C. J. Jolly (eds.), *Old World Monkeys*, pp. 29–56. Cambridge University Press, Cambridge.

Disotell, T. R., Honeycutt, R. L., and Ruvolo, M. 1992. Mitochondrial DNA phylogeny of the Old World monkey tribe Papionini. *Mol. Biol. Evol.* **9**:1–13.

Doyle, J. J., Doyle, J. L., and Brown, A. H. 1999. Incongruence in the diploid B-genome species complex of glycine (Leguminosae) revisited: histone H3-D alleles versus chloroplast haplotypes. *Mol. Biol. Evol.* **16**:354–362.

Goto, S. G., and Kimura, M. T. 2001. Phylogenetic utility of mitochondrial *COI* and nuclear *Gpdh* genes in *Drosophila*. *Mol. Phylogenet. Evol.* **18**:404–422.

Hanke, M., and Wink, M. 1994. Direct DNA sequencing of PCR-amplified vector inserts following enzymatic degradation of primer and dNTPs. *Biotechniques* **18**:858–860.

Harris, E. E., and Disotell, T. R. 1998. Nuclear gene trees and the phylogenetic relationships of the mangabeys (Primates: Papionini). *Mol. Biol. Evol.* **15**:892–900.

Kingdon, J. 1997. *The Kingdon Field Guide to African Mammals*. Academic Press, New York.

van der Kuyl, A. C., Kuiken, C. L., Dekker, J. T., and Goudsmit, J. 1995a. Phylogeny of African monkeys based upon mitochondrial 12s rRNA sequences. *J. Mol. Evol.* **40**:173–180.

van der Kuyl, A. C., Kuiken, C. L., Dekker, J. T., Perizonius, W. R., and Goudsmit, J. 1995b. Nuclear counterparts of the cytoplasmic mitochondrial 12s rRNA genes: a problem of ancient DNA and molecular phylogenies. *J. Mol. Evol.* **40**:652–657.

van der Kuyl, A. C., van Gennep, D. R., Dekker, J. T., and Goudsmit, J. 2000a. Routine DNA analysis based on 12s rRNA gene sequencing as a tool in the management of captive primates. *J. Med. Primatol.* **29**:307–315.

van der Kuyl, A. C., Dekker, J. T., and Goudsmit, J. 2000b. Primate genus *Miopithecus*: evidence for the existence of species and subspecies of dwarf guenons based on cellular and endogenous viral sequences. *Mol. Phylogenet. Evol.* **14**:403–413.

Leakey, M. 1988. Fossil evidence for the evolution of the guenons. In: A. Gautier-Hion, F. Bourlière, J.-P. Gautier, and J. Kingdon (eds.), *A Primate Radiation: Evolutionary Biology of the African Guenons*, pp. 7–12. Cambridge University Press, Cambridge.

Ledbetter, D. H. 1981. *Chromosomal Evolution and Speciation in the Genus Cercopithecus (Primates, Cercopithecinae)*. Ph.D. Thesis, The University of Texas at Austin, Austin, Texas.

Muleris, M., Couturier, J., and Dutrillaux, B. 1986. Phylogénie chromosomique des Cercopithecoidea. *Mammalia* **50**:38–52.

Page, S. L., and Goodman, M. 2001. Catarrhine phylogeny: noncoding DNA revidence for a diphyletic origin of the mangabeys and for a human-chimpanzee clade. *Mol. Phylogenet. Evol.* **18**:14–25.

Ruvolo, M. 1988. Genetic evolution in the African guenons. In: A. Gautier-Hion, F. Bourlière, J.-P. Gautier, and J. Kingdon (eds.), *A Primate Radiation: Evolutionary Biology of the African Guenons*, pp. 127–139. Cambridge University Press, Cambridge.

Ruvolo, M. 1997. Molecular phylogeny of the hominoids: inferences from multiple independent DNA data sets. *Mol. Biol. Evol.* **14**:248–265.

Satta, Y., Klein, J., and Takahata, N. 2000. DNA archives and our nearest relative: the trichotomy problem revisited. *Mol. Phylogenet. Evol.* **14**:259–275.

Schätzl, H. M., Da Costa, M., Taylor, L., Cohen, F. E., and Prusiner, S. B. 1995. Prion protein gene variation among primates. *J. Mol. Biol.* **245**:362–374.

Shultz, A. H. 1970. The comparative uniformity of the Cercopithecoidea. In: J. R. Napier, and P. H. Napier (eds.), *Old World Monkeys*, pp. 39–51. Academic Press, New York.

Sota, T., and Vogler, A. P. 2001. Incongruence of mitochondrial and nuclear gene trees in the carabid beetles *Ohomopterus*. *Syst. Biol.* **50**:39–59.

Swofford, D. L. 2001. PAUP*. Phylogenetic analysis using parsimony (*and other methods). Version 4. Sinauer Associates, Sunderland, Massachusetts.

Szalay, F. S., and Delson, E. 1979. *Evolutionary History of the Primates*. Academic Press, New York.

Tosi, A. J., Morales, J. C., and Melnick, D. J. 2000. Comparison of Y chromosome and mtDNA phylogenies leads to unique inferences of macaque evolutionary history. *Mol. Phylogenet. Evol.* **17**:133–144.

Tosi, A. J., Buzzard, P. J., Morales, J. C., and Melnick, D. J. 2002. Y-chromosomal window onto the history of terrestrial adaptation in the Cercopithecini. In: M. E. Glenn, and M. Cords (eds.).

The Guenons: Diversity and Adaptation in African Monkeys, pp. 15–26. Kluwer Academic Publishers, New York.

Wildman, D. E. 2000. *Mammalian Zoogeography of the Arabian Peninsula and Horn of Africa with a Focus on the Cladistic Phylogeography of Hamadryas Baboons (Primates: Papio hamadryas)*. Ph.D. Thesis, New York University, New York.

Yoder, A. D., Vilgalys, R., and Ruvolo, M. 1996. Molecular evolutionary dynamics of cytochrome b in strepsirrhine primates: the phylogenetic significance of third-position transversions. *Mol. Biol. Evol.* **13**:1339–1350.

Phylogeny of the *Cercopithecus lhoesti* Group Revisited: Combining Multiple Character Sets

4

JEAN-PIERRE GAUTIER, RÈGINE VERCAUTEREN DRUBBEL, and PIERRE DELEPORTE

Introduction

Previous phylogenies of cercopithecines (Gautier-Hion *et al.*, 1988) suggested that they are organized into two groups of species. The larger group was uniquely composed of forest forms, mainly arboreal, while the second group contained both savanna and forest forms, all terrestrial except *Miopithecus*. Among this second group, the relative phylogenetic position of the three species composing the *Cercopithecus lhoesti* group (*C. lhoesti* Sclater, 1899; *C. preussi* Matschie, 1898; *C. solatus* Harrison, 1988) is still questioned on the basis of their morphological and behavioral characteristics or their original disjointed distribution in Africa (Gautier-Hion *et al.*, 1999). The three species have gray fur, a white beard and a more or less extended brown saddle on

JEAN-PIERRE GAUTIER and PIERRE DELEPORTE ● UMR 6552, CNRS – Université de Rennes 1, Station Biologique 35380 Paimpont, France. RÈGINE VERCAUTEREN DRUBBEL ● Université Libre de Bruxelles – Anthropologie, CP 192 – B 1050 Bruxelles, Belgium.

The Guenons: Diversity and Adaptation in African Monkeys, edited by Glenn and Cords. Kluwer Academic/Plenum Publishers, New York, 2002.

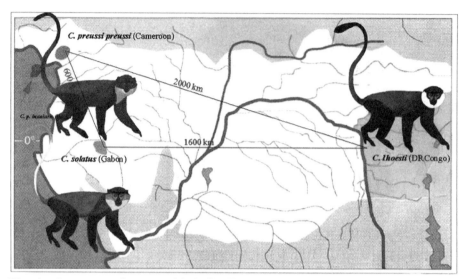

Fig. 1. Patterns and geographical distribution of the three species: *Cercopithecus preussi*, *C. solatus* and *C. lhoesti*.

the back. Only the sun-tailed monkey, *Cercopithecus solatus*, has a bright orange tail tip. Geographically, *Cercopithecus preussi* in Cameroon is separated by 2000 km from *Cercopithecus lhoesti* in Eastern Africa and by 600 km from *C. solatus* in Gabon. *Cercopithecus solatus* and *C. lhoesti* are separated by 1600 km (Fig. 1).

The present paper focuses on four questions: (1) Is the *Cercopithecus lhoesti* group monophyletic? If so, (2) where does it fit within the guenon phylogenetic tree? (3) What are the reciprocal affinities between the three species? and (4) What are the implications of this phylogeny for the interpretation of the evolution of the visual and vocal communication systems and the evolutionary biogeography of the guenons?

Methods

Unlike previous studies, the present phylogenetic analysis took into account several character sets in a single analysis. We simultaneously used and combined in a single data matrix karyological, morphological, behavioral and acoustical characters (Appendix). Among acoustical characters, we included loud calls of the first category defined by Gautier and Gautier-Hion (1977), as the highly species-specific calls displayed by the leader adult male only. We cladistically analyzed the data by Hennig 86 software (Farris, 1988), under option *ie* (finding all most parsimonious solutions) and with unordered characters (all transitions between character states are equally plausible). For multistate characters, we used conventional coding: we treated presence/absence of a given morpho-anatomical trait as a first character (i.e., a given line of cells in the data matrix). We then treated the different qualities of the trait as a second character (i.e., another line with neutral question marks in the cells corresponding to the taxa where the trait was absent) (Hawkins *et al.*, 1997). This method corresponds to the logical principle of using the total available relevant evidence (Kluge, 1989) and to the procedure of simultaneous analysis (Nixon and Carpenter, 1996). Introduction of all heritable characters into the data matrix supporting phylogenetic inference is a logical requirement, and does not impede the use of this same phylogeny for optimizing and testing evolutionary scenarios for the characters (Deleporte, 1993; Grandcolas *et al.*, 2001).

We used species for which we had a sufficient sampling of characters in the relevant categories. We included the following 12 species: *Erythrocebus patas, Cercopithecus solatus, C. preussi preussi, C. lhoesti, C. aethiops pygerythrus, C. pogonias grayi, C. cephus cephus, C. nictitans nictitans, C. neglectus, C. hamlyni hamlyni, C. diana diana, Allenopithecus nigroviridis*. We used a 13th species, *Lophocebus albigena albigena*, as an outgroup for rooting the tree. Our sample represents five arboreal and eight terrestrial species.

Results

The general data matrix for the 13 species is composed of 88 characters (Appendix), including 55 chromosomal mutations (rows 1–55), coded after the work of Dutrillaux *et al.* (1982, 1988a,b), 14 morphological and behavioral characters (rows 56–69), after Kingdon (1980, 1997), Henry (1989) and this study, and 19 acoustical characters (rows 70–88), after Strushaker (1970), Strushaker and Gartlan (1970), Gautier (1988), and this study. With *Lophocebus albigena albigena* being forced as the outgroup, the analysis gave two equally parsimonious rooted cladograms (Fig. 2).

On these cladograms, 13 non-ambiguous synapomorphies support the monophyly of cercopithecines (six karyological characters, five morphological and behavioral, and two acoustical ones). *Allenopithecus nigroviridis* (Allen's swamp monkey) stands as a sister-group to the other Cercopithecini, because the latter are gathered on the basis of five character states (59:9, 68:0, 78:1, 81:1, 87:1), notably the absence of a sexual skin and of a gobble loud call of the first category.

The 11 other species are split in two sister groups. One comprises six forest dwelling species (four arboreal: *Cercopithecus diana, C. cephus, C. nictitans, C. pogonias* and two semi-terrestrial: *C. neglectus* and *C. hamlyni*), and the other includes five species (two savanna forms: *Erythrocebus patas* and *C. aethiops,* and three forest dwelling terrestrial species: *C. solatus, C. lhoesti* and *C. preussi*). Eight characters support the former group: six karyological and two

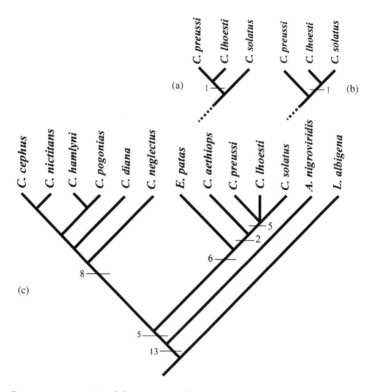

Fig. 2. Consensus tree (c) of the two equally parsimonious trees [(a) and (b)] found in the present study. Numbers on branches: number of non-ambiguous synapomorphies supporting the different clades.

acoustical. Six other characters support the latter group: four karyological, one morphological (presence of a particular color of the neonate), and one acoustical (absence of loud call I). Two karyological characters cluster the species of the second group with the exception of the patas monkey.

The *Cercopithecus lhoesti* group thus appears monophyletic. It is supported by a set of five synapomorphies (characters 53, 54, 63, 66, and 85), including two karyological mutations, one acoustical character, one morphological character (the vivid blue color of the scrotum), and one behavioral character (a tail display; Fig. 3).

In the consensus tree (Fig. 2), no particular affinity is shown between any two out of these three forms. This is explained by the fact that one karyological character (centromeric shift/3) supports the affiliation of *Cercopithecus lhoesti* with *C. preussi*, while one acoustical character (presence of atonality in the low pitch register, 71:1, Fig. 3) sustains the alternative affiliation of *C. lhoesti* with *C. solatus*. Other characters show an autapomorphic state in one of the three species: two centromeric shifts and the color of the tail are specific

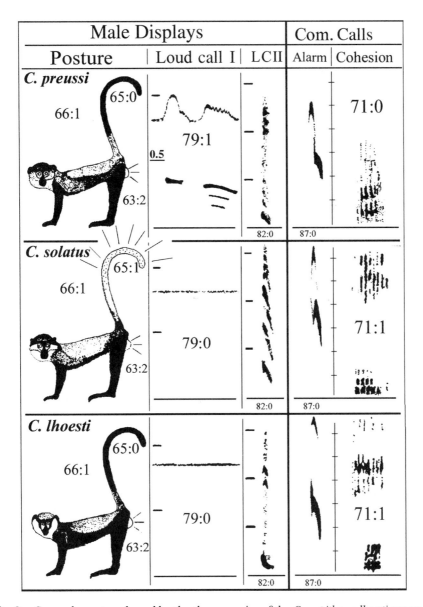

| Male Displays | | | Com. Calls | |
| Posture | Loud call I | LCII | Alarm | Cohesion |

C. preussi
65:0
66:1
79:1
0.5
63:2
82:0 87:0
71:0

C. solatus
65:1
66:1
79:0
63:2
82:0 87:0
71:1

C. lhoesti
65:0
66:1
79:0
63:2
82:0 87:0
71:1

Fig. 3. Some characters shared by the three species of the *Cercopithecus lhoesti* group, or by two of them. Morphological and behavioral characters: 65:0, black color of the tail in *Cercopithecus lhoesti* and *C. preussi*, instead of 65:1, the striking color differentiation in the sun-tailed monkey; 63:2, the vivid blue color of the scrotum; 66:1, posture and display of the tail. Structure of calls: 71:1, tonal and atonal structures of cohesion calls in *Cercopithecus lhoesti* and *C. solatus* instead of 71:0, only tonal structure in *C. preussi*; 82:0, loud call of the second category (L.C.II); 87:0, unitary warning chirp; 79:0, absence of a first category loud call instead of 79:1, presence of a particular first category loud call in *C. preussi*.

autapomorphies of *Cercopithecus solatus* while four acoustical characters are unique derived features in *C. preussi*, like loud call I (Fig. 3). Notably, the third possible phylogenetic combination (affiliating *Cercopithecus preusi* with *C. solatus*) is not supported by any character in our data set. Recent genetic analyses (Bibolet, pers. com.; Tosi *et al.*, 2002) confirmed the karyological analysis of Dutrillaux *et al.* (1988a,b) however, without resolving phylogenetic affinities between the three species.

Discussion

Our simultaneous analysis of multiple traits confirms some of the general phylogenetic trends already drawn from separate analyses (Gautier-Hion *et al.*, 1988).

Allenopithecus is the sister-group of all other guenons in our analysis. This relationship is not supported by the results of Tosi *et al.* (2002), which cannot specify the position of *Allenopithecus* and *Miopithecus*. The clustering of the lhoesti group with *Erythrocebus patas* and *Cercopithecus aethiops* within a clade of terrestrial forms is well supported, and this is confirmed by the analysis of the Y chromosome (Tosi *et al.*, 2002).

No particular affinity is shown between any two out of the three forms of the *lhoesti* group, which allows for possible alternative histories for the evolution of some characters in the *Cercopithecus lhoesti* group. Among the characters shared by *Cercopithecus lhoesti* and *C. preussi* is uniform color of the tail, while *C. solatus* possesses a differentiated tail color. Among the four acoustical characters shared by *Cercopithecus lhoesti* and *C. solatus*, the complex tonal-and-atonal structure of cohesion calls is an original synapomorphy of both species. A second shared character is the great acoustic complexity of calls within the low pitch register which are highly variable in duration, and which comprise tonal, atonal and noisy components. The latter character is symplesiomorphic at the level of the *Cercopithecus lhoesti* group, also being shared with the two closely related savanna species *C. aethiops* and *Erythrocebus patas*. *Cercopithecus preussi*, however, shows a moderately complex structure of its low pitch register and gives cohesion calls analogous to those of *C. nictitans* and *C. cephus*. The most conspicuous acoustical character differentiating *Cercopithecus preussi* from *C. lhoesti* and *C. solatus* is the presence of a loud call of first category in *C. preussi*. Like *Erythrocebus patas* and *Cercopithecus aethiops*, *C. lhoesti* and *C. solatus* never give this type of call. All other forest forms use such loud, species-specific calls, which play a role in spacing mechanisms, while in monkey species living in open habitats, visual displays play this role (Gautier and Gautier-Hion, 1977). Several species belonging to the papionines (e.g., the genera *Lophocebus* or *Cercocebus*) exhibit visual displays even in forest forms, but they are used in association with loud calls. In the three species of the *Cercopithecus lhoesti* group, the visual display is a ritualized posture of the body with a peculiar position of the tail (Fig. 3), which is reinforced by the vivid blue color of the scrotum. A bluish coloration of the scrotum is otherwise mainly observed in semi-terrestrial forest species like *Cercopithecus hamlyni* and *C. neglectus* and in the two savanna forms *Erythrocebus patas* and *C. aethiops*. However, each species in the *Cercopithecus lhoesti* group shows certain particularities. In addition to the coloration of the scrotum and the tail posture displayed by *Cercopithecus lhoesti*, *C. solatus* has a specifically colored tail, while *C. preussi* has a loud call. Moreover, the structure of this call differs from loud calls of all other cercopithecines or colobines (Gautier, 1988, Fig. 12.8). This structural difference is consistent with the autapomorphic and convergent status of this call in *Cercopithecus preussi* as shown on our phylogeny.

No character supports the phylogenetic arrangement in which *Cercopithecus preussi* is a sister species to *C. solatus* despite the fact that these two species show the more proximal geographical distribution, which indicates a complex historical biogeographic scenario for speciation in the group. Strong and lasting geographical barriers linked with fluctuating climatic episodes (Colyn *et al.*, 1991; Gautier *et al.*, 1999) are likely required to explain this paradoxical situation.

Summary

The phylogeny of the forest living, semi terrestrial *Cercopithecus lhoesti* group is revisited, using morphological, behavioral, acoustical and karyological characters in a simultaneous analysis. Results support monophyly of the group, with either *Cercopithecus lhoesti* and *C. preussi*, or *C. lhoesti* and *C. solatus* clustering first. A cluster *Cercopithecus preussi–C. solatus* is not supported, despite their much closer geographic distribution. Concerning the evolution of communication in this group, and despite their forest living habits, their common visual display associating vivid blue scrotum and question mark-shaped posture of the tail recall those of open habitat species. However, *Cercopithecus solatus* and *C. preussi* further evolved a divergent reinforcement of communication signals. An orange tail tip is an original feature of *Cercopithecus solatus*, while the presence of peculiar loud calls in *C. preussi* is functionally convergent with other forest living cercopithecines.

ACKNOWLEDGMENTS

We are much indebted to Dominique Mongy for help in elaborating the different illustrations of this paper, and to Annie Gautier-Hion for helpful criticism of the manuscript.

Appendix. General Matrix Describing Nature and Coding of the 88 Characters for 13 Taxa. Chromosomal mutations: 1–55; morphological and behavioral characters: 56–69; acoustical characters: 70–88

No.	Characteristics of karyo-mutation (1–55)/morpho-behavior (56–69)/ call (70–88)	L. albigena	A. nigroviridis	C. diana	C. hamlyni	C. neglectus	C. nictitans	C. cephus	C. pogonias	C. aethiops	E. patas	C. lhoesti	C. preussi	C. solatus
1	Robertsonian translocation/20;21	0	1	1	1	1	1	1	1	1	1	1	1	1
2	Inversion/6	0	1	1	1	1	1	1	1	1	1	1	1	1
3	Fission/3	0	1	1	1	1	1	1	1	1	1	1	1	1
4	Fission/5	0	1	1	1	1	1	1	1	0	0	0	0	0
5	Inversion/4	0	1	1	1	1	1	1	1	0	0	0	0	0
6	Inversion/1	0	1	0	0	1	0	0	0	0	0	0	0	0
7	Fission/6	0	1	0	0	1	0	0	0	0	0	0	0	0
8	Fission/1	0	0	1	1	1	1	1	1	0	0	0	0	0
9	Fission/1	0	0	1	1	1	1	1	1	0	0	0	0	0
10	Fission/2	0	0	1	1	1	1	1	1	0	0	0	0	0
11	Fission/6	0	0	1	1	1	1	1	1	0	0	0	0	0
12	Fission/8	0	0	1	1	1	1	1	1	0	0	0	0	0
13	Centromeric shift/5	0	0	1	1	1	0	0	0	0	0	0	0	0
14	Centromeric shift/1	0	0	1	0	1	0	0	0	0	0	0	0	0
15	Fission/3	0	0	0	1	1	1	1	1	0	0	0	0	0
16	Centromeric shift/3	0	0	0	0	1	0	0	1	0	0	0	0	0
17	Fission/1	0	0	0	0	1	0	0	0	0	0	0	0	0
18	Inversion/14	0	0	0	0	0	0	0	0	0	0	0	0	0
19	Fission/2	0	0	0	0	1	0	0	0	0	0	0	0	0
20	Other trans-location/1;13	0	0	0	0	1	0	1	0	0	0	0	0	0
21	Inversion/7	0	0	0	0	0	0	1	1	0	0	0	0	0
22	Centromeric shift/2	0	0	0	1	0	0	1	1	0	0	0	0	0
23	Fission/12	0	0	0	1	0	1	1	1	0	0	0	0	0
24	Fission/13	0	0	0	1	0	1	1	1	0	0	0	0	0
25	Fission/4	0	0	0	1	0	1	0	1	0	0	0	0	0
26	Fission/7	0	0	0	1	0	0	0	0	0	0	0	0	0
27	Centromeric shift/1	0	0	0	1	0	0	0	0	0	0	0	0	0

No.	Characteristics of karyo-mutation (1–55)/morpho-behavior (56–69)/ call (70–88)	L. albigena	A. nigroviridis	C. diana	C. hamlyni	C. neglectus	C. nictitans	C. cephus	C. pogonias	C. aethiops	E. patas	C. lhoesti	C. preussi	C. solatus
29	Robertsonian translocation/6;3	0	0	0	1	0	0	0	0	0	0	0	0	0
30	Inversion/7	0	0	0	0	0	0	1	1	0	0	0	0	0
31	Fission/7	0	0	0	0	0	0	0	1	0	0	0	0	0
32	Fission/6	0	0	0	0	0	1	1	1	0	0	0	0	0
33	Centromeric shift/1	0	0	0	0	0	0	0	0	0	0	0	0	1
34	Centromeric shift/5	0	0	0	0	0	0	0	1	0	0	0	0	0
35	Fission/7	0	0	0	0	0	1	0	0	0	0	0	0	0
36	Other trans-location/7;21	1	0	0	0	0	0	0	0	0	0	0	0	0
37	Robertsonian trans-location/20+22	1	0	0	0	0	0	0	0	0	0	0	0	0
38	Centromeric shift/20+21	0	0	0	0	0	0	0	0	1	1	1	1	1
39	Fission/2	0	0	0	0	0	0	0	0	1	1	1	1	1
40	Fission/1	0	0	0	0	0	0	0	0	1	0	1	1	1
41	Centromeric shift/1	0	0	0	0	0	0	0	0	0	0	0	0	0
42	Fission/7	0	0	0	0	0	0	0	0	0	0	0	0	0
43	Fission/6	0	0	0	0	0	0	0	0	0	0	0	0	0
44	Centromeric shift/2	0	0	0	0	0	0	0	0	0	0	0	0	0
45	Fission/6	0	0	0	0	0	0	0	0	1	1	1	1	1
46	Fission/7	0	0	0	0	0	0	0	0	0	1	1	1	1
47	Fission/2	0	0	0	0	0	0	0	0	1	0	1	1	1
48	Centromeric shift/3 A	0	0	0	0	0	0	0	0	0	1	0	0	1
49	Inversion/4	0	0	0	0	0	0	0	0	0	1	0	0	0
50	Centromeric shift/6	0	0	0	0	0	0	0	0	0	1	0	0	0
51	Fission/4	0	0	0	0	0	0	0	0	0	1	0	0	0
52	Centromeric shift/20+21	0	0	0	0	0	0	0	0	1	0	0	0	0
53	Fission/1	0	0	0	0	0	0	0	0	0	0	1	1	1
54	Inversion/9	0	0	0	0	0	0	0	0	0	0	1	1	1
55	Centromeric shift/3 B	0	1	0	0	0	0	0	0	0	0	1	1	0
56	Sexual dimorphism: 0 week/1 high	0	0	0	1	1	0	0	0	0	1	1	1	1

No.	Characteristics of karyo-mutation (1–55)/morpho-behavior (56–69)/call (70–88)	L. albigena	A. nigroviridis	C. diana	C. hamlyni	C. neglectus	C. nictitans	C. cephus	C. pogonias	C. aethiops	E. patas	C. lhoesti	C. preussi	C. solatus
57	Speckled fur: 0/1	0	1	1	1	1	1	1	1	1	1	1	1	1
58	Particular color neonates: 0/1	0	0	0	1	1	0	0	0	1	1	1	1	1
59	Catamenial skin: 0/1	1	1	0	0	0	0	0	0	0	0	0	0	0
60	General difference bottom part: 0/1	0	1	1	1	1	0	0	0	1	1	1	1	1
61	Naked perineal area: 0/1	1	0	0	0	0	0	0	0	0	0	0	0	0
62	Particular color bottom part: 0 no/1 fur/2 skin	0	2	1	2	1	0	0	0	0	1	0	0	0
63	Colored scrotum: 0 no/1 blue/2 vivid blue	0	0	0	2	1	0	0	0	1	1	2	2	2
64	Color of the penis: 0 no difference/1 red	0	1	0	0	0	0	0	0	1	0	0	0	0
65	Color difference tail: 0/1	0	0	0	0	0	0	1	0	0	0	0	0	1
66	Display of the tail: 0 no/1 tip up post./2 tip up ant.	2	0	0	0	0	0	0	0	0	0	1	1	1
67	Size of the naked face: 0 small/1 large	1	0	0	0	0	0	0	0	0	0	0	0	0
68	Earshape: 0 round/1 sharp	1	1	0	0	0	0	0	0	0	0	0	0	0
69	Scent marking: 0/1	0	1	0	1	1	0	0	0	1	0	0	0	0
70	Low-pitched register structuration: 0 low/1 medium/2 high	2	1	1	0	0	1	1	1	2	2	2	1	2
71	LPR: 0 tonal/1 tonal & atonal	0	0	0	0	0	0	0	0	0	0	1	0	1
72	High-pitched register structuration: 0 no/	0	1	2	0	0	1	1	2	0	0	0	0	0

No.	Characteristics of karyo-mutation (1–55)/morpho-behavior (56–69)/call (70–88)	L. albigena	A. nigroviridis	C. diana	C. hamlyni	C. neglectus	C. nictitans	C. cephus	C. pogonias	C. aethiops	E. patas	C. lhoesti	C. preussi	C. solatus
73	LPR: 0 unit/1 multiunit	1	0	0	0	0	0	0	0	0	0	0	0	0
74	LPR: 0 noquav./1 quav./2 noquav. Young & quav. Adult	0	1	1	2	0	1	1	0	1	0	1	1	1
75	LPR different from isolation call: 0/1	0	1	1	1	1	1	1	1	1	1	1	1	1
76	HPR: 0 noquav/1 quav./2 noquav & quav	NS	1	2	2	0	1	1	1	NS	2	1	?	1
77	L&HPR: 0 no/1 assoc./2 merged	0	1	2	0	0	1	1	2	0	0	0	0	0
78	HPR: True contact call: 0 abs/1 pres. Young/2 pres. Y.& Ad.	0	0	2	1	1	2	2	2	0	1	2	?	1
79	Loud call I: 0/1	1	1	1	1	1	1	1	1	0	0	0	1	0
80	LC I: 0 boom/1 near boom/2 boom & barks/3 barks/4 gobble/5 roar	4	4	5	0	2	2	3	2	NS	NS	NS	1	NS
81	LC II: 0/1	0	0	1	1	1	1	1	1	1	1	1	1	1
82	LC II: 0 unit/1 bin/2 multiforme	NS	0	2	0	0	0	0	1	2	1	0	0	0
83	LC II: 0 no ryth. series/1 Low RS/2 High RS	0	NS	2	0	1	2	1	1	2	0	0	0	0
84	LC III: 0/1	0	0	1	?	1	1	1	1	1	?	0	0	0
85	High-pitched warning call: 0 absent/1 present, young/2 present young & adult	1	2	1	0	0	2	2	1	1	1	2	2	2
86	HPWC: 0 whistles/1 chirps	1	1	0	NS	NS	1	1	1	1	?	1	1	1
87	HPWC: 0 unit/1 binary/2 multiunit	2	0	1	NS	NS	0	1	1	0	1	0	0	0
88	HPWC: 0 no series/1 series	0	0	0	0	NS	1	1	0	0	0	0	1	0

References

Colyn, M., Gautier-Hion, A., and Verheyen, W. 1991. A re-appraisal of paleoenvironmental history in Central Africa: Evidence for a major fluvial refuge in the Zaire Basin. *J. Biogeogr.* **18**:403–407.

Deleporte, P. 1993. Characters, attributes and tests of evolutionary scenarios. *Cladistics* **9**:427–432.

Dutrillaux, B., Couturier, J., Muleris, M., Lombard, M., and Chauvier, G. 1982. Chromosomal phylogeny of forty-two species or subspecies of Cercopithecoids (Primates Catarrhini). *Ann. Génét.* **25**:96–109.

Dutrillaux, B., Dutrillaux, A. M., Lombard, M., Gautier, J.-P., Cooper, R., Moysan, F., and Lernoud, J. M. 1988a. The karyotype of *Cercopithecus solatus* Harrison 1988, a new species belonging to *C. lhoesti*, and its phylogenetic relationships with other monkeys. *J. Zool. Lond.* **215**:611–617.

Dutrillaux, B., Muleris, M., and Couturier, J. 1988b. Chromosomal evolution of cercopithecinae. In: A. Gautier-Hion, F. Bourlière, J.-P. Gautier, and J. Kingdon (eds.), *A Primate Radiation: Evolutionary Biology of the African Guenons*, pp. 151–159. Cambridge University Press, Cambridge.

Farris, J. S. 1988. Hennig86, version 1.5. Published by the author, Port Jefferson, NY.

Gautier, J.-P. 1988. Interspecific affinities among guenons as deduced from vocalizations. In: A. Gautier-Hion, F. Bourlière, J.-P. Gautier, and J. Kingdon (eds.), *A Primate Radiation: Evolutionary Biology of the African Guenons*, pp. 194–226. Cambridge University Press, Cambridge.

Gautier, J.-P., and Gautier-Hion, A. 1977. Communication in Old World monkeys. In: S. Sebeok (ed.), *How Animals Communicate*, pp. 890–964. Indiana University Press, Bloomington and London.

Gautier-Hion, A., Bourlière, F., Gautier, J.-P., and Kingdon, J. 1988. *A Primate Radiation: Evolutionary Biology of the African Guenons.* 567p. Cambridge University Press, Cambridge.

Gautier-Hion, A., Colyn, M., and Gautier, J.-P. 1999. *Histoire Naturelle des Primates d'Afrique Centrale.* 162 p. Ecofac, Libreville, Gabon.

Grandcolas, P., Deleporte, P., Desutter-Grandcolas, L., and Daugeron, C. 2001. Phylogenetics and ecology: as many characters as possible should be included in the cladistic analysis. *Cladistics* **17**:104–110.

Harrison, M. J. S. 1988. A new species of guenon (genus Cercopithecus) from Gabon. *J. Zool. Lond.* **215**:561–575.

Hawkins, J. A., Hughes, C. E., and Scotland, R. W. 1997. Primary homology assessment, characters and character states. *Cladistics* **13**:275–283.

Henry, L. 1989. Nature et fonction de la coloration de la peau et du pelage (cas des primates simiens de l'ancien monde). Unpub. DEA Biol. Pop. Eco-Ethologie–Univ. Rennes 1, 21 pp.

Kingdon, J. 1980. The role of visual signals and face patterns in African forest monkeys (guenons). *Trans. Zool. Soc.* **35**:425–475

Kingdon, J. 1997. Lhoest's monkeys *Cercopithecus (l'hoesti)*. *The Kingdon field guide to African Mammals*, AP Natural World, pp. 63–66. Academic Press, New York.

Kluge, A. G. 1989. A concern for evidence and a phylogenetic hypothesis of relationships among Epicrates (Boidae, Serpentes). *Syst. Zool.* **38**:7–25.

Nixon, K. C., and Carpenter, J. M. 1996. On simultaneous analysis. *Cladistics.* **12**:221–241.

Strushaker, T. T. 1970. Phylogenetic implications of some vocalizations of *Cercopithecus* monkeys. In: J. R. Napier, and P. H. Napier (eds.), *Old World Monkeys, Evolution, Systematics and Behaviour*, pp. 367–444. Academic Press, New York.

Strushaker, T. T., and Gartlan, J. S. 1970. Observation on the behavior and ecology of the Patas monkey (*Erythrocebus patas*) in the Waza reserve, Cameroon. *J. Zool. Lond.* **161**:49–63.

Tosi, A. J., Buzzard, P. J., Morales, J. C., and Melnick, D. J. 2002. Y-chromosomal window onto the history of terrestrial adaptation in the Cercopithecini. In: M. E. Glenn, and M. Cords (eds.), *The Guenons: Diversity and Adaptation in African Monkeys*, pp. 15–26. Kluwer Academic Publishers, New York.

Terrestriality and the Maintenance of the Disjunct Geographical Distribution in the *lhoesti* Group

5

BETH A. KAPLIN

Introduction

Cercopithecus monkeys, or guenons, are generally considered forest-dwelling, arboreal and mainly frugivorous primates of African tropical forests (Cords, 1986; Gautier-Hion, 1988). The three species comprising the *lhoesti* group, the l'Hoest's monkey, *Cercopithecus lhoesti*, the Preuss's monkey, *C. preussi* and the sun-tailed monkey, *C. solatus*, although not well studied, are considered semiterrestrial (Gautier-Hion, 1988) or fully terrestrial (Gebo and Sargis, 1994), with behavioral and ecological characteristics divergent from other guenon species. These three species share a close phylogenetic relationship based on morphological and cytogenetic studies (Dutrillaux *et al.*, 1988; Harrison, 1988; Lernould, 1988; Gautier *et al.*, 2002; Tosi *et al.*, 2002) yet are geographically isolated from each other.

All three species (*Cercopithecus lhoesti*, *C. preussi* and *C. solatus*) are restricted mainly to montane and mature lowland forest (Wolfheim, 1983; Lernould,

BETH A. KAPLIN • Department of Environmental Studies, Antioch New England Graduate School, 40 Avon Street, Keene, NH 03431, USA.
The Guenons: Diversity and Adaptation in African Monkeys, edited by Glenn and Cords. Kluwer Academic/Plenum Publishers, New York, 2002.

1988) within a region in which Pleistocene climatic events are believed to have caused successive expansions and contractions of forest cover, leaving some areas as refugia during arid periods (Hamilton, 1981; Colyn *et al.*, 1991). Although a disjunct distribution pattern across the Congo Forest basin is common among other forest organisms (Kingdon, 1974; Diamond and Hamilton, 1980), the mechanism behind this disjunction is not well understood. In particular, the most widely accepted model of African forest refugia does not fully explain the disjunct distribution exhibited today by the *lhoesti* group (Oates, 1988; Haffer, 1990). As climate warmed following deglaciation and forests were able to expand, l'Hoest's, Preuss's and sun-tailed monkey populations, which were presumably in contact at one time, remained isolated and did not spread across the Congo basin as did other guenon species.

Because of their behavioral and ecological distinctions and their disjunct, restricted distribution, the *lhoesti* group is an interesting one in which to examine guenon biogeography and evolutionary history. Why are the three members of this group isolated today given their close phylogenetic relationship? What role does their ecology play in maintaining this isolation? Each guenon species has a unique range of environmental tolerances as a result of a distinct evolutionary history, and these differential tolerances produce different geographical ranges (Oates, 1988). These adaptations and tolerances, intimately linked to aspects of their natural history and ecology, are crucial to understanding the current geographical distribution of the *lhoesti* group. In this chapter, I address the following questions: What factors were responsible for the creation of the disjunct distribution in the *lhoesti* group? What guenons have maintained their disjunct distribution, in contrast to their arboreal congeners with wider geographic distributions? What role does the evolution of a terrestrial lifestyle play in maintaining the disjunct distribution? I present a hypothesis describing the mechanisms maintaining the distribution of the *lhoesti* group using an ecological and paleoecological framework. This is the first attempt at combining recent ecological field data on members of the *lhoesti* group with biogeographical and paleoecological data.

Methods

Biogeographical information for each of these species was collected from various published sources. Forest history and paleoclimatic data were collected from a review of the literature, focusing on the period from the late Miocene (23–5 mya) when the guenons are believed to have begun their evolutionary radiation, through the Pliocene (5–1.9 mya), the Pleistocene (1.9 mya–10 kya), to the Holocene (10 kya–present).

Data on the foraging ecology and habitat use patterns of *Cercopithecus lhoesti* were taken from a study conducted by the author from 1990 to 1992 in the Nyungwe Forest Reserve, Rwanda (Kaplin and Moermond, 2000). Data were collected during five to eight consecutive all-day follows each month of a habituated group of *Cercopithecus lhoesti* using 5 minute scan samples every 15 minutes for activity budget analysis and three 3 minute scan samples for diet composition every 15 minutes. A sympatric group of blue monkeys (*Cercopithecus mitis doggetti*) was studied using the same methods across the same months for comparative purposes. Data for *Cercopithecus lhoesti* were also taken from a three-month study conducted by Tashiro (unpub. data) in 1998 in the Kalinzu Forest, Uganda using scan sampling techniques and all-day follows of a habituated group. A sympatric group of *Cercopithecus mitis* was also studied concurrently for comparative purposes. Data for *Cercopithecus preussi preussi* are from a 1999 survey conducted by Seymour (unpub. data) in Nigeria, and data for *C. solatus* are reported in Brugière *et al.* (1998), Gautier-Hion *et al.* (1999), and Peignot *et al.* (1999).

Results

Geographic Distribution of the lhoesti Group

Cercopithecus lhoesti is separated by 2000 km from *C. preussi* and by 1600 km from *C. solatus*, while *C. preussi* and *C. solatus* are separated by 600 km (Gautier *et al.*, 2002). *Cercopithecus lhoesti* inhabit the East Central forest block between the Lualaba River and the Central African Rift, in the upper eastern Congo Basin (Colyn, 1988). This is medium-altitude and montane forest of the Albertine Rift region, straddling the Democratic Republic of Congo, Rwanda, Burundi, and southwestern Uganda (Lernould, 1988). In the lowland forest of the Congo basin *Cercopithecus lhoesti* is found only on one side of the Zaire River, and the Lindi River may act as a barrier to them downstream (Colyn, 1988). Colyn (1988) considers *Cercopithecus lhoesti* as endemic to this East Central forest region.

Rain forest at medium and high elevation appears to be the preferred habitat of the Preuss's monkey (Oates, 1988). *Cercopithecus preussi preussi* are narrowly restricted to southwestern Cameroon and eastern Nigeria (Oates, 1988). In Cameroon, *Cercopithecus preussi preussi* live on the northwestern side of Mt Cameroon, and in Nigeria they are restricted to the Obudu Plateau and the neighboring forests of Okwangwo Division, Cross River National Park, which reaches an elevation of 1500 m (Oates, pers. com.; Seymour, unpub. report). On the Obudu Plateau, *Cercopithecus preussi preussi* appear to be restricted to gallery forests in gullies and valleys, usually surrounded by grasslands. Although they occasionally move into grassland areas to forage, apparently they do not travel between widely separated forests across grasslands (Seymour, unpub. report). The Obudu Plateau, which receives high annual rainfall, was once covered in montane forest that today has been severely degraded by tree cutting and frequent burning (Oates, 1999). *Cercopithecus preussi insularis* is restricted to the highest elevations of the Gran Caldera of Bioko Island (Oates, 1988), which is only *ca.* 11,000 years old (Moreau, 1966).

Cercopithecus solatus also have a very restricted range centered in dense rain forest in the central Gabon lowlands, in the Forêt des Abeilles and extending into the Lopé Reserve (Harrison, 1988; Gautier-Hion *et al.*, 1999). The Forêt des Abeilles is a primary forest dominated by small valleys and ridgelines; maximum altitude in the north is 400 m and in the south the ridges reach 900 m (Gautier-Hion *et al.*, 1999). Brugière *et al.* (1998) suggested that *Cercopithecus solatus* is associated with dense understory vegetation, which may partly explain its distribution. The Ogooué River probably limits the distribution of *Cercopithecus solatus* in the north and east, and savannah in the northwest (Brugière *et al.*, 1998). Furthermore, Gautier *et al.* (1992) and Brugière *et al.* (1998) believe that *Cercopithecus solatus* avoids high elevation (above 500–550 m asl), and that altitude is the key factor limiting its distribution.

Forest History and the Formation of the Disjunct Distribution

During the late Miocene (11–5 mya), subtropical western and eastern African vegetation was characterized by annually wet lowland rain forest supported by a warm, humid climate (Hamilton, 1988). Conversely, glacial periods in Africa were cool and arid due to a decrease in sea surface temperatures, lower evaporation rates, and decreased southerly monsoons (Hamilton, 1981, 1988; van Zinderen Bakker SR, 1982). Forests retracted to higher locations and possibly to low, wet areas while savanna vegetation replaced the forest cover (Hamilton, 1988), although Colinvaux *et al.* (1996, 2000) and Willis and Whittaker (2000), working in the neotropics, offer recent evidence that low latitude lowland tropical forests were not replaced by savanna vegetation during glacial periods but dominated throughout. For the African continent, the current and well accepted theory is that during glacial phases, remnant forest islands served as refugia for floral and faunal populations, which subsequently expanded when the climate warmed (Booth, 1958; Hamilton, 1982; Oates, 1988).

Plio-Pleistocene environments were characterized by dry, tropical scrub and grassland with limited gallery forest along drainages. Climate became progressively cooler and drier after 2.8 mya as a result of glaciation, with mid-Pliocene vegetation shifting from closed canopy to open savanna vegetation, and arid-adapted flora and fauna expanded as glaciation proceeded (deMenocal, 1995). There have been about 20 cycles of major forest expansion and retreat in Africa during the last 2.3 mya (Hamilton, 1988; deMenocal,

1995). Evidence from pollen core data and the current distribution patterns of forest organisms suggest periods of massive forest reduction in tropical Africa from *ca*. 35,000–8000 years ago (Sowunmi, 1981; Hamilton, 1988).

Pollen data suggest that forests receded to higher elevations and remained during glacial phases as forest islands surrounded by drier regions of savannah or grassland vegetation. On either side of the Congo basin are major refugia that are believed to have persisted during glacial periods. One is in Cameroon and Gabon and another in eastern Democratic Republic of Congo, southwestern Uganda, and Rwanda (Hamilton, 1988). Colyn *et al.* (1991) propose the existence of another refuge, the Major Fluvial Refuge, in the basin of the Zaire River along its tributaries including the left and right banks of the Zaire/Lualaba River system.

During deglaciation phases, pollen records indicate that forest expanded across equatorial Africa connecting eastern and western Africa with continuous forest cover (Hamilton, 1982). The most recent forest expansion began at about 14,000 years ago and was extensive by 11,000 years ago with an increase in rainfall at about 12,500 BP (Hamilton, 1988). However, fossil data, deep sea cores, and climatic models suggest that there has been an overall trend towards greater aridity in Africa in the last 10,000 years (Hamilton, 1988).

The disjunct distribution of the *lhoesti* group could have been created by a secondary range disjunction, in which ranges were once connected during an interglacial phase but became separated in forest islands during glaciation events described above (Diamond and Hamilton, 1980; Hamilton, 1988; Haffer, 1990). Another possible explanation for the wide geographic separation of the *lhoesti* group is a primary range disjunction in which the original form dispersed to an isolated area by crossing a preexisting barrier (Diamond and Hamilton, 1980; Haffer, 1990). During interglacial phases, pioneering dispersal may have been possible by primate populations (Grubb, 1990). Divergence times of the members of the *lhoesti* group are needed to distinguish between a primary or a secondary range disjuncture. At this time, it is not possible to determine if divergence occurred before 70,000 BP, the approximate date of the last interglacial when wet forest was extensive across the Congo basin, or at an earlier time, during the full glacial phase or an earlier interglacial episode. However, the forest refuge model does not provide a complete explanation for the maintenance of the disjunct distribution in the group (Oates, 1988; Haffer, 1990). Following deglaciation the *lhoesti* group did not follow forest expansions and extend their ranges. The present distribution can be understood by examining ecological and behavioral correlates in this group.

Lhoesti *Group Foraging Ecology and Habitat Use*

Available field studies indicate members of the *lhoesti* group spend a considerable amount of time on the ground. Tashiro (unpub. data) found that

Table I.   Summary of Recent Studies on the Foraging Ecology of the *lhoesti* Group. Proportions are based on means across all months.

Species	Location of study	Months of study	Hours of observation	% Time on ground	% Fruit in diet	% Terrestrial herbs in diet
C. lhoesti	Nyungwe Forest,[a] Rwanda	10	574	38	24.5 (9.8)	35.2 (10.2)
C. lhoesti	Kalinzu Forest,[b] Uganda	3	280	?	18.4	10.2
C. solatus	CIRMF, Gabon[c]	14	>200	85.5	19.2	27.5
C. solatus	Forêt des Abeilles,[d] Gabon	?	?	40	?	?

[a]Kaplin and Moermond (2000).
[b]Tashiro, unpub. data.
[c]Peignot *et al.* (1999).
[d]Gautier-Hion *et al.* (1999).

the *Cercopithecus lhoesti* in her study group spent most of their time foraging 0–5 m above the forest floor for invertebrates, terrestrial herbs, and fruits (Table I). *Cercopithecus lhoesti* in the Nyungwe Forest, Rwanda spent 38% of observed activity time on the ground, and foraging was the most common terrestrial activity (Kaplin and Moermond, 2000). Similarly, *Cercopithecus solatus* spend a considerable amount of time on the ground (Table I) (Brugière *et al.*, 1998; Gautier-Hion *et al.*, 1999; Peignot *et al.*, 1999).

Terrestrial herbaceous vegetation is an important component of the *lhoesti* group diet. The diet of *Cercopithecus lhoesti* in the Nyungwe Forest was composed of a mean of 35% terrestrial herbs over the course of the 10 month study (Table I), and terrestrial herbs ranked first in terms of proportion of total diet in every month except two. The majority of the terrestrial herbs consumed by *Cercopithecus lhoesti* were taken from undisturbed forest habitat, or forest that was not recently logged, cleared, or along road edge (Kaplin and Moermond, 2000). Furthermore, terrestrial herbs were the determinant in the use of ground in undisturbed habitats by *Cercopithecus lhoesti*, where terrestrial herbs composed nearly 78% of the diet when the animals were in this habitat type. The *Cercopithecus mitis* in this study were never observed to consume terrestrial herbs and rarely descended to the ground. *Cercopithecus lhoesti* in the Kalinzu Forest, Uganda spent up to 15% of their feeding time consuming terrestrial herbs in one month of the study, while *C. mitis* were never observed to consume terrestrial herbs and rarely descended to the ground (Tashiro, unpub. data). In the Forêt des Abeilles, *Cercopithecus solatus* diet was composed mainly of terrestrial herbs, fruits, and invertebrates that the animals search for in leaves on the ground (Gautier-Hion *et al.*, 1999).

Discussion

Forest retreat during glaciation formed the initial geographic disjunction in the *lhoesti* group. I believe the disjunct distribution is maintained by their behavioral and ecological distinctions, including a terrestrial lifestyle and the consumption of terrestrial herbs as an important component of the diet, and the geography of the region. Within the *Cercopithecus* genus, there are several species that occupy a relatively widespread area across the Congo Basin. For example, the *Cercopithecus ascanius* group occurs in a nearly contiguous distribution from southern Sudan and Central African Republic south to Zambia, west to Angola and east into Kenya. Similarly, the *mitis/albogularis* group occurs in isolated populations but collectively is distributed across central, eastern and southern Africa in a variety of habitats. In contrast, the *lhoesti* group has a limited, disjunct geographic distribution. The central Congo basin, lying between the current population centers of *Cercopithecus solatus* and *C. preussi* to the west, and *C. lhoesti* to the east has no mountains, and during glaciation it was likely an extended dry grassland.

The *lhoesti* group, adapted to exploit the terrestrial herb layer found in some forests, would have been barred from dispersing out of their refugia by the existence of grasslands and riverine or gallery forest. The *lhoesti* group are unique among the *Cercopithecus* monkeys in their use of the forest floor for terrestrial herbs. They possess morphological features indicating adaptations to a terrestrial habitat (Kingdon, 1988; Gebo and Sargis, 1994) and their dental morphology suggests they are much more folivorous than other forest guenons (Kay and Hylander, 1978). It is likely that species with disjunct distributions in forest refugia have a lesser ability to deal with open grasslands and arid environments. Other *Cercopithecus* species, with greater arboreal mobility, had the ability to migrate using gallery and riverine forest corridors where fruits and tree leaves are available. Tree leaves have been shown to be an important component in the diets of some *Cercopithecus* species, especially where fruits are scarce or rare (Gautier-Hion, 1983; Lawes, 1991). But riverine and gallery forest may not support the density or species composition of terrestrial herbs the *lhoesti* group members rely on. African rain forests may have a higher density of edible terrestrial herbs than other continents, and the occurrence of this food resource in the ground layer may have influenced the evolution of terrestriality (Wrangham *et al.*, 1992). Montane forests in particular are characterized by open areas associated with the steep terrain where increased sunlight penetration allows growth of a dense layer of herbaceous vegetation, but terrestrial herbs are not distributed homogeneously throughout all forests, and may be a limiting factor in the distribution of the *lhoesti* group.

Wrangham and Peterson (1996) proposed a similar mechanism for the evolution of the bonobo, a species adapted to exploit the rich terrestrial herb layer of certain forests. They hypothesize that chimpanzees and gorillas were once distributed in the lowland forests of the Congo Basin, but during glaciation these forests could not support terrestrial herbs, and gorillas left for remnant forests in the mountains to the east and west (Wrangham and Peterson, 1996). When the glaciers retreated, the chimpanzees, which endured the arid phase, exploited without competition from gorillas the rich herb layer that developed, leading to the evolution of the bonobo (Wrangham and Peterson, 1996).

The foraging behavior of the *lhoesti* group differs from that of terrestrial primate species believed to be adapted to disturbed zones. Such species are considered forest-edge or savanna dwellers preadapted to open, meadow-like communities created by clear-cutting, grazing, and crop planting. The rhesus macaques (*Macaca mulatta*) in northwest Pakistan are an example; they are terrestrial herb feeders (mainly grasses and clover) that actively select disturbed sites for foraging and are believed to have adapted to disturbed zones (Goldstein and Richard, 1989; Richard *et al.*, 1989). The terrestrial herbs upon which the *lhoesti* group focus are forest herbs, not grassland herbs from human-dominated landscapes, found both in the forest interior (gaps and openings) and along road edges and other disturbed areas embedded in a forest matrix (Tashiro, pers. com.; Butynski, 1985; Kaplin and Moermond, 2000).

It is as yet unclear if terrestriality is the derived or ancestral condition of the guenons, but recent work by Tosi *et al.* (2002) suggests that terrestriality may be derived. Through adoption of a more terrestrial lifestyle and adaptations for consuming terrestrial foliage, the *lhoesti* group has become very successful at exploiting the open patches in montane and some lowland forest areas for the nutritious herbs, but they have become restricted to areas that support this resource. Additional comparative studies are needed on the behavioral ecology of the *lhoesti* group and botanical studies of the forests to assess this hypothesis. An understanding of the biogeography and ecology of the *lhoesti* group will help us to predict the impacts of environmental change and to plan for the conservation needs of this group.

Summary

This chapter presents a hypothesis for the maintenance of the current geographical distribution of the *lhoesti* group. I have synthesized recent studies of the behavioral ecology of members of the *lhoesti* group with biogeographical information and forest history data. The geography of the Congo Basin and the history of glaciation in the region created the disjunct distribution in this group. The *lhoesti* group retreated along with their forests to montane regions during cool, arid phases. I hypothesize that their behavioral and ecological distinctions have maintained the disjunct distribution pattern. Specifically, their adaptation to a terrestrial lifestyle and reliance on terrestrial herbs found in forests serves as a limiting factor in their distribution. Additional data are needed to develop this hypothesis, and an understanding of the biogeography and ecology of the *lhoesti* group will help us to predict the impacts of environmental change and to plan for the conservation needs of this group.

ACKNOWLEDGMENTS

I would like to thank Mary Glenn, Marina Cords and Annie Gautier-Hion for inviting me to join the symposium that led to this book, and Nicole Gross-Camp for assisting me in accessing unpublished data and information on these guenons. Comments from Marina Cords, Mary Glenn, James Jordan, and Russell Tuttle greatly improved an earlier version of this manuscript.

References

Booth, A. H. 1958. The Niger, the Volta and the Dahomey Gap as geographical barriers. *Evolution* **12**:48–62.

Brugière, D., Gautier, J.-P., and Lahm, S. 1998. Additional data on the distribution of *Cercopithecus (lhoesti) solatus*. *Folia Primatol.* **69**:331–336.

Butynski, T. M. 1985. Primates and their conservation in the Impenetrable (Bwindi) Forest, Uganda. *Primate Conserv.* **6**:68–72.

Colinvaux, P. A., De Oliveira, P. E., Moreno, J. E., Miller, M. C., and Bush, M. B. 1996. A long pollen record from lowland Amazonia: Forest and cooling in glacial times. *Science* **274**:85–88.

Colinvaux, P. A., De Oliveira, P. E., and Bush, M. B. 2000. Amazonian and neotropical plant communities on glacial time-scales: The failure of the aridity and refuge hypotheses. *Quat. Sci. Rev.* **19**:141.

Colyn, M. M. 1988. Distribution of guenons in the Zaire-Lualaba-Lomami river system. In: A. Gautier-Hion, F. Bourlière, J.-P. Gautier, and J. Kingdon (eds.), *A Primate Radiation: Evolutionary Biology of the African Guenons*, pp. 104–124. Cambridge University Press, Cambridge.

Colyn, M., Gautier-Hion, A., and Verheyen, W. 1991. A re-appraisal of paleoenvironmental history in Central Africa: Evidence for a major fluvial refuge in the Zaire Basin. *J. Biogeogr.* **18**:403–407.

Cords, M. 1986. Interspecific and intraspecific variation in diet of two forest guenons, *Cercopithecus ascanius* and *C. mitis. J. Anim. Ecol.* **55**:811–827.

deMenocal, P. B. 1995. Plio-Pleistocene African climate. *Science* **270**:53–55.

Diamond, A. W., and Hamilton, A. C. 1980. The distribution of forest Passerine birds and quaternary climatic change in tropical Africa. *J. Zool. Lond.* **191**:379–402.

Dutrillaux, B., Muleris, M., and Couturier, J. 1988. Chromosomal evolution of Cercopithecinae. In: A. Gautier-Hion, F. Bourlière, J.-P. Gautier, and J. Kingdon (eds.), *A Primate Radiation: Evolutionary Biology of The African Guenons*, pp. 150–159. Cambridge University Press, Cambridge.

Gautier, J.-P., Moysan, F., Feistner, A. T. C., Loireau, J. N., and Cooper, R. 1992. The distribution of *Cercopithecus (lhoesti) solatus*: An endemic guenon of Gabon. *Terre Vie* **47**:367–381.

Gautier, J.-P., Vercauteran Drubbel, R., and Deleporte, P. 2002. Phylogeny of the *Cercopithecus lhoesti* group revisited: Combining multiple character sets. In: M. E. Glenn, and M. Cords (eds.), *The Guenons: Diversity and Adaptation in African Monkeys*, pp. 37–48. Kluwer Academic Publishers, New York.

Gautier-Hion, A. 1983. Leaf consumption by monkeys in western and eastern Africa: A comparison. *Afr. J. Ecol.* **21**:107–113.

Gautier-Hion, A. 1988. The diet and dietary habits of forest guenons. In: A. Gautier-Hion, F. Bourlière, J.-P. Gautier, and J. Kingdon (eds.), *A Primate Radiation: Evolutionary Biology of the African Guenons*, pp. 257–283. Cambridge University Press, Cambridge.

Gautier-Hion, A., Colyn, M., and Gautier, J.-P. 1999. *Histoire Naturelle des Primates d'Afrique Centrale*. Multipress-Gabon, Libreville.

Gebo, D. L., and Sargis, E. J. 1994. Terrestrial adaptations in the postcranial skeletons of guenons. *Am. J. Phys. Anthropol.* **93**:341–371.

Goldstein, S. J., and Richard, A. F. 1989. Ecology of Rhesus Macaques (*Macaca mulatta*) in Northwest Pakistan. *Int. J. Primatol.* **10**:531–567.

Grubb, P. 1990. Primate geography in the Afro-tropical forest biome. In: G. Peters and R. Hutterer (eds.), *Vertebrates in the Tropics*, pp. 187–214. Alexander Koenig Zoological Research Institute and Zoological Museum, Bonn.

Haffer, J. 1990. Geoscientific aspects of allopatric speciation. In: G. Peters, and R. Hutterer (eds.), *Vertebrates in the Tropics*, pp. 45–61. Alexander Koenig Zoological Research Institute and Zoological Museum, Bonn.

Hamilton, A. C. 1981. The quaternary history of African forests: Its relevance to conservation. *Afr. J. Ecol.* **19**:1–6.

Hamilton, A. C. 1982. *Environmental History of East Africa*. Academic Press, New York.

Hamilton, A. C. 1988. Guenon evolution and forest history. In: A. Gautier-Hion, F. Bourlière, J.-P. Gautier and J. Kingdon (eds.), *A Primate Radiation: Evolutionary Biology of the African Guenons*, pp. 13–34. Cambridge University Press, Cambridge.

Harrison, M. J. S. 1988. A new species of guenon (genus *Cercopithecus*) from Gabon. *J. Zool.* **215**:561–575.

Kaplin, B. A., and Moermond, T. C. 2000. Foraging ecology of the mountain monkey (*Cercopithecus lhoesti*): Implications for its evolutionary history and use of disturbed forest. *Am. J. Primatol.* **50**:227–246.

Kay, R. F., and Hylander, W. L. 1978. The dental structure of mammalian folivores with special reference to primates and phalangeroidea (Marsupialia). In: G. G. Montgomery (ed.), *The Ecology of Arboreal Folivores*, pp. 46–51. Smithsonian Institute Press, Washington, DC.

Kingdon, J. 1974. *East African Mammals Volume I: An Evolution in Africa*. Academic Press, Chicago.

Kingdon, J. 1988. Comparative morphology of hands and feet in the genus *Cercopithecus*. In: A. Gautier-Hion, F. Bourlière, J.-P. Gautier, and J. Kingdon (eds.), *A Primate Radiation: Evolutionary Biology of the African Guenons*, pp. 184–193. Cambridge University Press, Cambridge.

Lawes, M. J. 1991. Diet of samango monkeys (*Cercopithecus mitis erythrarchus*) in the Cape Vidal dune forest, South Africa. *J. Zool.* **224**:149–173.

Lernould, J.-M. 1988. Classification and geographical distribution of guenons: A review. In: A. Gautier-Hion, F. Bourlière, J.-P. Gautier, and J. Kingdon (eds.), *A Primate Radiation: Evolutionary Biology of the African Guenons*, pp. 54–77. Cambridge University Press, Cambridge.

Moreau, R. E. 1966. *The Bird Faunas of Africa and its Islands*. Academic Press, London.

Oates, J. F. 1988. The distribution of Cercopithecus monkeys in West African forests. In: A. Gautier-Hion, F. Bourlière, J.-P. Gautier, and J. Kingdon (eds.), *A Primate Radiation: Evolutionary Biology of the African Guenons*, pp. 79–103. Cambridge University Press, Cambridge.

Oates, J. F. 1999. *Myth and Reality in the Rain Forest*. University of California Press, Berkeley.

Peignot, P., Fontaine, B., and Wickings, E. J. 1999. Habitat exploitation, diet and some data on reproductive behavior in semi-free-ranging colony of *Cercopithecus lhoesti solatus*, a guenon species recently discovered in Gabon. *Folia Primatol.* **70**:29–36.

Richard, A. F., Goldstein, S. J., and Dewar, R. E. 1989. Weed macaques: The evolutionary implications of macaque feeding ecology. *Int. J. Primatol.* **10**:569–594.

Sowunmi, M. A. 1981. Aspects of late quaternary vegetational changes in west Africa. *J. Biogeogr.* **8**:457–474.

Tosi, A. J., Buzzard, P. J., Morales, J. C., and Melnick, D. J. 2002. Y-chromosomal window onto the history of terrestrial adaptation in the *Cercopithecini*. In: M. E. Glenn and M. Cords (eds.), *The Guenons: Diversity and Adaptation in African Monkeys*, pp. 15–26. Kluwer Academic Publishers, New York.

van Zinderen Bakker SR, E. M. 1982. African palaeoenvironments 18,000 yrs BP. In: J. A. Coetzee and E. M. van Zinderen Bakker SR (eds.), *Palaeoecology of Africa and the Surrounding Islands*, pp. 77–99. A.A. Balkema, Rotterdam.

Willis, K. J., and Whittaker, R. J. 2000. The refugial debate. *Science* **287**:1406–1407.

Wolfheim, J. H. 1983. *Primates of the World: Distribution, Abundance, and Conservation*. University of Washington Press, Seattle.

Wrangham, R., and Peterson, D. 1996. *Demonic Males*. Houghton Mifflin Company, Boston.

Wrangham, R. W., Conklin, N. L., Chapman, C. A., and Hunt, K.D. 1992. The significance of fibrous foods for Kibale Forest chimpanzees. In: A. Whiten, and E. M. Widdowson (eds.), *Foraging Strategies and Natural Diets of Monkeys, Apes, and Humans*, pp. 11–18. Clarendon Press, Oxford.

Biogeographic Analysis of Central African Forest Guenons

6

MARC COLYN and PIERRE DELEPORTE

Introduction

With 37 taxa at the subspecific level, guenons constitute half of the central African forest primate fauna. Their taxonomy and distribution is fairly well documented (Dandelot, 1971; Colyn, 1988; Lernould, 1988; Oates, 1988; Grubb, 1990; Gautier-Hion *et al.*, 1999), though new discoveries are expected, as indicated by recent descriptions of new taxa in guenons (Grubb *et al.*, 1999; Colyn, 1999; Kingdon, 1997), red colobus (Grubb and Powell, 1999) and new validations in apes (Gonder *et al.*, 1997; Butynski, 2001). Nevertheless, the number of studies and the generally clear morphological distinctiveness of the taxa make them one of the best documented mammalian models for biogeographic studies in central African forests.

Previous biogeographic studies of African primates presented country species lists and regional communities (Oates, 1986, 1996), or distribution grids for a limited set of taxa (Oates, 1988; Grubb, 1990; Colyn, 1991). Other studies focused on the importance of geographical barriers, refuges, and intergradation zones in limited regions (Kingdon, 1980, 1990; Colyn, 1987,

MARC COLYN and PIERRE DELEPORTE • UMR 6552, CNRS – Université de Rennes 1, Station Biologique, 35380 Paimpont, France.
The Guenons: Diversity and Adaptation in African Monkeys, edited by Glenn and Cords. Kluwer Academic/Plenum Publishers, New York, 2002.

1988, 1991, 1993; Grubb, 1990; Colyn *et al.*, 1991). Most evolutionary interpretations focused on different taxa independently (Grubb, 1990; Kingdon,
1990). Gooder (1991) performed a vicariance analysis with *Colobus*, the
Cercopithecus mona + *C. pogonias* group, *C. cephus*, the *C. lhoesti* group, and
C. nictitans, but he did not systematically consider subspecific taxa and his
study was based on questionable phylogenetic relationships. In this study, we
use a large locality sampling to analyze biogeographic patterns of all Central
African primates, and then of guenons vs. other primates taken separately.
These simultaneous and separate analyses permit a reassessment of the
delineation of faunal regions for forest animals. We also question the possible
historical influence of lowland forest extension changes and of riverine
barriers on faunal distribution. From a methodological point of view, we
discuss the importance of considering different ecological categories within a
taxon for biogeographical analysis.

Methods

Data

We selected 74 localities (forest areas whose simian fauna is well docu-
mented) covering the whole range of central African lowland forests from the
Sanaga River in Cameroon to Western Rift forests in Eastern Democratic
Republic of Congo, Uganda, Rwanda, Burundi and Tanzania (Fig. 1, Map 1;
Appendix 1). Two western Cameroon localities were also included in the
analysis because they contain hybrids with taxa from neighboring central
African localities. We made the choice of localities independent of traditionally
defined faunal regions, but including all blocks of forest between major rivers.
Names like Edea do not refer to the city, but are convenient terms to designate
the neighboring forested area.

The 78 catarrhine taxa (species or subspecies) are listed in Appendix 2.
The sample includes 37 guenon taxa, together with 41 other taxa: eight
mangabeys, 21 colobus, five baboons and mandrills, and seven apes. In cases
of identified zones of secondary intergradation between taxa, we coded the
presence of hybrids in a given locality as the presence of all parent taxa
(Appendix 3). The faunistic list of each locality includes between four and 15
(mean 9.3) taxa (Appendix 1).

Analysis

When classic vicariance biogeography approaches are not possible for lack
of sufficient phylogenetic resolution at the required level (here semi-species
and subspecies), Parsimony Analysis of Endemicity (PAE) (Rosen, 1988) is a

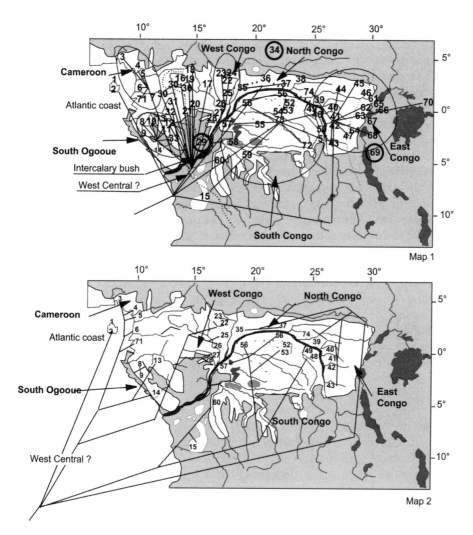

Fig. 1.   Number code for the localities (full description in Appendix 1). Map 1: PAE Analysis 1, 78 taxa, schematic consensus tree of mapped on the 74 localities. Identified faunal areas are named and circled. The three individually circled localities (29, 34, and 69) branch directly at the basal node of the tree. Non-circled localities inside West Central constitute the unresolved intercalary bush. Map 2: Analysis 2, 78 taxa, detailed consensus tree mapped on the 33 downstream localities retained by the geographical filter.

useful preliminary approach to identify areas of endemicity (Morrone and Crisci, 1995) or areas including several localities that share particular taxa. The method makes no use of phylogenetic information, though it attempts to group localities via classic parsimony algorithms of phylogenetic inference software. Unlike the standard phylogenetic data matrix, localities in the PAE biogeographic data matrix take the place of taxa in the rows, and the taxa in these localities take the place of characters in the columns (Rosen, 1988). We analyzed the data matrix using Hennig 86 software (Farris, 1988).

A frequently encountered source of bias in biogeographic studies is that the working biogeographic regions are more or less arbitrarily defined before the analysis (Hovenkamp, 1997). An advantage of our approach to endemicity is that elementary unit localities are used: thus, faunal regions are not pre-defined but will be allowed to emerge from the analysis itself (Rosen, 1988). Another possible source of arbitrariness is the rooting of the PAE cladogram. Rosen (1988) proposed to introduce an arbitrary outgroup locality in the form of an all zeros row (no species present) in the data matrix, without explicit justification. He otherwise suggested using approaches favoring the identification of areas of endemicity on the basis of presence, rather than absence, of taxa. We consider that this recommendation justifies the general use of an all zeros rooting, which should induce the grouping of localities on the basis of the presence of original taxa on stem branches supporting the groups for simple data sets (Deleporte and Colyn, 1999). Accordingly, our analyses using the all zeros rooting scheme effectively produced overwhelming support for faunal areas per presence of taxa rather than their absence.

Following the PAE using the whole data set, we performed additional analyses of subsets of the data. We used geographical filters or ecological filters to discard either a series of localities on the basis of geographical criteria or a series of taxa on the basis of ecological criteria.

We tested the hypothesis that rivers are major geographical barriers influencing the distribution of primates and thus the constitution of faunal regions (Colyn, 1987; Colyn et al., 1991; da Silva and Oren, 1996; Peres et al., 1996). Contacts and dispersal between faunas of adjacent inter-river forest blocks should be more likely in the upstream zones in periods of forest range extension, as attested for African lowland forest primates by the frequent presence of intergradation zones in upstream localities (Colyn, 1991, 1993). The river barriers hypothesis thus predicts that the faunas will be less well differentiated between upstream than downstream localities. We consequently analyzed a subset of the localities by applying a geographical filter to discard all localities situated in upstream zones of the river basins and retain the 33 downstream localities in Appendix 1.

Likewise, the hypothesis of a major influence of riverine barriers is supposed to apply to the fauna effectively arrested by such barriers. We therefore tested the effect of excluding from the analysis the seven typically good-swimming, riparian taxa (*Cercopithecus neglectus*, *Allenopithecus nigroviridis*, *Miopithecus talapoin*, *M. ogoouensis*, *Cercocebus agilis agilis*, *C. a. chrysogaster*, *C. torquatus*). Further, we included *Papio hamadryas tessellatus* in the first analyses because of its presence in forested areas in northeastern Democratic Republic of Congo. However, it is a largely savanna-dwelling species and thus unlikely to respond as forest species would to the biogeographic features we investigated. Accordingly, we excluded it from the ecological filter analysis. Finally, the ubiquitous *Colobus guereza occidentalis*, dwelling in forest, forest edges, savanna and forest patches, was also excluded. In sum, we excluded a total of seven riparian and two ubiquitous taxa.

We conducted seven separate analyses. Analysis 1 included all primates from all localities: 78 taxa, among which 69 are from Central African lowland forests and nine are from Western Rift forests. Analyses 2–7 were performed with a geographical filter, and thus limited to the 33 downstream localities of Central African lowland forests (without the Western Rift). Analysis 2 included all primates. Analysis 3 included only guenons and Analysis 4 included only non-guenon primates. In Analyses 5–7 we applied an ecological filter, excluding the nine riparian or ubiquitous taxa. Analysis 5 included all other species in our sample, while Analysis 6 was limited to guenons and Analysis 7 to non-guenon primates.

We present results as a strict consensus tree of the different most parsimonious trees obtained (*nelsen* option of Hennig 86). When the analysis could not be performed by exhaustive search (option *ie**), we used the best possible heuristic approach (Analysis 1: option *mhennig* with 100 repetitions, randomly changing the order of taxa in the data matrix for each repetition; Analysis 3: option *mhennig-, bb*; Analysis 6 and 7: option *ie-, bb*).

Results

Analysis 1 (all data) gave more than 1200 MPTs (most parsimonious trees) with a consistency index (CI) = 50, and retention index (RI) = 86 (consensus schematically represented in Fig. 1, Map 1). Besides the trivial separation of Bioko Island, it shows four major biogeographic areas, grouping localities on the basis of the shared presence of particular taxa (South Congo, East Congo, West Central and North Congo). The taxa characterizing the stem branch leading to each area are designated below as basally supporting taxa (BST).

First, a South Congo area (left bank of Congo River) is basally supported by the presence of BST *Colobus angolensis angolensis, Lophocebus aterrimus aterrimus, Pan paniscus, Cercopithecus ascanius katangae* and *C. wolfi wolfi*; also, *Cercopithecus ascanius whitesidei* and *Procolobus badius tholloni* support a large subunit in this area, located between the Congo, Kasaï and Lomami Rivers (localities 52, 53, 56, and 57). Second, an East Congo area (right bank) is supported by (BST) *Cercopithecus ascanius schmidti, C. mitis stuhlmanni, C. wolfi denti, C. lhoesti, Papio hamadryas tessellatus, Lophocebus albigena johnstoni* and *Pan troglodytes schweinfurthi*. In this region, a northern subarea may be distinguished by BST *Colobus angolensis cottoni, Cercocebus agilis agilis* and *Procolobus badius brunneus* in contiguous localities 74, 39, 44, 45, and 46. Third, a West Central area from the Atlantic Coast to the right bank of Congo and Ubangi Rivers is clearly identified, with BST *Cercopithecus nictitans nictitans, C. pogonias grayi, Lophocebus*

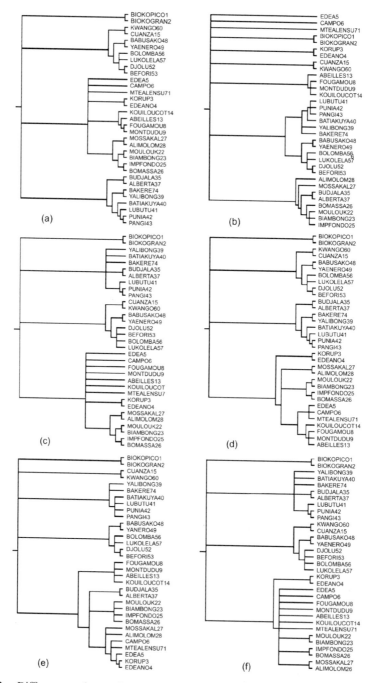

Fig. 2. Different analyses of 33 downstream localities (geographical filter): (a) Analysis 2, all primates; (b) Analysis 3, guenons only; (c) Analysis 4, non guenon primates. The three following analyses were performed with ecological filter: (d) Analysis 5, all primates; (e) Analysis 6, guenons only; (f) Analysis 7, non guenon primates.

albigena albigena, Pan troglodytes troglodytes and *Gorilla gorilla gorilla*. This vast western area presents three subunits (South Ogooue, West Congo and Cameroon) separated by a zone in which biogeographical affinities between localities are completely unresolved and which we call intercalary bush (Fig. 1, Map 1). Two of the subunits are coastal Atlantic areas sharing BST *Colobus satanas anthracinus* and *Mandrillus sphinx*. The first one is Cameroon (BST *Cercocebus torquatus*) and the second one is South Ogooue (BST *Cercopithecus pogonias nigripes, C. cephus cephodes*). The third subunit is the continental West Congo (west from Ubangi River, BST *Cercopithecus cephus ngottoensis*). The fourth major area is North Congo, in the northern Congo Basin (right bank, between Ubangi and Itimbiri Rivers, thus standing between the West Congo and East Congo areas, localities 35–38 with BST *Cercopithecus ascanius schmidti, C. nictitans nictitans, C. pogonias grayi, Lophocebus albigena johnstoni, Procolobus badius schubotzi* and *Pan troglodytes schweinfurthi*). Finally, three localities connect directly at the root of the cladogram, including two peripheral localities (34 and 69) and one containing a limited number of non-informative taxa (29).

Analysis 2 (all primates, downstream) gave seven MPTs [CI = 73, RI = 89, consensus in Fig. 1, Map 2; Fig. 2(a)]. This analysis identified the same four major areas as in Analysis 1, including the West Central area (same BST as in Analysis 1) and subunits South Ogooue and West Congo. Cameroon is not well differentiated, but there is otherwise more resolution of affinities among areas, namely North Congo clustering with East Congo (BST *Cercopithecus ascanius schmidti, Lophocebus albigena johnstoni, Papio hamadryas tessellatus, Pan troglodytes schweinfurthi*).

Analysis 3 (guenons only, downstream) gave three MPTs [CI = 71, RI = 88, consensus in Fig. 2(b)]. Despite the limited number of taxa (four to seven per locality), the cladogram is partly compatible with that of Analysis 2: West Central is not supported, South Ogooue is unchanged (BST *Cercopithecus cephus cephodes, C. pogonias nigripes*), but Cameroon is limited to Korup and Edeano (localities 3 and 4), and West Congo joins with North Congo (BST *Allenopithecus nigroviridis*). South Congo is well supported (BST *Cercopithecus wolfi wolfi, C. ascanius whitesidei, Allenopithecus nigroviridis*) as well as East Congo (BST *Cercopithecus ascanius schmidti, C. mitis stuhlmanni, C. wolfi denti*), but they join together on a weak and very doubtful basis (BST: the very widely distributed *Cercopithecus neglectus*).

Analysis 4 (non guenon primates, downstream) gave 12 MPTs [CI = 83, RI = 94, consensus in Fig. 2(c)]. There are two to eight taxa per locality, and the result is much like that of Analysis 2. West Central is clearly identified (BST *Gorilla gorilla gorilla, Lophocebus albigena albigena, Colobus guereza occidentalis, Pan troglodytes troglodytes*), but Cameroon and South Ogooue are not differentiated. South Congo is well supported (BST *Colobus angolensis angolensis, Lophocebus aterrimus aterrimus, Pan paniscus*), while North Congo joins East Congo (BST *Lophocebus albigena johnstoni, Papio hamadryas tessellatus, Pantroglodytes schweinfurthi*).

Analysis 5 (all primates, downstream, ecological filter) gave three MPTs [CI = 84, RI = 94, consensus in Fig. 2(d)]. The South Central, East Congo and North Congo areas are identified exactly as in Analysis 2 (same BST). West Central is also present (BST *Gorilla gorilla gorilla*) but the subunits are partly different: the Cameroon area is not supported in this analysis because Korup and Edeano separate from all other localities, and the group of remaining localities is supported by BST *Pan troglodytes troglodytes*, *Cercopithecus nictitans nictitans*, *C. pogonias grayi* and *Lophocebus albigena albigena*. West Congo separates into two subunits without clear affinities.

Analysis 6 (guenons, downstream, ecological filter) gave 8 MPTs [CI = 85, RI = 94, consensus in Fig. 2(f)]. As in Analysis 3, South Congo and East Congo areas are still identified (same BST), but they no longer cluster together, because of the elimination of *Cercopithecus neglectus*, the doubtful BST in Analysis 3. West Central is supported as a consistent area (BST *Cercopithecus nictitans*), including its three subunits (South Ogooue, Cameroon and West Congo: same BST as in Analysis 3), joined with North Congo.

Analysis 7 (non guenons primates, downstream, ecological filter) gave 28 MPTs [CI = 90, RI = 96, consensus in Fig. 2(f)]. The result is much like that of Analysis 4, with BST *Gorilla gorilla gorilla* for West Central, and also *Pan troglodytes troglodytes* except for Korup and Edeano.

Discussion

In our analyses (1 and 2) involving all primates, we found major primate biogeographic regions previously identified, notably by Oates (1986, 1996). The classic West Central region appears in our study as two well identified subunits: South Ogooue and West Congo, together with a less well supported Cameroon, these three areas being connected by a large set of localities without structured affinities. East from this, South Congo is clearly identified, as well as East Congo clustering the upper right bank of the river with Western Rift. North Congo, between Ubangi and Itimbiri Rivers, alternatively joins West Central or East Congo. The faunal coherence of these areas was supported by the presence of taxa with large geographic distribution covering a whole region or some of its subunits, particularly in ecological filter analyses (e.g., for West Central: *Cercopithecus nictitans nictitans*, *Gorilla gorilla gorilla* and *Pan troglodytes troglodytes*). North Congo has no exclusive endemic taxon and joins alternatively with West Congo or East Congo in Analyses 3–7. West Congo is identified west from Ubangi River with one endemic, while it has been classically considered to be an intergradation zone (Kingdon, 1980).

These results confirm the importance of riverine barriers for the biogeography of Central African forest primate communities. The more important rivers like Congo, Ubangi and Ogooue obviously constitute major biogeographic barriers. Accordingly, the whole central part of West Central has localities with unresolved affinities, which are effectively situated in the upstream part of river basins, around the line of ridges separating the coastal river basins from the continental Congo Basin. Presently, the primate communities in the continuous forest area between coastal and continental units contain numerous taxa but no exclusive endemic, while hybrids are frequent (Appendix 3). The present communities in the localities of the intercalary bush zone (Fig. 1, Map 1) would be the result of a recent (Holocene) extension of the forest, and ensuing dispersal movements of primates from the three areas Cameroon, South Ogooue and West Congo after an episode of separation of these units by savanna corridors. For example, *Mandrillus sphinx* and *Colobus satanas anthracinus* would have recently dispersed from the Dja River region (in Cameroon) toward West Congo. Otherwise, *Cercopithecus pogonias* hybrids between *C. pogonias* sspl (from Cameroon; see Appendix 2), *C. pogonias grayi* (from West Congo) and *C. pogonias nigripes* (from South Ogooue) also could have resulted from similar dispersal movements (Gautier-Hion *et al.*, 1999).

Our results argue in favor of a geographical filter approach. Some upstream zones are already known as intergradation zones with hybrids (Colyn, 1999). Accordingly, the exclusion of upstream localities (Analysis 2) enhanced the resolution of biogeographic affinities for a fauna influenced by riverine barriers, particularly on the right bank of the Congo River. In the all-primates analyses, the same areas were retrieved with and without an ecological filter, but the consistency of the result was better in the former case (with ecological filter, CI improved from 73 to 84, and RI from 89 to 94). Similarly, the separate analyses of guenons only to the exclusion of the four riparian taxa permitted identification of a West Central area.

On the whole, analyses of data from guenons revealed only some biogeographic patterns similar to those including other primates (e.g., the South Congo and East Congo regions). The West Central area, however, was not clearly identified by guenons alone. This was partly due to the confusing influence of riparian *Cercopithecus neglectus* and *Allenopithecus nigroviridis*. The best resolution of affinities within the West Central area came from the combination of data from all primates. The ecological filter applied to all primates provided a slightly different result inside West Central [Fig. 2(d)], with the two western-most localities (Korup and Edeano, west from Sanaga River) separating from all other ones. We must confess our lasting doubts concerning the coherence of the West Central area, not only because of the questionable status of the western-most localities, but also because of the large amount of unresolved affinities between subunits and intercalary localities. These doubts are reinforced by external information from non-primate forest mammals, also pointing to the complexity of the biogeographic history of this large zone (Colyn, unpub. data).

These analyses suggest several complementary approaches. First, we do not propose to definitely specify biogeographic areas on the basis of primates only; other biological taxa should be considered as well. Second, the analyses would benefit if the database were extended to all of tropical African forest primates: in this context, the westernmost localities like Korup and Edeano might show affinities with immediately neighboring western regions, at least up to the Niger River (with a question for the affinities of Bioko Island). The logic of geographical and ecological filters should be applied to these analyses in accordance with the biogeographic hypotheses being tested, e.g., riverine barriers or past savanna extensions. Third, improving the resolution of phylogenies at a low level would certainly contribute to understanding biogeographic affinities (Hovenkamp, 1997). Finally, basic taxonomy and distribution maps may certainly be improved. Recent descriptions of new primate taxa question the reliability and completeness of current knowledge of the composition of communities. For example, does the Ubangi River strictly separate *Lophocebus albigena albigena* from *L. a. johnstoni*, and *Pan troglodytes troglodytes* from *P. t. schweinfurthi*? There are also indications that *Miopithecus* occurs in northern Central African Republic (Lobão Tello, pers. com.), and three subspecies of guenons and red colobus monkeys remain to be described (Appendix 2). More studies in taxonomy, distribution maps, ecology and phylogeny, and an extension of the study to Western Africa, are needed to improve understanding of faunal biogeographic affinities already indicated by central African primates.

Summary

We analyzed biogeographic patterns of central African forest primates using a large locality sampling. Parsimony analyses of endemicity were performed simultaneously and separately on guenons and other primate fauna, on all localities and taxa, or by applying geographical and ecological filters. Results allow a reassessment of the delineation of forest faunal regions. Classic faunal regions are supported by this new and independent analysis, notably South Congo and East Congo. However, the West Central region reveals an unsuspected complexity, with two subunits South Ogooue and West Congo mixed with an otherwise largely unresolved intergradation zone.

ACKNOWLEDGMENTS

This work was supported by UMR CNRS 6552, University of Rennes 1. We thank John Oates and Annie Gautier-Hion for helpful suggestions, and Mary Glenn, Marina Cords and Russ Tuttle for valuable comments and precious help in improving the English.

Appendix 1. List of Localities with their Codes, Localities, Number of Taxa, Presence of Hybrids and Bibliographic References

Locality codes	Localities	Country	Downstream or upstream localities	Number of taxa	Presence of hybrids	Major bibliographic references
1. BIOKOPICO	Pico Basile, north Bioko	EG	D	4		1
2. BIOKOGRAN	Gran Caldera, southwestern Bioko	EG	D	7		1
3. KORUP	Korup, National Park	Cameroon	D	12	H	2,3,4,5,6,27,28
4. EDEANO	Edea (north)	Cameroon	D	13		2,3,4,5,6,27,28
5. EDEA	Edea (south)	Cameroon	D	14		2,3,4,5,6,[a]
6. CAMPO	Campo	Cameroon	D	12	H	4,6,7,[a]
7. MTALENNO	Moka, north Monte Alen,	EG	U	8	H	4,6,8,[a]
8. FOUGAMOU	Fougamou	Gabon	D	10		4,6,9,[a]
9. MONTDUDU	Mont Dudu	Gabon	D	10		4,6,9,[a]
10. MBIGOU	Mbigou, Eteke, Iboundji	Gabon	U	9		9,10,[a]
11. MOUNANA	Mounana	Gabon	U	8		10
12. KOULAMOU	Koulamoutou	Gabon	U	10		10
13. ABEILLES	South Lopé and Makandé	Gabon	D	9		6,10,11,[a]
14. KOUILCOT	Kouilou River (downstream)	RC	D	9	H	4,6,12,[a]
15. CUANZA	North of Cuanza River	Angola	D	5		2,3,4,13
16. EKOMDJA	Dja Reserve	Cameroon	U	11	H	4,6,[a]
17. YOKADOUM	Yokadouma	Cameroon	U	9		[a]
18. ABONGMBA	Abong-Mbang	Cameroon	U	9		[a]
19. LOMIEDJA	Lomié	Cameroon	U	12	H	[a]
20. ODZALPN	Odzala National Park, south	RC	U	10		4,6,[a]
21. ODZALECO	Mbandza, south eastern Odzala	RC	U	11		14,[a]
22. MOULOUK	Mouloukou, Lobaye River	CAR	D	10		6,[a]
23. BAMBIONG	Bambio, Mbaéré River	CAR	D	10		6,[a]
24. MBAIKING	Mbaïki	CAR	U	9		6,[a]
25. IMPFONDO	Impfondo	RC	D	10		4,6,[a]
26. BOMASSA	Bomassa	RC	D	11		4,6,15,[a]
27. MOSSAKAL	Mossaka, south of Likouala River	RC	D	10		4,6,[a]

(Cont.)

Locality codes	Localities	Country	Downstream or upstream localities	Number of taxa	Presence of hybrids	Major bibliographic references
28. ALIMOLOM	Olombo, south of Alima River,	RC	D	9		4,6,[a]
29. LEFINI	Lefini River	RC	U	4		[a]
30. MINKEBE	Minkébé, Djoum, Mintom II	Gabon	U	13	H	4,6,9,15
31. BELINGA	Bélinga	Gabon	U	11	H	4,6,9,16
32. MOUYABI	Mouyabi	Gabon	U	9	H	10,17
33. KOUILMOS	Kouilou River (upstream), Mossendjo	RC	U	8	H	4,6,12,[a]
34. MANOVO	Manovo St Floris National Park	CAR	U	7		18,[a]
35. BUDJALA	Budjala	DRC	D	9		19,20,21,[a]
36. BUSINGA	Businga	DRC	U	10		19,20,21,[a]
37. ALBERTA	Alberta	DRC	D	9		19,20,21,[a]
38. BONDO	Bondo	DRC	D	10		19,20,21,[a]
39. YALIBONGO	Yalibongo	DRC	D	10		19,20,21,[a]
40. BATIAKUY	Batiakuya	DRC	D	10		19,20,21,[a]
41. LUBUTU	Lubutu	DRC	D	9		19,20,21,[a]
42. PUNIA	Punia	DRC	D	10		19,20,21,[a]
43. PANGI	Pangi	DRC	D	11		19,20,21,[a]
44. NIAPU	Niapu	DRC	U	11		19,20,21,[a]
45. MUNGBERE	Mungbéré	DRC	U	13		19,20,21,[a]
46. MAMBASA	Mambasa	DRC	U	15	H	19,20,21,[a]
47. SHABUNDA	Shabunda	DRC	U	12	H	19,20,21,[a]
48. BABUSAKO	Babusako	DRC	D	8		19,20,21,[a]
49. YAENERO	Yaénéro	DRC	D	8		19,20,21,[a]
50. KASUKU	Kasuku, Lokando	DRC	U	6		19,20,21,[a]
51. LUEKI	Luéki, Kindu	DRC	U	5		19,20,21,[a]
52. DJOLU	Djolu	DRC	D	9		19,20,21,[a]
53. BEFORI	Béfori	DRC	D	9		19,20,21,[a]
54. BOTSIMA	Botsima	DRC	U	7		19,20,21,[a]
55. WAFANIA	Wafania	DRC	U	8		19,20,21,[a]
56. BOLOMBA	Bolomba, Elisabetha	DRC	D	8		19,20,21,[a]
57. LUKOLELA	Lukolela	DRC	D	9		19,20,21,[a]
58. LUKENIE	South of Lukenie River	DRC	U	8		19,20,21,[a]

Locality codes	Localities	Country	Downstream or upstream localities	Number of taxa	Presence of hybrids	Major bibliographic references
59. SANKURU	South of Sankuru River	DRC	U	6		19,20,21,[a]
60. KWANGO	Kwango Region	DRC	D	6		19,20,21,[a]
61. SEMLIKI	Semliki River Forest	DRC, Uganda	U	12	H	22
62. NORDPNV	North of Virunga National Park	DRC	U	10		19,20,21,[a]
63. SUDPNV	South of Virunga National Park	DRC	U	10		19,20,21,[a]
64. KAHUZI	Kahuzi-Biega, National Park	DRC	U	13	H	19,20,21,[a]
65. BWAMBA	Bwamba Forest	Uganda	U	11		23
66. KIBALE	Kibale Forest	Uganda	U	8		24
67. VISOKE	Visoke	Rwanda	U	11		23
68. NYUNGWE	Nyungwe	Rwanda	U	11		[a]
69. TAKATAFO	Takata Forest	Tanzania	U	6		25
70. KAKAMEGA	Kakamega Forest	Kenya	U	5		26
71. MTALENSU	Monte Alen, Bibe River, south	EG	D	11	H	4,6,8,[a]
72. LUBEFU	Lubefu	DRC	U	8		19,20,21,[a]
73. BESOIPNS	Besoï, Salonga National Park	DRC	U	8		19,20,21,[a]
74. BAKERE	Bakéré	DRC	D	10		19,20,21,[a]

[a]This study (field data and museum collections); (1) Butynski & Koster, 1994; (2) Oates, 1996; (4) Lernould, 1988; (5) Gartlan and Struhsaker, 1972; (6) Gautier-Hion et al., 1999; (7) Mitani, 1991; (8) Garcia and Mba, 1997; (9) Blom et al., 1992; (10) Gautier et al.,1992; (11) Brugière and Gautier, 1999; (12) Dowsett and Granjon, 1991; (13) Machado de Barros, 1969; (14) Colyn and Perpète, 1995; (15) Mitani, 1990; (16) Gautier and Gautier-Hion, 1969; (17) Gautier, pers. com.; (18) Fay, 1988; (19) Colyn, 1988; (20) Colyn, 1991; (21) Colyn, 1993; (22) Haddow et al., 1947; (23) Kingdon, 1971; (24) Rudran, 1978; (25) Kano, 1971; (26) Cords, 1987; (27) Gonder et al. 1997; (28) Gagneux et al., 2001.

Appendix 2.   List of 78 Taxa Considered in this Analysis

Central African lowland forests (from the Cross River to the Western Rift in Eastern DRC)

1. *Cercopithecus erythrotis erythrotis* Waterhouse, 1838
2. *Cercopithecus erythrotis camerunensis* Hayman, 1940
3. *Cercopithecus cephus cephus* Linnaeus, 1758
4. *Cercopithecus cephus cephodes* Pocock, 1907
5. *Cercopithecus cephus ngottoensis* Colyn, 1999
6. *Cercopithecus ascanius ascanius* (Audebert, 1799)
7. *Cercopithecus ascanius katangae* Lönnberg, 1919
8. *Cercopithecus ascanius schmidti* Matschie, 1892
9. *Cercopithecus ascanius whitesidei* Thomas, 1909
10. *Cercopithecus neglectus* Schlegel, 1876
11. *Allenopithecus nigroviridis* (Pocock, 1907)
12. *Cercopithecus hamlyni hamlyni* Pocock, 1907
13. *Miopithecus talapoin* (Schreber, 1774)
14. *Miopithecus ogoouensis* Kingdon, 1997
15. *Cercopithecus dryas* Schwarz,1932
16. *Cercopithecus lhoesti* Sclater, 1899
17. *Cercopithecus preussi insularis* Thomas, 1910
18. *Cercopithecus preussi preussi* Matschie, 1898
19. *Cercopithecus solatus* Harrison, 1988
20. *Cercopithecus mitis mitis* Wolf, 1822
21. *Cercopithecus mitis stulhmanni* Matschie, 1893
22. *Cercopithecus mitis heymansi* Colyn et Verheyen, 1987
23. *Cercopithecus nictitans nictitans* Linnaeus, 1766
24. *Cercopithecus nictitans martini* Waterhouse, 1838
25. *Cercopithecus nictitans* ssp1[a]
26. *Cercopithecus mona* (Schreber, 1774)
27. *Cercopithecus pogonias pogonias* Bennett, 1833
28. *Cercopithecus pogonias* ssp1[a]
29. *Cercopithecus pogonias nigripes* Chaillu, 1860
30. *Cercopithecus pogonias grayi* Fraser, 1850
31. *Cercopithecus (pogonias) wolfi wolfi* Meyer, 1891
32. *Cercopithecus (pogonias) wolfi denti* Thomas, 1907
33. *Cercopithecus (pogonias) wolfi elegans* Dubois et Matschie, 1912
34. *Cercopithecus (pogonias) wolfi pyrogaster* Lönnberg, 1919
35. *Cercocebus agilis agilis* Rivière, 1886
36. *Cercocebus agilis chrysogaster* Lydekker, 1900
37. *Cercocebus torquatus* (Kerr, 1792)
38. *Lophocebus albigena albigena* (Gray, 1850)
39. *Lophocebus albigena johnstoni* (Lydekker, 1900)
40. *Lophocebus albigena osmani* Groves, 1978
41. *Lophocebus aterrimus aterrimus* Oudemans, 1890
42. *Lophocebus aterrimus opdenboschi* Schouteden, 1944
43. *Colobus guereza occidentalis* (Rochebrune, 1886–87)
44. *Colobus satanas satanas* Waterhouse, 1838
45. *Colobus satanas anthracinus* Leconte, 1858
46. *Colobus angolensis angolensis* Sclater, 1860
47. *Colobus angolensis cordieri* Rahm, 1959
48. *Colobus angolensis cottoni* Lydekker, 1905

(Cont.)

49. *Procolobus badius pennantii* Waterhouse, 1838
50. *Procolobus badius preussi* Matschie, 1900
51. *Procolobus badius bouvieri* (Rochebrune, 1886–87)
52. *Procolobus badius oustaleti* Trouessart, 1906
53. *Procolobus badius* ssp1[a]
54. *Procolobus badius schubotzi* Matschie, 1914
55. *Procolobus badius brunneus* Lönnberg, 1919
56. *Procolobus badius langi* J.A. Allen, 1925
57. *Procolobus badius lulindicus* Matschie, 1914
58. *Procolobus badius parmentierorum* Colyn et Verheyen, 1987
59. *Procolobus badius tholloni* Milne Edwards, 1886
60. *Mandrillus leucophaeus leucophaeus* (F. Cuvier, 1807)
61. *Mandrillus leucophaeus poensis* Zukowsky, 1922
62. *Mandrillus sphinx* (Linnaeus, 1758)
63. *Papio hamadryas tessellatus* Elliot, 1909
64. *Pan troglodytes troglodytes* Blumenbach, 1779
65. *Pan troglodytes schweinfurthi* Giglioli, 1872
66. *Pan troglodytes vellerosus* Gray, 1862
67. *Pan paniscus* Schwarz, 1929
68. *Gorilla gorilla gorilla* Savage et Wyman, 1847
69. *Gorilla gorilla graueri* Matschie, 1914

Western Rift forests (taxa used only for global Analysis 1 with all localities)
70. *Procolobus badius semlikiensis* Colyn, 1991
71. *Procolobus badius foai* Pousargues, 1899
72. *Procolobus badius tephrosceles* Elliot, 1907
73. *Colobus angolensis ruwenzorii* Thomas, 1901
74. *Cercopithecus hamlyni kahuziensis* Colyn et Rahm, 1987
75. *Cercopithecus mitis kandti* Matschie, 1906
76. *Cercopithecus mitis doggetti* Pocock, 1907
77. *Papio hamadryas kindae* Lönnberg, 1919
78. *Gorilla gorilla beringei* Matschie, 1923

[a]Remarks: *Cercopithecus nictitans* ssp1 from Bioko Island, *Cercopithecus pogonias* ssp1 from the region between Cross and Sanaga Rivers, and *Procolobus badius* ssp1 from Lobayi River basin in C.A.R. are undescribed taxa.

Appendix 3. List of Localities with Hybrids Present [this Study (Field Data and Museum Collections), Lernould (1988) and Gautier-Hion et al. (1999)]

Code	Locality	Hybrid identity
3	Korup	*Cercopithecus mona* × *C. pogonias* ssp1[a]
5	Edea	*C. mona* × *C. pogonias grayi*
5	Edea	*C. p.* ssp1 × *C. p. grayi*
5	Edea	*C. erythrotis camerunensis* × *C. c. cephus*
6	Campo	*C. p.* ssp1 × *C. p. grayi*
7	Monte Alen north	*C. p.* ssp1 × *C. p. grayi*
14	Kouilou coast	*C. p.* ssp1 × *C. p. grayi*
16	Ekom Dja	*Lophocebus a. albigena* × *L. a. osmani*
19	Lomié Dja	*L. a. albigena* × *L. a. osmani*
30	Minkébé	*C. p.* ssp1 × *C. p. grayi*
31	Belinga	*C. p.* ssp1 × *C. p. grayi*
32	Mouyabi	*C. p.* ssp1 × *C. p. grayi*
33	Kouillou Mossendjo	*C. p.* ssp1 × *C. p. grayi*
46	Mambasa	*Procolobus b. brunneus* × *P. b. langi* × *P. b. semlikiensis*
47	Shabunda	*P. b. lulindicus* × *P. b. foai*
61	Semliki	*Colobus angolensis cottoni* × *C. a. ruwenzorii*
64	Kahuzi	*P. b. lulindicus* × *P. b. foai*
71	Monte Alen south	*C. p.* ssp1 × *C. p. grayi*

[a] *C. pogonias* ssp1 from the region between Cross and Sanaga Rivers is an undescribed taxon.

References

Blom, A., Alers, M. P. T., Feistner, A. T. C., Barnes, R. F. W., and Barnes, K. L. 1992. Primates in Gabon: current status and distribution. *Oryx* **26**:223–234.

Brugière, D., and Gautier, J.-P. 1999. Status and conservation of the sun-tailed guenon *Cercopithecus solatus*, Gabon's endemic monkey. *Oryx* **33**:67–74.

Butynski, T. M. 2001. Africa's great apes. In: B. Beck, T. S. Stoinski, M. Hutchins, T. L. Maple, B. Norton, A. Rowan, E. F. Stevens, and A. Arluke (eds.), *Great Apes and Humans: The Ethics of Coexistence*, pp. 3–56. Smithsonian Institution Press, Washington, DC.

Butynski, T. M., and Koster, S. H. 1994. Distribution and conservation statuts of primates in Bioko Island, Equatorial Guinea. *Biodiversity Conserv.* **3**:893–809.

Colyn, M. 1987. Les Primates des forêts ombrophiles de la cuvette du Zaïre: interprétations zoogéographiques des modèles de distribution. *Rev. Zool. afr.* **101**:183–196.

Colyn, M. 1988. The Distribution of Guenon Monkeys in the Lowland Rain Forest of the Zaïre-Lualaba-Lomami River System. In: A. Gautier-Hion, F. Bourlière, J.-P. Gautier, and J. Kingdon (eds.), *A Primate Radiation: Evolutionary Biology of the African Guenons*, pp. 104–124. Cambridge University Press, Cambridge.

Colyn, M. 1991. L'importance géographique du bassin du fleuve Zaïre pour la spéciation: le cas des primates simiens. *Ann. Musée Roy. Afr. Centr. Tervuren*, vol. 264.

Colyn, M. 1993. Coat color polymorphysm of red colobus monkeys (*Colobus badius*, Primates, Colobinae) in Eastern Zaire: Taxonomic and biogeographic implications. *J. Afr. Zool.* **107**: 301–320.

Colyn, M. 1999. Etude populationnelle de la super espèce *Cercopithecus cephus* habitant l'enclave forestière Sangha—Oubangui et description de *C. c. ngottoensis* subsp. nov. *Mammalia* **63**: 137–147.

Colyn, M., Gautier-Hion, A., and Verheyen, W. 1991. A re-appraisal of palaeoenvironmental history in Central Africa: evidence for a major fluvial refuge in the Zaire basin. *J. Biogeogr.* **18**:403–407.

Colyn, M., and Perpète, O. 1995. Missions d'expertise zoologique—Cameroun—Réserve de Faune du Dja—Septembre–Novembre 1994. Rapport ECOFAC/CEE.

Cords, M. 1987. Mixed-species association of *Cercopithecus* monkeys in the Kakamega Forest, Kenya. *Univ. California: Publ. Zool.* **117**:1–109

Dandelot, P. 1971. Order Primates, part 3. In: J. Meester and H. W. Setzer (eds.), *Mammals of Africa, an Identification Manual*, pp. 1–45. Smithsonian Institution Press, Washington, DC.

Deleporte, P., and Colyn, M. 1999. Biogéographie et dynamique de la biodiversité: application de la PAE aux forêts planitiaires d'Afrique centrale. *Biosystema* **17**:37–43.

Dowsett, R. J., and Granjon, L. 1991. Liste préliminaire des mammifères du Congo. In: R. J. Dowsett, and F. Dowsett-Lemaire (eds.), *Flore et faune du bassin du Kouliou (Congo) et leur exploitation*, pp. 297–310. Tauraco Research Report No. 4. Tauraco Press, Belgique.

Farris, J. S. 1988. Hennig 86 version 1.5. Computer program and documentation. Port Jefferson, New York.

Fay, M. 1988. Forest monkey populations in the Central African Republic: the northern limits. A census in Manovo-Gounda-St. Floris National park. *Mammalia* **52**: 57–74.

Gagneux, P., Gonder, M. K., Goldberg, T. L., and Morin, P. A. 2001. Gene flow in wild chimpanzee populations: what genetic data tell us about chimpanzee movement over space and time. *Phil. Trans. R. Soc. Lond. B.*, **356**:889–897.

Garcia, J. E., and Mba, J. 1997. Distribution, status and conservation of primates in Monte Alen National Park, Equatorial Guinea. *Oryx* **31**:67–76.

Gartlan, J. S., and Struhsaker, T. T. 1972. Polyspecific associations and niche separation of rain-forest anthropoids in Cameroon, West Africa. *J. Zool. Lond.* **168**:221–266.

Gautier, J.-P., and Gautier-Hion, A. 1969. Les associations polyspécifiques chez les Cercopithecidae du Gabon. *Rev. Ecol. (Terre Vie)* **2**:164–201.

Gautier, J.-P., Moysan, F., Feistner, A. T. C., Loireau, J. N., and Cooper, R. W. 1992. The distribution of *Cercopithecus (Lhoesti) solatus.* An endemic guenon of Gabon. *Rev. Ecol. (Terre Vie)* **47**:367–381.

Gautier-Hion, A., Colyn, M., and Gautier, J. P. 1999. *Histoire Naturelle des Primates d'Afrique Centrale.* ECOFAC, Libreville, Gabon.

Gonder, M. K., Oates, J. F., Disotell, T. R., Forstner, M. R., Morales, J. C., and Melnick, D. J. 1997. A new West African chimpanzee subspecies? *Nature* **388**:337.

Gooder, S. J. 1991. *A Phylogenetic and Vicariance Analysis of Some African Forest Mammals.* Ph.D. Dissertation, University of Liverpool.

Grubb, P. 1990. Primate geography in the Afro-tropical forest biome. In: G. Peters and R. Hutterer (eds.), *Vertebrates in the Tropics*, pp. 187–214. Museum Alexander Koenig, Bonn.

Grubb, P., and Powell., C. B. 1999. Discovery of red colobus monkeys (*Procolobus badius*) in the Niger Delta, with the description of a new and geographically isolated subspecies. *J. Zool. Lond.* **248**:67–73.

Grubb, P., Lernould, J. M., and Oates, J. F. 1999. Validation of *Cercopithecus erythrogaster pococki* as the name for the nigerian white-throated guenon. *Mammalia* **63**:389–392.

Haddow, A. J., Smithburn, K. C., Mahaffy, A. F., and Bugher, J. C. 1947. Monkeys in relation to yellow fever in Bwamba county, Uganda. *Trans. R. Soc. Trop. Med. Hyg.* **40**:677–700.

Hovenkamp, P. 1997. Vicariance events, not areas, should be used in biogeographical analysis. *Cladistics* **13**:67–80.

Kano, T. 1971. Distribution of the primates on the Eastern shore of Lake Tanganyika. *Primates* **12**:281–304.

Kingdon, J. 1971. *East African Mammals: An Atlas of Evolution in Africa, vol. 1.* Academic Press, London.

Kingdon, J. 1980. The role of visual signals and face patterns in African forest monkeys of the genus *Cercopithecus*. *Trans. Zool. Soc. Lond.* **35**:425–475.

Kingdon, J. 1990. *Island Africa.* Collins, London.

Kingdon, J. 1997. *African Mammals.* Academic Press, London.

Lernould, J. M. 1988. Classification and geographical distribution of guenons: a review. In: A. Gautier-Hion, F. Bourlière, J.-P. Gautier, and J. Kingdon (eds.), *A Primate Radiation: Evolutionary Biology of the African Guenons*, pp. 54–78. Cambridge University Press, Cambridge.

Machado, A. de B. 1969. Mamíferos de angola ainda não citados ou pouca conhecidos. *Publçoes cult. Co. Diam. Angola* **46**:93–232.

Mitani, M. 1990. A note on the present situation of the primate fauna found from south-eastern Cameroon to northern Congo. *Primates* **31**:625–634.

Mitani, M. 1991. Present situation of primate fauna found in northern Congo and in southeastern Cameroon in the Central Africa. In: A. Ehara, and T. Kimura (eds.), *Primatology Today*, pp. 43–46. Elsevier Science Publishers, New York.

Morrone, J. J., and Crisci, J. V. 1995. Historical biogeography: introduction to methods. *Ann. Rev. Ecol. Syst.* **26**:373–401.

Oates, J. F. 1986. *African Plan for African Primate Conservation: 1986–90.* Stony Brook, New York: IUCN/SSC Primate Specialist Group.

Oates, J. F. 1988. The distribution of *Cercopithecus* monkeys in West African forests. In: A. Gautier-Hion, F. Bourlière, J.-P. Gautier, and J. Kingdon (eds.), *A Primate Radiation: Evolutionary Biology of the African Guenons*, pp. 79–103. Cambridge University Press, Cambridge.

Oates, J. F. 1996. *African Primates. Status Survey and Conservation Action Plan.* Revised Edition. IUCN, Gland, Switzerland.

Peres, C. A., Patton, J. L., and da Silva, M. N. 1996. Riverine barriers and gene flow in Amazonian saddle-back tamarins. *Folia Primatol.* **67**:113–124.

Rosen, B. R. 1988. From fossils to earth history: applied historical biogeography. In: A. A. Myers and P. S. Giller (eds.), *Analytical Biogeography: an Integrated Approach to the Study of Animal and Plant Distributions,* pp. 434–481. Chapman & Hall, London.

Rudran, R. 1978. Intergroup dietary comparisons and folivorous tendencies of two groups of blue monkeys. In: G. G. Montgomery (ed.), *The Ecology of Arboreal Folivores*, pp. 483–503. Smithsonian Institution Press, Washington, DC.

da Silva, J. M. C., and Oren, D. C. 1996. Application of PAE in Amazonian Biogeography. An example with primates. *Biol. J. Linn. Soc.* **59**:427–437.

Hybridization between Red-tailed Monkeys (*Cercopithecus ascanius*) and Blue Monkeys (*C. mitis*) in East African Forests

7

KATE M. DETWILER

Introduction

Natural hybridization can be an important source of evolutionary innovations (Harrison, 1990; Arnold, 1997; Grant and Grant, 1998) and may have played an active role in the speciation process of the genus *Cercopithecus* (Dutrillaux *et al.*, 1988; Struhsaker *et al.*, 1988; Disotell and Raaum, 2002). Among *Cercopithecus* monkeys, as in other primate taxa (Jolly, 2001), most cases of natural hybridization occur between neighboring allotaxa in their zones of contact (Grubb, 1999). Examples include crosses between *Cercopithecus erythrotis camerunensis* and *C. cephus cephus* (Struhsaker, 1970), *C. campbelli campbelli* and *C. c. lowei* (Booth, 1955; Kingdon, 1997), and *C. mitis*

KATE M. DETWILER • Department of Anthropology, New York University, 25 Waverly Place, New York, NY 10003, USA and New York Consortium in Evolutionary Primatology (NYCEP).

The Guenons: Diversity and Adaptation in African Monkeys, edited by Glenn and Cords. Kluwer Academic/Plenum Publishers, New York, 2002.

stuhlmanni and *C. m. klobi* (Kingdon, 1971). Less common are cases of hybridization between well-differentiated and ecologically distinct species that live sympatrically in many areas with no evidence of interbreeding (Jolly, 2001). Examples of this kind of hybridization raise interesting questions about conditions favoring the breakdown of reproductive barriers between species that normally do not hybridize. Among guenons, such hybridization occurs between red-tailed monkeys (*Cercopithcus ascanius*, a member of the *cephus* species group) and blue monkeys (*C. mitis*, a member of the *nictitans* species group) (Struhsaker *et al.*, 1988).

In spite of the wide sympatry of the parental species, hybrids between red-tailed and blue monkeys have been recorded from only four localities: Budongo, Kibale, and Itwara Forests in Uganda (*Cercopithecus ascanius schmidti* × *C. mitis stuhlmanni*), and Gombe National Park, Tanzania (*C. ascanius schmiditi* × *C. mitis doggetti*; Fig. 1 and Table I). At Budongo, Aldrich–Blake (1968) observed a female hybrid and her infant living in a group of blue monkeys. He suggested, on the basis of their appearance and association with a blue monkey group, that the female was the F1 offspring of a cross between a female blue monkey and a male red-tailed monkey, and that her infant resulted from a back-cross to a male blue monkey.

From 1978 to 1986, Struhsaker *et al.* (1988) observed two hybrid females and their offspring and one hybrid male living in red-tailed monkey social groups in the Ngogo study area in the Kibale Forest, and concluded that hybridization was due to blue × red-tailed monkey matings. Because the Ngogo blue monkey population included few heterosexual social groups and a high density of solitary males, Struhsaker *et al.* (1988) concluded that cross-breeding by male blue monkeys most likely reflected a shortage of conspecific females. Struhsaker *et al.* (1988) extended this explanation to the occurrence of hybrids at Gombe, previously reported by Clutton–Brock (1972), and to the occasional hybridization reported by Aldrich–Blake (1968) at Budongo, inferring in the latter case that a deficit of female *Cercopithecus ascanius* caused male red-tailed monkeys to seek heterospecific mates.

Hybrids appear to be much more frequent at Gombe than at the Ugandan sites, having been regularly observed throughout the 1970s, 1980s, and 1990s by primatologists whose primary focus was other species (Clutton–Brock, 1972; Goodall, 1986, 2000; Stanford, 1998; K. Detwiler, unpub. data). During their studies of red colobus monkeys, Clutton–Brock (1972) and Stanford (1998) frequently observed hybrid guenons in red-tailed monkey groups and in mixed groups of red-tailed and blue monkeys. The work reported here represents the first attempt to confirm these impressions by quantifying the incidence and distribution of hybrid guenons at Gombe, and comparing them with observations made at other East African sites. My findings support the view that hybridization occurs much more often at Gombe than elsewhere, and suggest that a more systematic study of the situation in this region could shed light upon factors that facilitate inter-species hybridization among forest guenons.

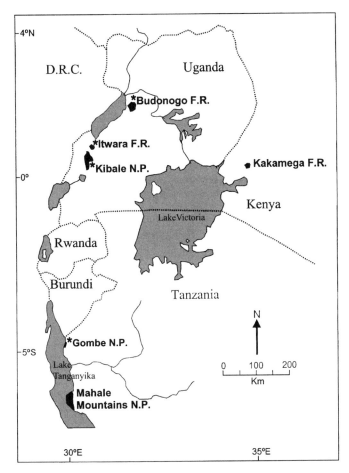

Fig. 1. Map of East Africa showing the location of six government forest reserves and parks where *Cercopithecus mitis* and *Cercopithecus ascanius* occur sympatrically. Forests marked with asterisks indicate sites where hybrid monkeys have been observed. Modified from Ghiglieri (1984).

Methods

From February to May 1996, I censused valley forests in Gombe National Park, Tanzania for hybrid monkeys. Gombe, comprising 13 steep-sided valleys, extends roughly 16 km north/south, covering an area approximately 35 km^2 along Lake Tanganyika (Clutton–Brock and Gillett, 1979; Goodall, 1986; Collins and McGrew, 1988; Stanford, 1998; Williams *et al.*, in press; Figs. 1 and 2). Heavy deforestation in the region over the last 50 years has left Gombe

Table I. Reported Sightings of Hybrid Monkeys

Study site	Date and duration of hybrid sighting	No. of hybrids[a]	Hybrid group association	Reference
Budongo F. R.	1968 (37 days)	2	Blue monkey group	Aldrich–Blake (1968)
Itwara F. R.	1971 (1 day)	2–3	Mixed species group of red-tailed and blue monkeys	Oates, unpub. data
Ngogo/ Kibale N. P.	1975–1986	9	Red-tailed monkey groups	Struhsaker et al. (1988)
Kanyawara/ Kibale N. P.	1983 (Aug–Nov)	1	Solitary	Struhsaker et al. (1988)
Gombe N. P.	1960–present	Sighted often	Red-tailed monkey groups, blue monkey groups, mixed species groups	Clutton–Brock (1972), Goodall (1986, 2000), Stanford (1998), this study

[a] Number of hybrids includes all individuals with intermediate, red-tailed-like hybrid, and blue-like hybrid phenotypes.

completely isolated from any other forest or woodland areas (Moreau, 1942; Goodall, 1986; Kamenya, 1997).

I considered as hybrids all individuals phenotypically intermediate between red-tailed and blue monkeys, and allocated such animals to one of three categories: H, RH, and BH. H individuals were most intermediate between the parental species in appearance, with white to light gray nose spots, light gray cheek hair with no narrow black border, dark pelage, and brown tail (Aldrich–Blake, 1968; Struhsaker et al., 1988). RH (red-tailed-like hybrid) and BH (blue-like hybrid) individuals tended, respectively, toward the red-tailed or blue parental phenotypes.

I used broad surveys as my main sampling technique to locate red-tailed and blue monkey groups, examine them for hybrid members, and collect data on group size and composition (National Research Council, 1981). Most of my broad surveys (85/98) were conducted in the northern half of the park, particularly in the Mkenke ($n = 41$) and Mitumba ($n = 26$) valley areas, where forest habitat is extensive and trails are well maintained by the Gombe Stream Research Center (Fig. 2). I generally conducted broad surveys during the first 6 and last 4 hours of the day, beginning at 0800 and 1500 h, respectively. When possible, I conducted two surveys per day. During surveys, I walked slowly uphill from the lakeshore, due east, along valley streams, or streambeds, to lookout points on valley ridges. Upon detecting monkeys, I actively followed them, maintaining contact as long as possible. I recorded

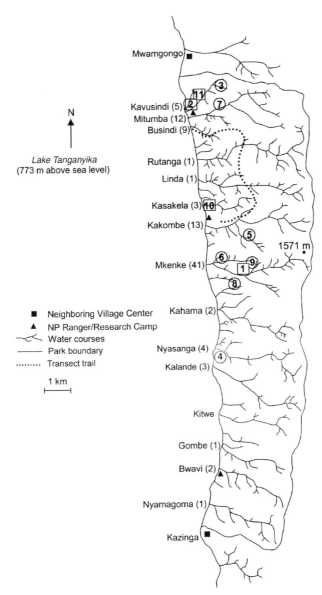

Fig. 2. Map of Gombe National Park, showing all major valleys and the approximate location of 11 unhabituated monkey groups for which group composition data are available. Circles refer to groups containing hybrid monkey(s) and squares refer to groups without hybrids. Low group numbers indicate groups comprising mostly red-tailed monkeys (1–3), while high group numbers indicate groups comprising mostly blue monkeys (8, 9, and 11). Intermediate numbers (4–7) represent mixed-species groups with hybrid monkeys. Numbers in parentheses next to valley names indicate the number of broad surveys conducted in that valley. Modified from Goodall (1986).

the date and time the group was detected, the time observation ended, location in valley, number of individuals, phenotypes of individuals, quality of count, and when possible age–sex class of individuals. If I encountered three or more monkeys of any combination of phenotypes engaging in affiliative interactions such as contact calling and grooming, traveling along the same route of progression, or within 50 m of one another, I scored them as a group (Glenn, 1997). I characterized groups by the phenotypes of all visible individuals, distinguishing six types: red-tailed monkey group, red-tailed monkey group with hybrid(s), blue monkey group, blue monkey group with hybrid(s), mixed-species group, and mixed-species group with hybrid(s).

I based my estimates of group members' age and sex on previous research experience with blue monkeys in a habituated population at Kakamega, and used the same classes and definitions (Table II; Cords, 1987). I scored the quality of group counts as incomplete or reliable. A count was reliable only if it was made when a group was traveling and I was close enough, generally ≤25 m, to identify each individual's phenotype accurately. Even reliable counts rarely yielded complete age–sex data, as it was difficult to classify all individuals, and I often missed the few individuals that lingered behind, left early, or used a different route that was out of sight. Nevertheless, reliable counts provided data on minimum group size and composition. I identified distinct groups on the basis of group composition data, the space and time in which I observed each group, and known travel rates, daily path lengths, and home range sizes of red-tailed and blue monkey groups at other intensively studied sites (Cords, 1987; Jones and Bush, 1988; Struhsaker and Leland, 1988; Butynski, 1990). I used reliable counts of clearly distinct groups to quantify the minimum number of groups with hybrids, and to demonstrate variation in group composition.

During the last month of my study (May 4–23), I employed a second sampling technique, line transect censuses, to collect systematic sighting

Table II. Age-sex Class[a] of all Individuals Identified in 11 Independent Groups Encountered during Broad Surveys[b]

Phenotype	AM	AF	BIG	LJ	MJ	SJ	TJ/IN	UNID	Total
Red-tailed	7	24	3	4	10	11	12	36	107
Red-tailed-like hybrid (RH)	0	1	0	0	1	2	4	0	8
Hybrid intermediate (H)	3	6	0	2	3	1	4	0	19
Blue-like hybrid (BH)	1	5	0	0	0	3	3	0	12
Blue	7	12	7	4	6	4	3	28	71
Total hybrids[c]	4	12	0	2	4	6	11	0	39

[a] *Abbreviations*: AM, adult male; AF, adult female; BIG, a large individual with sex and age class undetermined; LJ, large juvenile; MJ, medium juvenile; SJ, small juvenile; TJ/IN, tiny juvenile or infant; UNID, unidentified sex and age class.
[b] Groups 1–11 shown in Figure 2.
[c] The sum of all three hybrid phenotypes.

frequency data to calculate estimates of abundance—number of monkeys observed per km of census (Mitani *et al.*, 2000). I conducted 10 censuses that all began between 0730–0800 and lasted 5.3–6.8 hours (mean = 6.1 hours; SD = 0.5). Each census followed the same transect trail that was approximately 6.2 km long, marked every 50 m, and traversed five northern valleys (Fig. 2). All censuses started in Busindi valley and ended in Kakombe valley at the 6.1 km mark (the first four censuses) or 5.1 km mark (the last six censuses). Most of the transect trail passed through forest (~4 km) and woodland (~1.6 km) habitats. Only approximately 600 m of the trail passed through open grassland areas along ridge crests, and I deducted this latter section from the total trail length in my analyses. During each census, I walked slowly (approximately 1.1 km/hour) and paused often to look and listen for guenons. Upon encountering monkeys, I remained on the transect trail and stopped for ≤15 minutes (mean = 10 minutes; SD = 4.5; $n = 48$); on one occasion I extended my observation limit of 15 minutes to 30 minutes for an opportunistic group count. I recorded the time of sighting, transect position, number of individuals, phenotypes of individuals, time encounter ended, and when possible, age–sex class of individuals. I defined groups using the same criteria that I used during broad surveys. During transect censuses and broad surveys, I recorded *ad libitum* all observations related to reproduction, including inter- and intra-specific matings, hybrid females with clinging infants, and hybrid females with a nursing infant or juvenile.

I collected information about the presence of hybrid monkeys at four other East African sites where the parental species are sympatric: Budongo Forest Reserve and Kibale National Park (Uganda), Kakamega Forest Reserve (Kenya), and Mahale Mountains National Park (Tanzania). I questioned researchers and field assistants about sightings of hybrids at each of the sites and reviewed the literature for recent reports. Between August 1995 and August 1996 at each site except Budongo, I spent at least one week walking forest trails, and carefully examining red-tailed and blue monkey groups for hybrid individuals, concentrating upon areas within the site where hybrids had previously been observed.

Results

Frequency of Hybridization

I observed red-tailed and blue monkey hybrids only at Gombe National Park. No hybrid monkeys were detected among red-tailed and blue monkey groups examined at Ngogo, Kibale NP (seven red-tailed monkey and two blue monkey groups, July 2–12, 1996), at Isecheno, Kakamega FR (three red-tailed monkey and five blue monkey groups, September 15–October 9, 1995), and Kasoje, Mahale Mountains NP (six red-tailed monkey groups and one blue monkey group, June 4–9, 1996). In addition, apart from Stanford (1998), I found no recent published account of a hybrid monkey sighting, and none of the researchers and field assistants I surveyed reported recently observing a hybrid monkey (Budongo FR, A. J. Plumptre, pers. com., January 2001; Kibale NP, T. T. Struhsaker and T. L. Windfelder, pers. com., June 2000 and January 2001, respectively; Kakamega FR, M. Cords and P. J. Fashing, pers. com., January 2001; Mahale Mountains NP, T. Nishida, pers. com., December 1995; R. Kitopeni and M. Hamisi, pers. com., June 1996).

It is, however, much easier to prove the presence of hybrids at a study site than to prove their absence. Even though I searched the same forest areas that Struhsaker *et al.* (1998) used for their hybrid study and found no sign of hybrids, it is still possible that hybrids exist in the Kibale guenon population. However, if hybridization still occurs at Kibale, it is sufficiently rare to escape the notice of the many experienced observers who work there. Similarly, at Budongo Forest, no hybrid sighting has been reported by the several researchers (Fairgrieve, 1995; Plumptre, 2000; Mnason and Obua, 2001; Sheppard and Paterson, 2001) who have studied red-tailed and blue monkeys since Aldrich–Blake (1968) reported a hybrid female and her infant. At Kakamega Forest, guenon research has been ongoing since 1979 and there has been only one report of a possible hybrid sighting; T. Rowell observed an infant that looked like a hybrid, but this was not found in the group the next year (M. Cords, pers. com.). Therefore, as at Kibale, guenon hybridization at Budongo and Kakamega is probably rare. In the history of primate studies at Mahale, the one detailed census study that carefully examined red-tailed and blue monkey groups (Uehara and Ihobe, 1998) reported no confirmed sighting of a hybrid monkey.

During my surveys at Gombe, I observed red-tailed monkeys, blue monkeys and their hybrids in all valleys north of the Nyasanga–Kalande ridge. I did not see or hear a red-tailed monkey, blue monkey or hybrid monkey during broad surveys ($n = 8$) in the southern-most valleys. During the remaining 90 surveys, I encountered unhabituated monkeys 121 times (83 observations of bisexual groups, nine of solitary adult males, 29 incomplete observations) for a total of 177.8 contact hours. Of the nine solitary adult male monkey sightings, four were of blue monkeys, four were of red-tailed monkeys, and one was a hybrid. The mean duration of encounters of groups was two hours (SD $= 1.3$, min $= 0.06$, max $= 5.20$). While I observed all six group types, most (60% of sightings) were mixed-species groups of red-tailed monkeys, blue monkey(s), and hybrid monkey(s).

By eliminating all possible repeat sightings, I identified a minimum of 11 independent groups (Fig. 2), and used the most complete counts to order them by phenotypic variation (Fig. 3). Low numbered groups were comprised of mostly red-tailed monkeys, while high numbered groups consisted mostly of blue monkeys. Groups with intermediate numbers were mixed-species

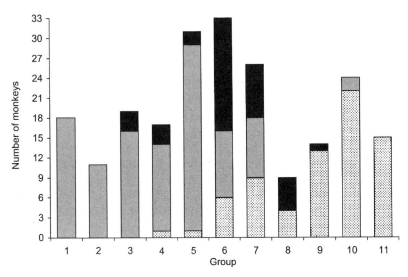

Fig. 3. The minimum number of red-tailed monkeys (gray), blue monkeys (stippled), and hybrid monkeys (black) counted in each of the 11 groups shown in Fig. 2.

groups with hybrid monkeys. Overall, 18% of the monkeys were hybrid (H, RH, and BH combined), 33% were blue monkeys, and 49% were red-tailed monkeys. Table II lists the sum of all age–sex classes of all individuals identified in the 11 groups.

Because I did not obtain confirmed repeat counts of groups and the groups counted were not the subject of long-term study, I was not able to clarify the relationship between monkeys counted and the behaviorally defined social group. This was especially true of mixed-species groups containing several hybrid monkeys. For these groups, it was unclear whether individuals were permanent members of a fused group of heterospecifics and hybrids, or had dual membership—each belonging to a smaller group of monkeys of like phenotype, as well as to the larger mixed-species group with hybrids. Thus, the following compositions of the 11 groups should be interpreted cautiously within this context.

Four of the seven groups containing hybrid monkeys were mixed-species groups with hybrid monkeys, but their compositions differed considerably. Groups 4 and 5 were similar, each containing one adult male red-tailed monkey, one adult male blue monkey, 13–28 adult female and juvenile red-tailed monkeys, and two to three nonadult male hybrid monkeys. In contrast, groups 6 and 7 each had ⩾5 nonadult male monkeys of each of the two parental species, and ⩾8 nonadult male hybrid monkeys. I counted three adult males in group 6 (one red-tailed monkey, one blue monkey, one hybrid) and two adult males in group 7 (one red-tailed monkey, one blue monkey). The composition of group 6 was the most diverse, with hybrid monkeys ($n = 17$) outnumbering the two parental species combined ($n = 6$ blue monkeys + 10 red-tailed monkeys).

The other three groups with hybrids were groups 3, 8, and 9. Group 3 was a red-tailed monkey group with hybrids. It contained one adult male red-tailed monkey, two adult male hybrid monkeys, 15 adult female and juvenile red-tailed monkeys, and one small juvenile hybrid monkey. In contrast, groups 8 and 9 were blue monkey groups with hybrid monkey(s): group 8 comprised one adult male blue monkey, four nonadult male blue monkeys, and six nonadult male hybrid monkeys, while group 9 contained one adult male hybrid monkey and 13 nonadult male blue monkeys. The male hybrid monkey in group 9 gave an unusual pyow-like call while in view, and although no adult male blue monkey was visible, I heard one other male giving loud calls characteristic of adult male blue monkeys from within the general area of the group. Fourteen days after counting group 9 and at the same location, I sighted an adult male hybrid monkey within a blue monkey group of approximately the same size and composition as group 9; the male however, had a fresh wound on the right side of his face. He gave an unusual pyow-like call while in view, and I heard trills and grunts from two adult female and two juvenile blue monkeys sitting within 20 m of him. Approximately 300 m from the blue monkey group with the male hybrid, I encountered an adult male and juvenile blue monkey traveling together. These preliminary observations suggest that the resident male of group 9 was a hybrid.

Of the four groups with no hybrid monkey observed, two were red-tailed monkey groups (groups 1 and 2) and two were blue monkey groups (groups 10 and 11). In each of these groups, I counted only one adult male. I observed a pair of red-tailed monkeys and a solitary red colobus monkey traveling with members of group 10. Groups 1 and 9, and groups 2 and 11, respectively, formed mixed species groups (Fig. 2).

Sighting frequency data collected during transect censuses provide estimates of relative abundance of red-tailed and blue monkeys and their hybrids. These estimates are, however, preliminary because they are based on only 10 censuses conducted over three weeks. The mean number of monkeys sighted per kilometer of census varied from 3.6 (SD = 1.2) red-tailed monkeys to 2.8 (SD = 2.0) blue monkeys to 0.8 (SD = 0.5) hybrid monkeys (H, RH, BH combined). Of the total groups recorded ($n = 48$), 42% were blue monkey groups, 27% were red-tailed monkey groups with hybrid monkey(s), 25% were red-tailed monkey groups with no hybrid monkey, and 6% were mixed-species groups with hybrid monkeys. Of the two types of group with hybrid monkeys, mixed species groups had three to five hybrid monkeys, eight to 13 red-tailed monkeys, and two to five blue monkeys; while the mean number of hybrid monkeys in red-tailed monkey groups was 1.8 (SD = 1.3), of which the most frequently recorded age–sex class was adult male ($n = 8$). The only group type that I did not encounter was a blue monkey group with hybrid monkey(s). I sighted members of red-tailed and blue monkey groups within larger mixed-species groups on six occasions, and on one occasion I encountered a solitary adult male blue monkey among members of a red-tailed monkey group.

All monkey sightings along the transect trail were at least 1 km from sighting locations of groups 1–11 of broad surveys. Therefore, I consider data from the two sampling periods to be independent and combined them in estimating the incidence of hybridization at Gombe. These data are, however, too incomplete to yield accurate density estimates of the two parental species. For example, the high number of blue monkey group sightings during transect censuses may be misleading because these groups occurred only along the first 1.8 km of the transect. This suggests that blue monkey density is particularly high in the Busindi valley area and that blue monkey density may vary significatly between valleys within the park.

Reproductive Behavior

Hybrid females with a clinging infant (encountered on 48 occasions) or nursing juvenile (five sightings) comprised at least 15 different individuals with offspring, presumably their own, strongly suggesting that hybrid females are fully fertile.

Of the 10 complete copulations (mounts with presumed ejaculatory pause) that I observed, one was between heterospecific individuals, one was between two hybrid individuals, three were between a hybrid individual and an individual of one of the parental species, and five were between conspecific individuals (Table III).

On several other occasions, I observed hybrids participating in mating behavior without copulation. For example, on March 6 within the general area—mid Mkenke valley—where I frequently encountered a mixed-species

Table III. Observed Copulations among Red-tailed Monkeys, Blue Monkeys, and Hybrid Monkeys

Date (1996)	Valley location[a]	Copulations observed
2/13	Mid Mkenke	Male blue × female blue
2/16	Mid Mkenke	Male red-tailed × female red-tailed
3/16	Mid Chihaga	Male hybrid × female blue
4/11	Mid Busindi	Male hybrid × female red-tailed-like hybrid
4/13	Mid Busindi	Male red-tailed × female red-tailed
4/15	Mid Busindi	Male red-tailed × female red-tailed
5/14	High Rutanga	Male red-tailed × female blue
5/14	High Busindi	Male blue × female blue
5/17	High Busindi	Male red-tailed × female red-tailed-like hybrid
5/19	Low Busindi	Male red-tailed × female hybrid

[a]The general location where copulations were observed: low, mid, and high refer to relative elevation in valley.

group with hybrid monkeys ($n = 20$ encounters, of which one included the count of group 6), I saw an adult female hybrid monkey present to an adult male blue monkey and then groom him. The male did not show interest in the female until she left, upon which he sniffed the area where she had been sitting and then pursued her.

In the middle of April in mid-Busindi valley, I frequently heard multiple males give loud calls and nasal screams, and over the course of five days (April 11–15) I counted seven different adult males in the area—three were red-tailed monkeys, three were hybrid monkeys, and one was a blue monkey. The attraction appeared to be several females receptive simultaneously within a red-tailed monkey group with hybrid monkeys. On the 11th, I saw a female hybrid (RH) monkey copulate with a male hybrid (H) monkey; the pair was within 25 m of another male hybrid (BH) monkey, a male red-tailed monkey, and a grooming pair of adult male and female blue monkeys. Two days later at the same location, I observed an adult female red-tailed monkey copulate with a conspecific adult male; just a few minutes before the copulation I saw the same female and an adult male hybrid (H) in the same tree. On the 15th, I observed a female red-tailed monkey copulate with a conspecific male. These data suggest that my study occurred during the breeding season of both parental species and their hybrids, and that I encountered at least one group during a promiscuous mating period.

Discussion

Hybridization between *Cercopithecus ascanius* and *C. mitis* seems to be at most rare and sporadic over most of their extensive shared range. By contrast, at Gombe hybrids are frequent, and have been so for several decades, at least. This qualifies Gombe—alone of the East African sites so far investigated—as a localized sympatric hybrid zone in the sense of Woodruff (1973). Such zones, whether between sympatric or parapatric taxa, are recognized as valuable natural laboratories for the study of processes involved in speciation and population differentiation, and potentially the source of new species.

Two related but distinct questions arise. First, what circumstances permit rare cases of hybridization, as observed at Kibale and Budongo? And second, under what conditions does hybridization progress to the situation seen at Gombe, where hybrids evidently comprise a substantial proportion of the breeding population?

The variety of hybrid phenotypes observed at Gombe strongly suggests that hybrids themselves breed successfully, so that intrinsic, post-zygotic barriers to gene flow between *Cercopithecus ascanius* and *C. mitis* are weak or absent. Conversely, the very rarity and sporadic occurrence of hybridization elsewhere in the extensive area of sympatry attests to the efficacy of pre-zygotic barriers under normal circumstances. A full understanding of the conditions under which these barriers fail, and why they appear to have failed so completely at Gombe, will require much more information than is presently available about mate-recognition rules in these species including what they are, how an individual acquires them in its ontogeny, and how much flexibility in mate choice they permit. Evolutionary theory and empirical data do, however, permit us to speculate about conditions under which pre-zygotic barriers might be overcome, and thus to construct hypothetical scenarios of hybridization.

Occasional hybridization between sympatric species has been related to a local scarcity of conspecific mates (Bernstein, 1966; Lehman *et al.*, 1991; Wirtz, 1999; Alberts and Altmann, 2001; Randler, 2002), and this explanation was advanced for hybridization at Kibale and Budongo (Struhsaker *et al.*, 1988). In both cases, a deficit of females was assumed (*Cercopithecus mitis* at Kibale, *C. ascanius* at Budongo), but only at Kibale was the deficit actually apparent in the field. This explanation assumes that males of both species are normally unattracted to females of the other but overcome this aversion when the prospects of successfully entering a conspecific female group are reduced beyond a critical threshold. Though untested, this scenario is plausible, but fails to explain why females of the commoner species do not always reject heterospecific matings. Because female guenons, like most mammals, invest more heavily in parenting than do males, theory predicts that they should be the more discriminating sex in mate choice (Trivers, 1972). A genotype that permits a female to accept heterospecific partners will be strongly disfavored by selection if genes invested in hybrid offspring fail to enter the female's own gene-pool. If observation shows that in practice females do not always avoid heterospecific matings, there are several possible explanations.

Although female guenons should in theory be "choosy", natural selection might in fact never have provided them with the species recognition mechanisms needed to make the choice. If males usually initiate mating (by joining a female social group), and normally attempt to join only homospecific groups, females would be shielded from strong selection for species discrimination. Though red-tailed and blue monkeys differ markedly in pelage coloration and face patterns, they have similar mating systems and courtship behavior. It is possible that female guenons lack a mate choice program that uses pattern to distinguish between conspecifics and closely related heterospecifics, or that they have a weak program that is easily overridden by other factors, such as insistent attention from a "lone" heterospecific male, or lack of interest by conspecific males.

Another possibility is that female selectively exists, but is occasionally overcome by forced copulation (Randler, 2002). There is, however, no evidence of intraspecific forced copulations in red-tailed and blue monkeys, and females of both species are known to effectively evade unwanted matings (Cords *et al.*, 1986; Cords, 1988; Henzi and Lawes, 1988; Rowell, 1994).

The female's age might also influence the probability of extraspecific mating. Young inexperienced females might make more mistakes in mate recognition (Grant and Grant, 1997), or be more likely to accept a heterospecific mate because conspecific adult males ignore or reject them (Dunbar and Dunbar, 1974). Grant and Grant (1997) found no support for the "inexperience" hypothesis in birds and it remains untested in guenons. Adult male blue monkeys, however, appear to prefer older females (Rowell, 1994), and resident males have been seen to physically reject young females that persistently solicit copulations (Rowell, 1994). A young receptive female that is being ignored by the resident male may perhaps seek, or at least tolerate, copulations by interested heterospecific males. On the other hand, because hybrids would be much more frequent if *all* young females showed a tendency to mate heterospecifically, this can be at most a contributing factor.

Alternatively, heterospecific mating may originate with a female if a "mutation" in her mate selection program leads her to exercise lower selectivity, or even to prefer heterospecific mates. Presumably, a female guenon acquires its program of mate-selection behaviors by the action of an appropriate social setting on a species-specific genotype, but the details of this process are unknown. In theory, either a genetic mutation or a developmental abnormality such as "misimprinting" (Grant and Grant, 1997, 1998), or both, might be involved.

There is, therefore, no lack of potential "explanations" for occasional deviation from the expected pattern of female homospecific mate choice, especially if the aberration occurs in the context of an excess of heterospecific potential mates. The level of hybridization seen at Gombe clearly requires additional explanation. At this site, obvious hybrids have been observed continuously for decades. This survey found recognizable hybrids comprising a majority in some social groups, and participating fully in reproduction. Hybrids of different degrees are present in both *Cercopithecus ascanius* and *C. mitis* groups, showing that introgression has occurred in both directions, unlike the unidirectional hybridization suggested for Budongo and Kibale.

These observations suggest that the rate of primary hybridization, producing F1 hybrids, has been higher at Gombe than at the other sites. This difference might be attributed at least in part to Gombe's topography, which concentrates guenon habitat into narrow, blind-ended valleys. In such a setting, each guenon social group tends to have relatively few conspecific neighbors with which to exchange migrants. This means that frequent, severe, sex-ratio biases and shortages of conspecific mates are to be expected at a very local level, regardless of the densities of the two species over the region as a whole.

While this mechanism may have initially contributed to abnormally frequent hybridization in Gombe (Clutton–Brock, 1972), it cannot entirely account for the situation in 1996: members of each species cross-mated, even when conspecific mates were apparently available. The interspecific mating that I observed was between a male red-tailed monkey and female blue monkey, contrary to predictions based on previous estimates of low blue monkey density, high red-tailed monkey density, and frequent sightings of solitary male blue monkeys (Clutton–Brock, 1972; Struhsaker *et al.*, 1988; Stanford, 1998).

Such observations, together with the observed mating success of hybrids, raises the possibility that "normal" patterns of mate recognition have been modified at Gombe, in response to an initially high frequency of primary hybridization. The striking differences between the parental species in pelage coloration and face pattern are believed to function in mate recognition by one or both sexes (Kingdon, 1988), and thus to act as pre-zygotic mechanisms isolating the gene-pools of the species. While plausible, this suggestion has yet to be tested experimentally, and the occurrence of hybridization shows that the "barrier" is not absolute. An effective barrier requires the stimuli represented by the distinctive pelage phenotypes to be complemented by an appropriate agenda of behavioral responses. The determinants of that agenda are unknown, but presumably comprise a mix of genetic and social–environmental factors. Either component could be modified in the direction favoring more hybridization in a deme where, by demographic accident, F1 hybrids have come to comprise a significant proportion of the breeding membership. Genes underlying a predisposition to indiscriminate mating would rapidly increase in social groups entered by heterospecific or hybrid males. On the environmental side, to the extent that animals acquire their mating cues from the appearance of their mother or their mother's consort, the presence of breeding hybrid animals would also tend to weaken discriminatory behavior. Meanwhile, as hybrids increased in frequency in the population at large, more animals would present potential mates with phenotypic stimuli that were less extreme, and therefore presumably less aversive, than those presented by "full-blooded" heterospecifics. Indeed, some females of either species are known to show strong preferences for nonresident males that may be unfamiliar (Struhsaker, 1988; Cords, 2000), and for such animals the novel appearance of a hybrid male might even add to his attractiveness. The current, anthropogenic isolation of forest guenon habitat at Gombe would presumably enhance the positive feedback among these elements, by preventing the immigration of genetically pure and behaviorally naive animals from neighboring populations.

Though the distinctiveness of the Gombe guenon populations is undoubted, the explanatory model presented here is obviously highly speculative, and much work at all levels from the genetic to the ecological and behavioral will be required to test its various components. If its main features are confirmed, it suggests that genetically and phenotypically novel guenon populations might arise by population fusion, as well as by cladogenesis, and that such events might have been especially influential in ecologically unstable mosaic habitats fringing major forest blocks.

Summary

Hybridization between *Cercopithecus ascanius* and *C. mitis* has been previously recorded at several localities in East Africa. However, recent sightings of red-tailed and blue monkey hybrids suggest that they are restricted to Gombe National Park, Tanzania. At Gombe, hybrid monkeys of various age and sex classes are commonly sighted. They are found within red-tailed monkey groups, blue monkey groups, and mixed-species groups, suggesting that introgression is bidirectional. Observations of hybrid females nursing infants and juveniles provide evidence of female hybrid fertility. The high incidence of hybridization at Gombe calls for its recognition as a localized sympatric hybrid zone between *Cercopithecus ascanius schmidti* and *C. mitis doggetti*. Gombe—a terrestrial island habitat—then becomes of special interest to evolutionary primatologists because of its possible implications for speciation in guenons.

ACKNOWLEDGMENTS

I thank the Thomas J. Watson Foundation for providing funds for all research and travel expenses. The Tanzanian Commission for Science and Technology, Tanzania National Parks, and Serengeti Wildife Research Institute granted permission to work in Gombe, and the director and staff of Gombe National Park provided much cooperation and assistance. D. A. Collins, S. M. Kamenya, and fellow members of the Gombe Stream Research Center provided valuable logistical and research support. Thanks to Mr. Ulimwengu Hilaly of Gombe Stream Research Center for his assistance and dedication in the field. J. F. Oates provided access to his field notes. I thank A. Tosi, R. Raaum, T. Disotell, A. Burrell, and P. Fashing for access to manuscripts and discussion. I thank T. T. Struhsaker for his assistance at Ngogo and helpful comments on a preliminary draft. I thank W. Staples for his generous technical instruction. I am especially thankful to C. J. Jolly for academic support, theoretical discussion, and assistance with revisions. I am grateful to M. E. Glenn and M. Cords for their patience, encouragement, and constructive criticisms. J. Gray and the Detwiler family have been, and continue to be, very supportive of this project. And special thanks to M. Cords for her instructorship and mentoring in the field and in New York.

References

Alberts, S. C., and Altmann, J. 2001. Immigration and hybridization patterns of yellow and anubis baboons in and around amboseli, Kenya. *Am. J. Primatol.* **53**:139–154.

Aldrich–Blake, F. P. G. 1968. A fertile hybrid between two *Cercopithecus* species in the Budongo Forest, Uganda. *Folia Primatol.* **9**:15–21.

Arnold, M. L. 1997. *Natural Hybridization and Evolution.* Oxford University Press, Oxford.

Bernstein, I. S. 1966. Naturally occurring primate hybrid. *Science* **154**:1559–1560.

Booth, A. H. 1955. Speciation in the Mona monkeys. *Journal of Mammalogy* **36**:434–449.

Butynski, T. M. 1990. Comparative ecology of blue monkeys (*Cercopithecus mitis*) in high-and low-density subpopulations. *Ecol. Monogr.* **60**:1–26.

Clutton–Brock, T. H. 1972. *Feeding and Ranging Behaviour in the Red Colobus.* Ph.D. Thesis. University of Cambridge, Cambridge.

Clutton–Brock, T. H., and Gillett, J. B. 1979. A survey of forest composition in the Gombe National Park, Tanzania. *Afr. J. Ecol.* **17**:131–158.

Collins, D. A., and McGrew, W. C. 1988. Habitats of three groups of chimpanzees (*Pan troglodytes*) in western Tanzania compared. *J. Human Evol.* **17**:553–574.

Cords, M. 1987. Mixed-species association of *Cercopithecus* monkeys in the Kakamega forest. *Univ. Calif. Pub. Zool.* **117**:1–109.

Cords, M. 1988. Mating systems of forest guenons: a preliminary review. In: A. Gautier–Hion, F. Bourlière, J.-P. Gautier, and J. Kingdon (eds.), *A Primate Radiation: Evolutionary Biology of the African Guenons*, pp. 323–329. Cambridge University Press, Cambridge.

Cords, M. 2000. The number of males in guenon groups. In: P. Kappeler (ed.), *Primate Males*, pp. 84–96. Cambridge University Press, Cambridge.

Cords, M., Mitchell, B. J., Tsingalia, H. M., and Rowell, T. E. 1986. Promiscuous mating among blue monkeys in the Kakamega Forest, Kenya. *Ethology* **72**:214–226.

Disotell, T. R., and Raaum, R. L. 2002. Molecular timescale and gene tree incongruence in the guenons. In: M. E. Glenn, and M. Cords (eds.), *The Guenons: Diversity and Adaptation in African Monkeys*, pp. 27–36. Kluwer Academic Publishers, New York.

Dunbar, R. I. M., and Dunbar, P. 1974. On hybridization between *Theropithecus gelada* and *Papio anubis* in the wild. *J. Hum. Evol.* **3**:187–192.

Dutrillaux, B., Muleris, M., and Conturier, J. 1988. Chromosomal evolution of Cercopithecinae. In: A. Gautier–Hion, F. Bourlière, J.-P. Gautier, and J. Kingdon (eds.), *A Primate Radiation: Evolutionary Biology of the African Guenons*, pp. 151–159. Cambridge University Press, Cambridge.

Fairgrieve, C. 1995. Infanticide and infant eating in the blue monkey (*Cercopithecus mitis stuhlmanni*) in the Budongo Forest Reserve, Uganda. *Folia Primatol.* **64**:69–72.

Ghiglieri, M. P. 1984. *The Chimpanzees of Kibale Forest: A Field Study of Ecology and Social Structure.* Columbia University Press, New York.

Glenn, M. E. 1997. Group size and group composition of the mona monkey (*Cercopithecus mona*) on the island of Grenada, West Indies. *Am. J. Primatol.* **43**:167–173.

Goodall, J. 1986. *The Chimpanzees of Gombe: Patterns of Behavior.* Harvard University Press, Cambridge.

Goodall, J. 2000. In: D. Peterson (ed.), *Africa in My Blood: An Autobiography in Letters, the Early Years*, Houghton Mifflin, Boston.

Grant, P. R., and Grant, B. R. 1997. Hybridization, sexual imprinting, and mate choice. *Am. Nat.* **149**:1–28.

Grant, B. R., and Grant, P. R. 1998. Hybridization and speciation in Darwin's finches. In: D. J. Howard, and S. H. Berlocher (eds.), *Endless Forms: Species and Speciation*, pp. 404–422. Oxford University Press, New York.

Grubb, P. 1999. Evolutionary processes implicit in distribution patterns of modern African mammals. In: T. G. Bromage, and F. Schrenk (eds.), *African Biogeography, Climate Charge, and Human Evolution*, pp. 150–164. Oxford University Press, New York.

Harrison, H. G. 1990. Hybrid zones: windows on evolutionary process. *Oxford Surv. Evol. Biol.* **7**:69–128.

Henzi, S. P., and Lawes, M. 1988. Strategic responses of male samango monkeys (*Cercopithecus mitis*) to a decline in the number of receptive females. *Int. J. Primatol.* **9**:479–495.

Jolly, C. J. 2001. A proper study for mankind: Analogies from the papionin monkeys and their implications for human evolution. *Yrbk. Phys. Anthropol.* **44**:177–204.

Jones, W. T., and Bush, B. B. 1988. Movement and reproductive behavior of solitary male redtail guenons *(Cercopithecus ascanius). Am. J. Primatol.* **14**:203–222.

Kamenya, S. M. 1997. *Changes in the Land Use Patterns and their Impacts on Red Colobus Monkey's Behavior: Implications for Primate Conservation in Gombe National Park, Tanzania.* Ph.D. Thesis, University of Colorado at Boulder, Colorado.

Kingdon, J. 1971. *East African Mammals: An Atlas of Evolution in Africa,* Vol. 1. Academic Press, London.

Kingdon, J. 1988. What are face patterns and do they contribute to reproductive isolation in guenons? In: A. Gautier–Hion, F. Bourlière, J.-P. Gautier, and J. Kingdon (eds.), *A Primate Radiation: Evolutionary Biology of the African Guenons,* pp. 227–245. Cambridge University Press, Cambridge.

Kingdon, J. 1997. *The Kingdon Field Guide to African Mammals.* Academic Press, New York.

Lehman, N., Eisenhawer A., Hansen, K. Mech, L. D., Peterson, R. O., Gogan, P. J. P., and Wayne, R. K. 1991. Introgression of coyote mitochondrial DNA into sympatric North American gray wolf populations. *Evolution* **45**:104–119.

Mitani, J. C., Struhsaker, T. T., and Lwanga, J. S. 2000. Primate community dynamics in old growth forest over 23.5 years at Ngogo, Kibale National Park, Uganda: Implications for conservation and census methods. *Int. J. Primatol.* **21**:269–286.

Mnason, T., and Obua, J. 2001. Feeding habits of chimpanzees *(Pan troglodytes),* red-tail monkeys *(Cercopithecus ascanius schmidti)* and blue monkeys *(Cercopithecus mitis stuhlmanii)* on figs in Budongo Forest Reserve, Uganda. *Afr. J. Ecol.* **39**:133–139.

Moreau, R. E. 1942. The distribution of the chimpanzee in Tanganyika Territory. *Tanganyika Notes and Records* **4**:52–55.

National Research Council. 1981. *Techniques for the Study of Primate Population Ecology.* National Academy Press, Washington, DC.

Plumptre, A. J. 2002. Monitoring mammal populations with line transect techniques in African Forests. *J. Appl. Ecol.* **37**:356–368.

Randler, C. 2002. Avian hybridization, mixed pairing and female choice. *Anim. Behav.* **63**: 103–119.

Rowell, T. E. 1994. Choosy or promiscuous: it depends on the time scale. In: J. J. Roeder, B. Thierry, J. R. Anderson, and N. Herrenschmidt (eds.), *Current Primatology: Social Development, Learning and Behaviour,* pp. 11–18. Université Louis Pasteur, Strasbourg.

Sheppard, D., and Paterson, J. D. 2001. *Cercopithecus ascanius* logged and unlogged diet and ranging comparisons in the Budongo Forest, Uganda. Abstract, *18th Congress of IPS,* Adelaide.

Stanford, C. 1998. *Chimpanzee and Red Colobus: The Ecology of Predator and Prey.* Harvard University Press, Cambridge.

Struhsaker, T. T. 1970. Phylogenetic implications of some vocalizations of *Cercopithecus* monkeys. In: J. R. Napier, and P. H. Napier (eds.), *Old World Monkeys: Evolution, Systematics and Behaviour,* pp. 365–444. Academic Press, New York.

Struhsaker, T. T. 1988. Male tenure, multi-male influxes, and reproductive success in redtail monkeys *(Cercopithecus ascanius).* In: A. Gautier–Hion, F. Bourlière, J.-P. Gautier, and J. Kingdon (eds.), *A Primate Radiation: Evolutionary Biology of the African Guenons,* pp. 340–363. Cambridge University Press, Cambridge.

Struhsaker, T. T., and Leland, L. 1988. Group fission in redtail monkeys *(Cercopithecus ascanius)* in the Kibale Forest, Uganda. In: A. Gautier–Hion, F. Bourlière, J.-P. Gautier, and J. Kingdon (eds.), *A Primate Radiation: Evolutionary Biology of the African Guenons,* pp. 364–388. Cambridge University Press, Cambridge.

Struhsaker, T. T., Butynski, T. M., and Lwanga, J. S. 1988. Hybridization between redtail *(Cercopithecus ascanius schmidti)* and blue *(C. mitis stuhlmanni)* monkeys in the Kibale Forest,

Uganda. In: A. Gautier–Hion, F. Bourlière, J.-P. Gautier, and J. Kingdon (eds.), *A Primate Radiation: Evolutionary Biology of the African Guenons*, pp. 477–497. Cambridge University Press, Cambridge.

Trivers, R. L. 1972. Parental investment and sexual selection. In: B. Campbell (ed.), *Sexual Selection and the Descent of Man 1871–1971*, pp. 136–179. Aldine, Chicago.

Uehara, S., and Ihobe, H. 1998. Distribution and abundance of diurnal mammals, especially monkeys, at Kasoje, Mahale Mountains, Tanzania. *Anthropol. Sci.* **106**:349–369.

Williams, J. M., Pusey, A. E., Carlis, J. V., Farm, B. P., Goodall, J. (in press). Female competition and male territorial behavior influence female chimpanzees' ranging patterns. *Anim. Behav.*

Wirtz, P. 1999. Mother species – father species: unidirectional hybridization in animals with female choice. *Anim. Behav.* **58**:1–12.

Woodruff, D. S. 1973. Natural hybridization and hybrid zones. *Syst.Zool.* **22**:213–218.

A Genetic Study of a Translocated Guenon: *Cercopithecus mona* on Grenada

8

K. ANN HORSBURGH, ELIZABETH MATISOO-SMITH, MARY E. GLENN, and KEITH J. BENSEN

Introduction

The endemic range of *Cercopithecus mona* is centered in Nigeria, extending west to the Volta River in Ghana and south across the Sanaga River in Cameroon including much of Togo and Benin (Lernould, 1988; Oates, 1988). *Cercopithecus mona* is now free-ranging on the West African islands of São Tomé and Príncipe, and Grenada, a small island in the Caribbean. The most likely explanation for the presence of mona monkeys on these islands is that they were transported by slave traders who had been working out of West Africa and shipping slaves and other cargo to the Americas. São Tomé was used

---

K. ANN HORSBURGH • Department of Anthropology, 701 East Kirkwood Avenue, Indiana University, Bloomington, IN 47405, USA. ELIZABETH MATISOO-SMITH • Department of Anthropology, The University of Auckland, Private Bag 92019, Auckland, New Zealand. MARY E. GLENN • Department of Anthropology, Humboldt State University, Arcata, CA 95521, USA. KEITH J. BENSEN • Windward Islands Research and Education Foundation, 11 East Main Street, Bayshore, NY 11706, USA.
The Guenons: Diversity and Adaptation in African Monkeys, edited by Glenn and Cords. Kluwer Academic/Plenum Publishers, New York, 2002.

as a staging post for transport of slaves from West Africa (Poiters, 1559, cited in Law 1991) and hundreds of thousands of slaves were shipped through São Tomé to Europe, the Americas and the Caribbean (Law, 1991). Grenada was the end point for many transport ships; thus, it is possible that there were multiple opportunities for mona monkeys to be introduced to both Grenada and São Tomé. Príncipe was not a stopping point en route to the slave markets, but it was a plantation island with its own populations of both African slaves and mona monkeys.

Methods

To determine the extent of the bottleneck experienced by the population of mona monkeys that colonized Grenada and to discover their geographical origins, we employed a molecular approach. The mitochondrial genome, and more specifically the control region, is a molecule particularly appropriate for the study of intraspecific variability (Brown *et al.*, 1979; Wilson *et al.*, 1985). We therefore focused on the mitochondrial control region to determine the history of the Grenada mona monkey population.

We collected mona monkey tissue samples during field seasons in Grenada between 1992 and 1994, and in São Tomé and Príncipe in 1998 (Glenn, 1996; Glenn *et al.*, unpub. report). Unfortunately, no fresh tissue samples were available for African mainland populations. However, skins had been collected in the early 1900s by naturalists, and these samples were stored and made available to us by the Powell-Cotton Museum, Kent, and the Museum of Natural History, London. In total, 45 samples from Grenada, 17 samples from São Tomé, four samples from Príncipe, ten skin snips from Cameroon, and two skin snips from Benin were available for genetic studies. Tissue samples were preserved in ethanol at ambient temperatures for extended periods; as a result many were degraded and no DNA was present. Skin snips were stored in unknown conditions in museums, and a few snips provided no extractable DNA. Table I presents the number of samples successfully analyzed in this study.

Table I. Sample Origins

Origin	Number	Obtained by
Grenada	21	Glenn and Bensen
São Tomé	12	Glenn and Bensen
Príncipe	4	Glenn and Bensen
Cameroon	9	Museum of Natural History, London
		Powell Cotton Museum, Kent
Benin	1	Museum of Natural History, London
Total	47	

We extracted whole genomic DNA from the modern tissue samples using a standard phenol/chloroform protocol (Sambrook *et al.*, 1989) and then PCR-amplified a 1 kb portion of the mitochondrial control region using universal mammalian primers (Kocher *et al.*, 1989). PCR products were directly sequenced at the Centre for Gene Technology, University of Auckland. The sequences obtained from the modern samples were aligned by eye. We used these to design an additional primer for use with the skin samples. Unfortunately, due to degradation of DNA post-mortem, DNA extracted from museum specimens (often referred to as ancient DNA, or aDNA) is often sheared. As a result, only small fragments of approximately 150–250 base pairs (bp) in length can be amplified from old or degraded material. We used the following primers for amplification from the skin extracts:

H 16419: 3'GCACTCTTGTGCGGGATATTG5'
Mona 1: 3'CCGAATATCAACCGAACAAC5'

The primer H 16419, previously designed for the human mitochondrial genome, was regularly in use at the Department of Anthropology, University of California, Davis, and was found to bind near a variable region identified in the aligned 1 kb fragment. The Mona 1 primer was designed to bind approximately 180 bp away, so that a 220 bp variable region of the d-loop was amplified (Horsburgh, 2000).

We extracted whole genomic DNA from the museum samples using a phenol/chloroform protocol (Sambrook *et al.*, 1989) and standard ancient DNA techniques to control for contamination (Hummel and Herrmann, 1994). All the samples were extracted and amplified in a physically isolated, dedicated ancient DNA lab, and negative extraction controls were run in parallel with all samples. The PCR products were purified using QIAquick silica columns according to the manufacturers' instructions, and were directly sequenced in both directions at the DNA Sequencing Facility, Division of Biological Sciences, University of California, Davis. The sequences obtained were then aligned with the modern sequences for phylogenetic analyses.

Data Analysis

This study includes the analysis of both modern and ancient samples. As such, some of the samples are represented by longer sequences than are others. Consequently, the bulk of the analyses described herein were undertaken on all the samples truncated to the length of the shortest sequences—137 bp in length. Running the analyses described below on the longer sequences showed that while the shorter sequences tended to result in lower indications of diversity, they did not alter the topology of the tree.

Genetic Diversity

The genetic characteristics of an island population are highly dependent on the characteristics of the colonizing generation, specifically on the size of the group. It is expected that the genetic diversity of an island population will generally increase with the size of the founding population and with any subsequent introductions to the population (Tajima, 1990). A measure of genetic diversity therefore gives a valuable indication of the number of introductions to a population. Consequently, we employed Nei's (1987, p. 179) gene diversity index (*h*) to compare the complexity of population histories of the various islands. Due to stochastic forces, some lineage extinction through time occurs in any population (Avise *et al.*, 1984). Consequently, there may have been more maternal lineages introduced to Grenada than are currently represented in the population.

Phylogeny

We constructed a phylogenetic tree using the computer program PAUP* 4.0 (Swofford, 1999) with a maximum likelihood optimality criterion. We used four different models of molecular evolution, chosen with assistance of the MODELTEST software (Posada and Crandall, 1998) to test whether there were particular biases in the data. All tests produced identical trees, and thus a simple evolutionary model could legitimately be applied (Horsburgh, 2000).

Results

The modern samples yielded an alignment of 499 bp, with 146 polymorphic sites and nine insertion/deletion events (indels). As expected in an analysis of variation within a species (Brown *et al.*, 1979), the number of indels was high and transitions outnumbered transversions.

The three haplotypes represented in the Grenada population are closely related to each other, differing by at most only 0.08%. It is possible that this level of diversification has occurred within the population since they were introduced to Grenada. If this is not the case, the founding females of the population had to have shared a relatively recent common ancestor. In contrast, the pairwise differences within the São Tomé population are much higher, reaching a maximum of 22.6%, which is as high as the comparisons between all island populations studied (Table II).

The shorter sequence alignment for 137 bp, which included the African samples from Cameroon, showed 48 polymorphic sites with ten indels. There were 43 substitution events, of which 29 were transitions and 14 were

Table II. Minimum and Maximum Percentage Pairwise Sequence Differences Calculated on the Basis of the Long Sequences (499 bp)

	Grenada		São Tomé		Príncipe	
	Min	Max	Min	Max	Min	Max
Grenada ($n = 21$)	0	0.08	0	22.6	1.05	3.95
São Tomé ($n = 12$)	0	22.60	0	22.6	1.05	22.60
Príncipe ($n = 4$)	1.05	3.59	1.05	22.6	0	2.52

transversions. The maximum pairwise divergence for the shorter sequences across all populations is 16.2%, which is also the maximum population pairwise difference within the São Tomé population (Table III).

To include the African sequences in the analysis, we had to reduce the amount of sequence studied to 137 bp per sample. To determine the effect of discarding a proportion of the modern sequence, we considered the relationship between the short and the long sequences. The modern sequences yield a maximum difference within São Tomé of 22.6% across the 499 bp. When the two most divergent sequences are compared in their truncated "ancient length" of 137 bp, this pairwise difference is 16.2%. It seems likely therefore that the maximum difference of 16.2% identified within Africa is an underestimate of the difference we would have obtained if we had been able to look at a full 499 bp for the African sequences.

Diversity Indices

The Grenada sample size is almost twice that of São Tomé, yet, within the long sequences that contain the most information, São Tomé has three times the number of haplotypes found in Grenada (Table IV). The diversity indices are high for each of the populations except Grenada. With a sample size of only four, the Príncipe results are unlikely to be a reliable figure for the true

Table III. Minimum and Maximum Percentage Pairwise Sequence Differences Calculated on the Basis of the Short Sequences (137 bp)

	Grenada		São Tomé		Príncipe		Africa	
	Min	Max	Min	Max	Min	Max	Min	Max
Grenada ($n = 21$)	0	0	0	15.32	0.09	2.70	3.60	11.71
São Tomé ($n = 12$)	0	15.32	0	15.32	0	16.22	2.07	16.22
Príncipe ($n = 4$)	0.09	2.70	0	16.22	0	1.80	0.09	12.61
Cameroon ($n = 10$)	3.60	11.71	2.07	16.22	0.09	12.61	0	16.22

Table IV. Nei's Diversity Index (*h*) and Number of Haplotypes for the Long (499 bp) and Short Sequences (137 bp)

Location	*n*	*h* (long)	Haplotypes long	*h* (short)	Haplotypes short
Grenada	21	0.1856	3	0	1
São Tomé	12	0.9697	9	0.7879	4
Príncipe	4	0.8333	3	0.8333	3
Cameroon	10	–	–	0.8667	6

level of diversity for the population as a whole. However, even with a sample size of four, Príncipe has, looking at the long sequences, the same number of haplotypes as Grenada, and the haplotypes are more divergent from each other than those on Grenada, implying a significantly higher level of diversity on Príncipe.

Phylogenetic Analyses

As noted previously, all of the models of molecular evolution tested resulted in only one tree (Fig. 1). It identifies two major clades (Fig. 1). Clade I contains the Grenada, the Príncipe and two of the São Tomé samples. Clade II contains the remaining São Tomé lineages and three of the African lineages. The remaining three African lineages are basal to both Clades I and II.

Discussion

The results of our analyses strongly suggest that there was a very limited introduction of founding females to Grenada. Mitochondrial DNA is maternally inherited and as such, it reflects the population movement of female monkeys only. Consequently, our results presented are uninformative as to the role of male mona monkeys in the history of these populations. The higher levels of mtDNA variation in Príncipe and more noticeably in São Tomé suggest a more complex history of introductions. Additional samples from both islands are needed, however, to more fully understand the true degree of variation.

The phylogenetic analyses strongly suggest that the founding population for Grenada mona monkeys came from São Tomé. It is also likely that the Príncipe population is the result of introductions from São Tomé. We would suggest, however, that either a larger number of individuals were involved or that introductions occurred over a period of time, as the Príncipe population is more variable than the Grenada population. The São Tomé population is

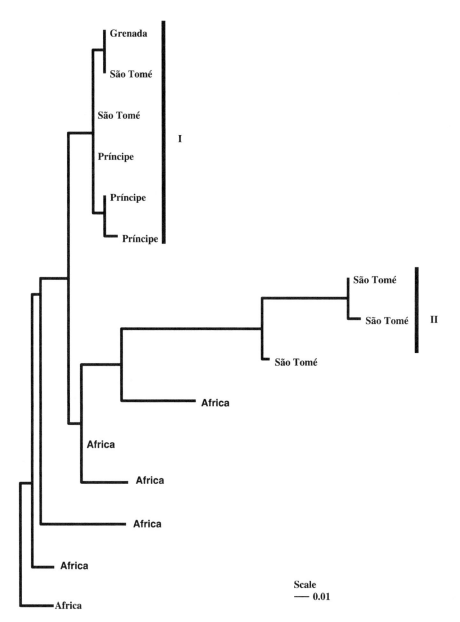

Fig. 1. Single best tree found by heuristic parsimony search ($n = 1000$ replicates) under a maximum likelihood criterion.

highly variable, even when compared with samples collected on at least two different expeditions in Cameroon, across a region significantly larger and more diverse than the island of São Tomé.

In previous studies of mtDNA control region variation, the highest pairwise differences reported were 18% in butterflyfishes (McMillan and Palumbi, 1997), and 6.6 and 9.5% in ocelots and margays, two South American cats (Eizirik *et al.*, 1998). Therefore the 22.6% maximum pairwise distance found in mona monkeys from São Tomé is remarkable. Without further study of the Cercopithecids it is difficult to determine the cause of this extreme variation. It is possible that our results indicate a hybridization event. Although the frequency of hybridization within the guenons is unknown, Struhsaker (1970) reported numerous hybrids in Cameroon guenons including *Cercopithecus mona–C. pogonias* crosses, *C. nictitans–C. mona* crosses and *C. erythrotis–C. cephus* hybrids. In addition, high numbers of *Cercopithecus ascanius–C. mitis* hybrids have been observed (Detwiler, 2002). While the F1 hybrids are apparently identifiable phenotypically, it is unknown how long the morphological characteristics would persist in future generations. It is certainly possible that a second species of guenon was taken to São Tomé and that it was present in such low numbers that there is now no trace of it phenotypically. With the further genetic study of guenons it may be possible to determine if such a hybridization event did take place, and if so, to identify the species involved.

If we were to reject the possibility of hybridization in São Tomé mona monkeys, we would have to suggest that the population was derived from at least two geographically diverse African populations. This would not be unlikely given the wide geographical range of the African mona monkeys and the widespread activity of the Portuguese slave traders. Clearly, given the range of *Cercopithecus mona* in Africa, our ten samples are unlikely to represent the total diversity of African mona monkeys. Perhaps after further African sampling, the Grenada and Príncipe samples that presently cluster with those from São Tomé would associate more closely with African lineages we have not yet identified.

Ultimately, our data, in conjunction with information about other guenons, may help to shed light on the processes of speciation. The guenons are a particularly interesting group with which to investigate speciation because they have undergone a rapid radiation into some 23 species within the last million years (Leakey, 1988). In addition, the population of mona monkeys on Grenada may offer a unique opportunity to follow the evolution of an isolated monkey population with limited genetic diversity.

A number of important questions for future research have been raised by our work. Among them is the degree to which the African mona monkey populations are genetically structured; that is, do they exhibit clinal genetic variation? Secondly, the divergent mitochondrial lineages on São Tomé may represent a series of ancient hybridization events between mona monkeys and another guenon species. Alternatively, the São Tomé lineages represent a portion of the African mona monkey population that has yet to be sampled. Both of these alternatives are likely the explanations for the high diversity on

São Tomé. More broadly, as a piece of the speciation puzzle, given that guenons are capable of extensive and successful hybridization, they are a valuable group in which to study the maintenance of species identity.

By further studying all facets of each of the populations of mona monkeys, and eventually other guenon species, an understanding of the mechanisms that turn populations of one species into two may be developed. There is a great deal to be gained from the multidisciplinary study of guenons, which diversified so extensively and so recently within and beyond the forests of West Africa.

Summary

At some point during the last 350 years, *Cercopithecus mona*, the mona monkey, was taken from West Africa to the islands of Grenada, in the Caribbean, and São Tomé and Príncipe, off the west-central coast of Africa. In this chapter, we present results of a mitochondrial DNA analysis of the Grenada monkeys that was undertaken to assess the degree of mtDNA variation in the population in order to estimate the number of founding mtDNA lineages and the origin(s) of those lineages. Historical records indicate that the mona monkeys on Grenada may have been introduced from São Tomé rather than directly from Africa. Mitochondrial DNA sequences from the Grenada monkeys were therefore compared with those obtained from extant populations on the islands of São Tomé and Príncipe, and with the sequences extracted from museum skin samples from West African monkeys collected in the early 1900s. In addition to answering the specific questions regarding the histories and origins of these island populations, this and future genetic studies of these isolated populations may shed light on broader evolutionary questions such as the relationship between isolation and speciation events, and the effects of reduced genetic variation on population biology and behavior.

ACKNOWLEDGMENTS

Funding for the research conducted by M.E.G. and K.J.B. was provided by ECOFAC, the Windward Islands Research and Education Foundation, The Rockefeller University, Yerkes Regional Primate Research Center, Wildlife Conservation Society, and the Center for Environmental Research and Conservation at Columbia University. We are grateful to the Grenada National Parks, the São Tomé and Príncipe ECOFAC staff, and the Governments of Grenada, and São Tomé and Príncipe for allowing us to conduct this research. Special thanks go to Heather Bruce and Kevin Jensen for their help in the field in Grenada, and São Tomé and Príncipe, respectively. We are also grateful to

Marina Cords, Thelma Rowell and Russell Tuttle for their helpful comments on this paper.

Funding for the research conducted by K.A.H. and E.M.-S. was provided by the Windward Islands Research and Education Foundation. We thank the Powell-Cotton Museum, Kent, and the Museum of Natural History, London for samples. We are also grateful to David G. Smith, at the University of California, Davis, for the provision of PCR primers and absorbing laboratory costs for the portion of this research conducted in his lab.

References

Avise, J. C., Neigel, J. E., and Arnold, J. 1984. Demographic influences on mitochondrial DNA lineage survivorship in animal populations. *J. Mol. Evol.* **20**:99–105.

Brown, W. M., George, M., and Wilson, A. C. 1979. Rapid evolution of animal mitochondrial DNA. *Proc. Natl. Acad. Sci. USA* **76**:1967–1971.

Detwiler, K. M. 2002. Hybridization between red-tailed monkeys (*Cercopithecus ascanius*) and blue monkeys (*C. mitis*) in East African forests. In: M. E. Glenn, and M. Cords (eds.), *The Guenons: Diversity and Adaptation in African Monkeys*, pp. 79–97. Kluwer Academic Press, New York.

Eizirik, E., Sandro, L., Bonatto, L., Johnson, W. E., Crawshaw Jr., P. G., Vié, J. C., Brousset, D. M., O'Brien, S. J., and Salzano, F. 1998. Phylogeographic patterns and evolution of the mitochondrial DNA control region in two neotropical cats (Mammalia, Felidae). *J. Mol. Evol.* **47**:613–624.

Glenn, M. E. 1996. *The Natural History and Ecology of the Mona Monkey* (Cercopithecus mona *Schreber 1774*) *on the Island of Grenada, West Indies*. Ph.D. dissertation, Northwestern University.

Glenn, M. E., Bensen, K. J., and Jensen, K. F. (unpub. report). Comparative Studies of Island and Mainland Dwelling Populations of Mona Monkeys: São Tomé and Príncipe, Research Report to AGRECO and ECOFAC Africa, May 1999, 37 pp.

Horsburgh, K. A. 2000. *A Genetic Study*: Cercopithecus mona *on Grenada*. MA thesis. University of Auckland.

Hummel, S., and Herrmann, B. 1994. General aspects of sample preparation. In: B. Herrmann, and S. Hummel (eds.), *Ancient DNA Recovery and Analysis of Genetic Material from Paleontological, Archaeological, Museum, Medical, and Forensic Specimens*. Springer, New York.

Kocher, T. D., Thomas, W. K., Meyer, A., Edwards, S., Pääbo, S., Villablanca, F. X., and Wilson, A. C. 1989. Dynamics of mitochondrial DNA evolution in animals: amplification and sequencing with conserved primers. *Proc. Natl. Acad. Sci. USA* **86**:6196–6200.

Law, R. 1991. *The Slave Coast of West Africa 1550–1750. The Impact of the Atlantic Slave Trade on an African Society*. Clarendon Press, Oxford.

Leakey, M. 1988. Fossil evidence for the evolution of guenons. In: A. Gautier-Hion, F. Boulière, J.-P. Gautier, and J. Kingdon (eds.), *A Primate Radiation: Evolutionary Biology of the African Guenons*, pp. 7–12. Cambridge University Press, Cambridge.

Lernould, J.-M. 1988. Classification and geographical distribution of guenons: a review. In: A. Gautier-Hion, F. Boulière, J.-P. Gautier, and J. Kingdon (eds.), *A Primate Radiation: Evolutionary Biology of the African Guenons*, pp. 57–78. Cambridge University Press, Cambridge.

McMillan, W. O., and Palumbi, S. R. 1997. Rapid rate of control-region evolution in Pacific butterflyfishes (Chaetodontidae). *J. Mol. Evol.* **45**(5):473.

Nei, M., 1987. *Molecular Evolutionary Genetics*. Columbia University Press, New York.

Oates, J. F. 1988. The distribution of *Cercopithecus* monkeys in West African forests. In: A. Gautier-Hion, F. Boulière, J.-P. Gautier, and J. Kingdon (eds.), *A Primate Radiation: Evolutionary Biology of the African Guenons*, pp. 79–103. Cambridge University Press, Cambridge.

Posada, D., and Crandall, K. A. 1998. Modeltest: testing the models of DNA substitution. *Bioinformatics* **14**:817–818.

Sambrook, J., Fritsch, E. F., and Maniatis, T. 1989. *Molecular Cloning: A Laboratory Manual, 2nd ed.*, Cold Spring Harbor Laboratory Press, Salem, MA.

Struhsaker, T. T. 1970. Phylogenetic implications of some vocalisations of *Cercopithecus* monkeys. In: J. R. Napier, and P. H. Napier (eds.), *Old World Monkeys: Evolution, Systematics and Behaviour*, pp. 365–411. Academic Press, London.

Swofford, D. L. 1999. *Phylogenetic Analysis Using Parsimony (PAUP*) Version 4.0.* Sinauer Associates Inc., MA.

Tajima, F. 1990. Relationship between migration and DNA polymorphism in a local population. *Genetics* **126**:231–234.

Wilson, A. C., Cann, R. L., Carr, S. M., George, M., Gyllensten, U. B., Helm-Bychowski, K. M., Higuchi, R. G., Palumbi, S. R., Prager, E. M., Sage, R. D., and Stoneking, M. 1985. Mitochondrial DNA and two perspectives on evolutionary genetics. *Biol. J. Linn. Soc.* **26**:375–400.

Part II

Behavior

Diversity of Guenon Positional Behavior

9

W. SCOTT McGRAW

Introduction

The guenons are among the most species-rich groups of primates yet there have been few studies of their positional behavior. They are routinely characterized as a postcranially uniform radiation consisting largely of generalized arboreal quadrupeds (Schultz, 1970; Fleagle, 1999). These are convenient and, in some cases, accurate labels; however, they also obscure variation that is useful for understanding broader aspects of guenon biology. For example, an examination of the amount of time some members of this radiation spend on the ground reveals interesting differences among species. Here I first briefly review the development of a few significant ideas about guenon positional behavior. One of the most important issues for primatologists concerns the evolution of terrestriality, so I next summarize current data on the ground-dwelling tendencies of extant guenons. Finally, I assess the diversity of locomotor and postural behavior in free-ranging populations and examine how positional diversity is associated with other aspects of guenon natural history including diet, body size and limb length.

W. SCOTT McGRAW • Department of Anthropology, The Ohio State University, 1680 University Drive, Mansfield, OH 44906, USA.

The Guenons: Diversity and Adaptation in African Monkeys, edited by Glenn and Cords. Kluwer Academic/Plenum Publishers, New York, 2002.

Studies of Guenon Positional Behavior and Anatomy: A Brief Review

The first field studies of guenons were concerned primarily with mapping their distributions (Booth, 1955, 1956, 1958) or with traditional aspects of behavioral ecology such as feeding, ranging, communication and social structure. Positional behavior *per se* was not a principal area of study, and information about guenon locomotion and posture accumulated sporadically from accounts of field biologists with other interests. Haddow (1952) described general movement patterns of red-tailed monkeys (*Cercopithecus ascanius*) while Tappen (1960) summarized guenon habitat use, particularly the occupation of forest strata, with occasional remarks on locomotion. The first researchers to specifically address issues of positional behavior were Napier (1962), and Ashton and Oxnard (1964a), who sought associations between locomotor categories and the occupation of specific strata. They characterized tree-dwelling guenons as generalized quadrupeds that differed primarily in the canopy levels in which they moved. Studies in the mid-1960s on patas monkeys (Hall, 1965) and vervets (Brain, 1965; Hall and Gartlan, 1965; Struhsaker, 1967; Gartlan and Brain, 1968; Fedigan and Fedigan, 1988) contained brief remarks on positional behavior as it related to intraspecific communication and vigilance. The study of *Cercopithecus campbelli* by Bourlière *et al.* (1970) is noteworthy for including the first detailed description of an arboreal guenon's movement. The major qualifier of this study was the highly altered habitat. Rose (1974, 1978, 1979) provided the first data on guenon positional behavior with a study of vervets. Over the next 20 years there was a significant increase in guenon field research (Gautier-Hion and Gautier, 1974; Gautier-Hion, 1978, 1988a,b; Struhsaker, 1978, 1980, 1981; Struhsaker and Leland, 1979; Cords, 1984, 1987; Galat and Galat-Luong, 1985; Oates, 1985; Gautier, 1988), but locomotion and posture of free-ranging guenons went largely unstudied. Only two other quantitative field studies of guenon positional behavior have been conducted since Rose's work: Gebo and Chapman's (1995a,b, 2000) study in Uganda's Kibale Forest and my study of cercopithecids in Ivory Coast's Taï Forest (McGraw, 1996, 1998a,b,c, 2000). These studies notwithstanding, guenons continued to be characterized as consisting primarily of locomotively equivalent tree dwellers and a few semiterrestrial species, most notably vervets (Rowe, 1996; Fleagle, 1999).

Studies of guenon anatomy and positional behavior have traditionally focused on two areas: (1) gait characteristics and kinematics, and (2) the relationship between limb morphology and stratal occupation, including the ground. Ashton and Oxnard (1964a,b) and Ashton *et al.* (1965) identified some skeletal correlates of gross locomotor tendencies focusing principally on the forelimb and shoulder girdle. Jolly (1965) assigned arboreality ratings to a number of guenons and related the degree of tree dwelling to differences in limb proportions (e.g., arm and forearm length increase with increasing terrestriality). Tuttle (1969) related patterns of digitigrady to gross habitat preference in a large sample of catarrhines including three guenon species. Manaster (1979) undertook a more comprehensive study of the guenon locomotor skeleton operating on the assumption that average height in the trees reflected degree of acrobatic locomotion. Her analysis yielded some features that separated the most terrestrial species (*Cercopithecus aethiops*) from the more arboreal species and, to a lesser extent, lower arboreal and ground-dwelling forms from upper arboreal species. She concluded that the ancestral guenon was most likely a generalized, middle-canopy tree-dweller.

Rollinson and Martin (1981) examined gait by studying films of six semi-free-ranging guenons (*Cercopithecus nictitans, C. cephus, C. pogonias, C. neglectus, C. aethiops* and *Miopithecus talapoin*). They argued that the *Cercopithecus* spp. were descended from a semiterrestrial cercopithecine ancestor and had become secondarily arboreal. Meldrum (1991) used these films to study foot kinematics of the six species and concluded that semiterrestrial and arboreal guenons were atypical in not showing the flexed abducted limbs that are common in arboreal quadrupeds. He suggested that this was evidence of a recent terrestrial past. Gait and limb mechanics of *Cercopithecus aethiops* have been particularly well studied in captivity (Larson and Stern, 1989; Vilensky and Gankiewicz, 1990; Whitehead and Larson, 1994; Larson, 1998; Schmitt, 1998). These studies have confirmed that certain aspects of quadrupedalism in primates are unique. Kingdon (1988) studied hands and feet of a small sample of guenons and concluded that a mosaic of features in guenons hinted at an ancestral condition somewhere between the morphology characterizing fine-branch arborealism and semiterrestriality.

The most recent study of guenon limb anatomy is that of Gebo and Sargis (1994) who identified characters associated with terrestriality in *Cercopithecus lhoesti* but found few features that distinguish *C. aethiops* from the arboreal forms or ones that differentiate arboreal species. They concluded that the ancestral guenon was most likely arboreal and that terrestriality in guenons evolved independently at least twice. In addition, various authors have included guenons in broader analyses of limb proportions, size, scaling and biomechanics (Jungers, 1984; Strasser, 1987, 1992, 1994; Jungers and Burr, 1994; Jungers *et al.*, 1998).

Most studies related to guenon positional behavior have been concerned principally with where it takes place, e.g., in the trees or on the ground or mid-canopy vs. high canopy. Considerable attention has been directed at identifying features related to terrestriality or semiterrestriality in extant guenons and what this information implies for the ancestral guenon's locomotor mode and substrate preference. Inferring the positional habits of the guenon ancestor is important for understanding the clade's evolution (Strasser, 1987; Leakey, 1988; Pickford and Senut, 1988; McCrossin *et al.*, 1998; Tosi *et al.*, 2002). I believe, however, that by concentrating on issues of guenon terrestriality we have ignored the diversity of their arboreal behavior. Moreover, basic information about the habitat preference of certain extant taxa may be inaccurate or entirely lacking, which only confounds attempts to identify the morphological correlates of ground movement. For example, in their attempt to associate skeletal anatomy and behavior, Ashton and Oxnard (1994a) assumed that *Cercopithecus diana* was a high canopy monkey that rarely descended to the understory or ground. In fact, diana monkeys frequently descend below the middle canopy and occasionally forage on the forest floor (McGraw, 1996, 1998a,b,c). Stratal occupation is a crucial component of positional behavior, but important differences also exist between similarly arboreal species. Although positional data on guenons are still scarce, there is enough information to begin looking at patterns and differences in behavior. In this chapter, I will summarize the major differences among species based on available data and examine how guenon locomotion and posture is correlated with other aspects of their biology.

Methods

The positional data that I discuss are from three quantitative field studies (Rose, 1974, 1978, 1979; Gebo and Chapman, 1995a,b, 2000; McGraw, 1996, 1998a,b,c, 2000). These data are from six out of approximately 23 guenon species. This is a small percentage, but they include at least one member from most of the major guenon species groups (Lernould, 1988; *sensu* Oates, 1988): *Cercopithecus cephus* and *C. ascanius* from the cephus group; *C. campbelli* from the mona group; *C. mitis* from the nictitans group; *C. diana* from the Diana/dryas group; *C. aethiops* from the vervet group. I assume that each group or superspecies consists of closely related species that are behaviorally more similar to one another than they are to members of other groups. The taxonomic breadth of the data allows investigation of the locomotor diversity among the arboreal forms. I will not discuss how differences in methods, definitions and protocols might render interspecific comparisons inappropriate, a topic discussed at length by Hunt *et al.* (1996), Dagasto and Gebo (1998), McGraw (1998b, 2000) and Moffett (2000). Different methods were used to collect the data that I discuss and interested readers should refer to the original papers for methodological details of each study. There are two caveats: McGraw (1996, 1998a,b,c, 2000) collected data on adult females only, while Gebo and Chapman (1995a,b, 2000) combined data from both sexes. Since most guenon groups contain more adult females than males, I assume that data from Kibale on *Cercopithecus ascanius* and *C. mitis* derive predominantly from female subjects. The correlations between positional behaviors and body size are based on bodily weights of adult females only (Smith and Jungers, 1997). Secondly, the locomotor frequencies in each correlation are for overall locomotion (or travel and feeding combined). Important variation exists between the two maintenance activities; however, I combined them because of differences across studies in how they are defined. I culled data on terrestriality from these studies and from additional literature.

Results

Terrestriality in Guenons

Most guenons are primarily tree dwellers (Gautier *et al.*, 2002). As more field data are collected it becomes increasingly clear that virtually all guenons spend some time on the ground (Gartlan and Brain, 1968; Kavanagh, 1980; Lawes, 1990; McGraw, 1994; Gonzalez-Kirchner, 1996). The critical issue for primatologists is how much terrestriality varies among taxa. Published data on terrestriality are presented in Table I. Most authors agree that the patas monkey is the most terrestrial guenon (Isbell *et al.*, 1998) yet no published data exist on the percentage of time they spend on the ground. Chism and Rowell (1988) report that one group spent nearly 70% of its time in open acacia woodland, and it can be inferred that this habitat preference probably required much ground movement. A recent study on semifree-ranging *Cercopithecus solatus* (Peignot *et al.*, 1999) indicates that the sun-tailed guenon may rival the patas as the most terrestrial species, spending nearly 86% of the time on the forest floor (Harrison, 1988). This figure is more than twice that reported previously (Gautier, Gautier-Hion and Loireau, unpub. data reported in Gautier-Hion, 1988b). Additional

Table I. Frequency of Terrestriality in 13 Species of Guenons

Species	% Terrestriality	Site	Reference
C. solatus	86	(CIRMF) Gabon	Peignot *et al.* (1999)
C. lhoesti	38	Nyungwe Forest, Rawanda	Kaplin and Moermond (2000)
C. preussi	35	Cameroon	Gartlan and Struhsaker (1972) and Gautier-Hion (1988b)
C. neglectus	30	Gabon	Quris (1976) and Gautier-Hion and Gautier (1978)
C. aethiops	20	Lake Navasha, Kenya	Rose (1974)
C. campbelli	15.2	Taï, Ivory Coast	McGraw (1998a)
C. cephus	11	Makokou, Gabon	Gautier-Hion and Gautier (1974)
C. ascanius	10	Ituri Forest, DRC	Thomas (1991)
C. mitis	5	Ituri Forest, DRC	Thomas (1991)
M. talapoin	5	Gabon	Gautier-Hion (1971)
C. pogonias	<2	Makokou, Gabon	Gautier-Hion and Gautier (1974)
C. diana	<2	Taï, Ivory Coast	McGraw (1996, 1998a)
C. petaurista	<2	Taï, Ivory Coast	McGraw (2000)

studies on sun-tailed guenons in Gabon under completely natural conditions are needed to confirm this result (Gautier *et al.*, 1992; Brugiere *et al.*, 1998). After these two apparently extreme taxa, there is a significant decline in the degree of ground dwelling among guenons. Struhsaker (1969) was the first to note the terrestrial tendencies of *Cercopithecus lhoesti*. Kaplin and Moermond (2000) report that l'Hoest's monkey in Rwanda spends 38% of time traveling and foraging among dense vegetation on the ground, a figure slightly greater than that seen in the closely related *Cercopithecus preussi* (Kaplin, 2002). Vervets (*Cercopithecus aethiops*) are frequently thought of as particularly ground adapted, yet the only published figures available (Rose, 1974) indicate that they spend only about 20% of the time terrestrially. This figure is 10% less than that for two groups of *Cercopithecus neglectus* (Quiris, 1976; Gautier-Hion and Gautier, 1978) and <5% more than that of *C. campbelli* (McGraw, 1996, 1998a), a true forest dweller. Two members of the cephus group, *Cercopithecus cephus* and *C. ascanius*, are similar in the degree of terrestriality, 11 and 10% respectively. The remaining guenons, including talapoin monkeys, are reported to spend ≤ 5% of time on the ground.

Locomotor Diversity of Arboreal Guenons

Locomotor frequencies for six guenon species are presented in Table II. Most guenons are predominantly quadrupedal, spending ≥50% of time either walking or running. The average for the six species is 55%. The possible

Table II. Overall Locomotion (Travel and Feeding Combined) of Six Guenon Species

Species	Site	Quad-rupedalism	Leap-ing	Climb-ing	Reference	N
C. ascanius	Kibale, Uganda	39.0	15.0	43.0	Gebo and Chapman (1995a,b)	3653
C. aethiops	Lake Naivasha, Kenya	53.9	9.6	29.5	Rose (1979)	Not available
C. mitis	Kibale, Uganda	54.0	11.0	35.0	Gebo and Chapman (1995a,b)	2182
C. diana	Taï, Ivory Coast	70.1	10.4	19.4	McGraw (1996, 1998a)	1553
C. petaurista	Taï, Ivory Coast	71.1	10.1	18.8	McGraw (1996, 1998a)	512
C. campbelli	Taï, Ivory Coast	80.3	5.2	14.4	McGraw (1996, 1998a)	596

NB: "Other" locomotor behaviors are not included. Data from McGraw (1996, 1998a) were collected using time point sampling at 3 minute intervals. Data from Gebo and Chapman (1995a,b) were collected using the bout sample method. Differences in sample size among species are due in part to differences in sampling protocol.

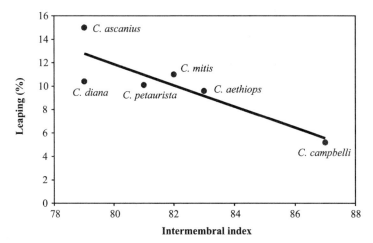

Fig 1. Leaping frequencies compared with intermembral index in five species of guenons.

exception is red-tailed monkeys in Kibale Forest, Uganda, which are quadrupedal only 39% of the time (Gebo and Chapman, 1995a). The range in climbing frequencies is considerable: *Cercopithecus campbelli* climbs 14% of time and *C. ascanius* 43% of time. The average for the group is 27%. All guenons are capable leapers, but none leap >15% of time. Leaping is the least variable locomotor behavior, ranging between 5 and 15%. Based on these numbers, the average guenon is predominantly quadrupedal. The amount of climbing varies significantly and most guenons leap only about 10% of time.

Locomotion, Bodily Size and Limb Proportions

Previous studies of primate locomotion demonstrated predictable relationships between leaping, climbing and bodily size. In some primate groups, climbing increases with bodily size (Fleagle and Mittermeier, 1980). Among guenons, there is no relationship between the two variables ($r = 0.08, p = 0.7$). Similarly, unlike among other primate groups, smaller guenons do not leap more than larger ones ($r = -0.06, p = 0.66$). The relationship between leaping and limb proportions is stronger though not statistically significant (Fig. 1); guenons with relatively long hind limbs leap more than do those with shorter hindlimbs ($r = -0.684, p = 0.027$).

Locomotion and Diet

Fleagle (1984) found few predictive relationships between locomotion and diet among primates. Other authors have used diet to explain differences in

Table III.   Diet of Five Species of Guenons

Species	Leaves	Fruit	Insects	Other	N	Reference
C. ascanius	15.0	43.7	21.8	19.5	1,327	Struhsaker (1978)
C. ascanius	7.2	61.3	25.1	6.4	9,009	Cords (1986)
C. mitis	20.6	45.1	19.8	14.5	2,566	Struhsaker (1978)
C. mitis	18.9	54.6	16.8	9.7	10,167	Cords (1986)
C. petaurista	5.5	77.2	7.3	10.0	110	Galat and Galat-Luong (1985)
C. campbelli	0.0	78.3	15.2	6.5	46	Galat and Galat-Luong (1985)
C. diana	3.5	52.0	37.8	6.7	4,415	Wachter *et al.* (1997)

the extent that specific behaviors were used by two taxa. For example, Struhsaker (1979, 1980) argued that red-tailed monkeys in Kibale climbed more than red colobus monkeys because of increased reliance on insects by the guenons. McGraw (2000) used the same logic when comparing climbing in *Cercopithecus petaurista* and *C. ascanius*: the more insectivorous red-tailed monkeys at Kibale climbed more than did the lesser spot-nosed monkeys at Taï. Dietary data for five guenons are presented in Table III. Among this larger sample, the relationship breaks down. Increased climbing is not significantly associated with high degrees of insectivory ($r = 0.11$, $p = 0.86$). The amount of climbing, however, appears to be negatively correlated with the amount of fruit intake ($r = 0.83$, $p = 0.08$, Fig. 2). It must be noted that the diets of at least two species, *Cercopithecus ascanius* and *C. mitis*, have been shown to vary at different localities (Struhsaker, 1978; Cords, 1986; Lambert, 2002). In the absence of positional data from all sites where feeding data are available, it is most appropriate to use the dietary profiles on groups at localities where locomotor data were gathered. This is done to minimize confounding

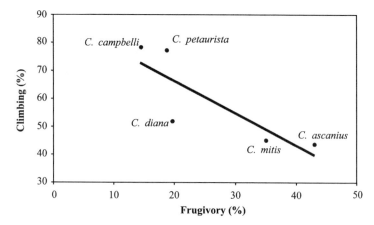

Fig. 2.   Climbing frequencies compared with percent fruit intake in five species of guenons.

variation between sites (by comparing locomotion of a species at one site with the diet of that species at another site); however, it is not yet known whether guenon locomotion is as variable between sites as is diet.

Postural Diversity in Arboreal Guenons

Postural data for five guenons are presented in Table IV. Two behaviors account for ≥75% of all postures among the species. Sitting is the most common posture representing 50–75% of all postures in all species. Quadrupedal standing is the second most common posture in all species with the greatest frequency (34.8%, *Cercopithecus campbelli*), twice that of the least (17.4%, *C. petaurista*). With one exception, no other behavior contributes >8% to the postural profile of any guenon: reclining and suspension are rarely performed behaviors. The single exception is the comparatively high (19.9%) frequency with which *Cercopithecus diana* employs a stationary climb (SC) posture (Hunt *et al.*, 1996; McGraw, 1998c).

Posture and Diet

McGraw (1998c) argued that postural differences among Taï colobines and cercopithecines could be explained by the spatial distribution of preferred food items: insect- and fruit-eating promotes less permanent (e.g., sitting, bipedal stand, stationary climb) postures than did leaf eating (Fleagle, 1984). Among the larger sample of guenons, insect eating can explain in part the distribution of postures: as insect eating increases, the amount of sitting tends to decrease (Fig. 3, $r = 0.67$, $p = 0.2$). Diana monkeys are the most insectivorous guenons in the sample and they sit the least. Quadrupedal postures account for just under 26% of their postural profile. Transitional postures that allow more efficient movement to the next feeding patch, i.e., a bug, account for almost 20% of diana monkey posture, nearly three times the next highest total. The least insectivorous guenon, *Cercopithecus petaurista*, sits the most, stands the least, and uses transitional postures just over 6% of total feeding time.

Discussion

Guenons show considerable variation in the extent to which they frequent the ground. With the exception of patas and perhaps sun-tailed monkeys, all guenons are predominantly arboreal, yet the degree of ground movement varies from <5% (*Cercopithecus petaurista*, *C. diana*, *Miopithecus*

Table IV. Postures during Feeding of Five Guenon Species

Species	Site	Sit	Quadrupedal stand	Bipedal stand/ stationary climb	Recline	Other	N	Reference
C. ascanius	Kibale, Uganda	69.0	25.0	5.0	<1	<1.0	2309	Gebo and Chapman (1995a,b)
C. mitis	Kibale, Uganda	65.0	29.0	3.0	2	<1.0	2355	Gebo and Chapman (1995a,b)
C. petaurista	Taï, Ivory Coast	74.6	17.4	6.2	0	1.8	780	McGraw (2000)
C. campbelli	Taï, Ivory Coast	54.1	34.8	7.9	0	3.2	508	McGraw (1998b, 2000)
C. diana	Taï, Ivory Coast	51.5	25.9	19.9	0	2.7	1468	McGraw (1998b, 2000)

NB: Data from McGraw (1998b, 2000) were collected using time point sampling at 3 minute intervals. Data from Gebo and Chapman (1995a,b) were collected using the bout sampling method.

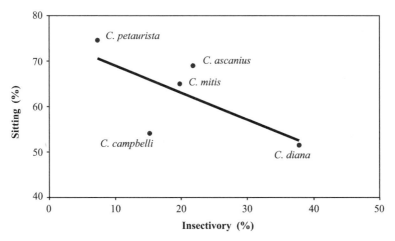

Fig. 3. Sitting frequencies compared with percent insect intake in five species of guenons.

talapoin) to nearly 40% (*C. lhoesti, C. preussi*). Vervets are usually regarded as ground-adapted, yet published data indicate that they may be significantly more arboreal than *Cercopithecus neglectus* for example. This surprising conclusion is probably due to lack of information on vervet monkeys in other areas of Africa. Additional data are needed on how often other populations use the ground. If the frequencies published by Rose (1974), however, are representative of the whole group, this may help to explain why morphological studies have revealed few terrestrial features in postcrania of *Cercopithecus aethiops* (Ashton and Oxnard, 1964a,b; Jolly, 1965; Manaster, 1979; Meldrum, 1989, 1991; Gebo and Sargis, 1994).

The range of ground-dwelling within this purported arboreal radiation is significant and may require a reassessment of what it means to be semi-terrestrial as well as which species fit the category. Nevertheless, our inability to even specify the terrestriality of patas monkeys highlights the gaps in our knowledge and underscores the need for more field data. Only by identifying how much time guenon species actually spend on the ground will we be able to successfully correlate morphology, locomotion and habitat use in extant as well as extinct taxa.

Locomotor data are available for only six guenon species: five predominantly forest-dwelling species and vervet monkeys. With one exception, running and walking dominate their locomotion; the frequency of quadrupedalism varies between 54 and 80%. Leaping is the least common locomotor behavior in all six guenon species, and increased leaping is associated with a low intermembral index. The most variable behavior is climbing, a behavior not related to differences in bodily size but that may be associated with interspecific dietary requirements. For example, the amount of climbing decreases as fruit intake increases. If locomotion during feeding is being driven by the spatial distribution of preferred foods, however, one would expect climbing to be correlated positively with degree of insectivory. It is not, and additional data are needed to evaluate the complex relationship between locomotion, diet and bodily size. This is especially true with regard to climbing, a critically important but difficult behavior to define consistently in the field (Tuttle, 1975; Fleagle *et al.*, 1981; Gebo, 1996; McGraw, 2000).

A stronger case can be made for the relationship between diet and posture. In the sampled taxa, posture varies less than locomotion, but species that rely more on patchily distributed (and mobile) foods such as insects, sit less, stand more, and employ more transitional postures. These postural tendencies may be adaptations for more efficient capture of foods that are relatively scarce. Unlike colobines, guenons do not employ reclining postures often, if at all. The sprawling postures of colobines are probably related to the requirements of leaf processing and are enabled by a more mobile hip and shoulder joint.

Guenons appear to violate some relationships among positional behavior, habitat use and bodily size that characterize other primate groups. Unlike platyrrhines (Fleagle and Mittermeier, 1980), neither climbing nor leaping frequencies change systematically with increasing bodily size among guenons (Gebo and Chapman, 1995b). Data on additional taxa are needed, but current findings suggest that as a group, guenons are capable of a high degree of behavioral flexibility that may not correspond closely with their anatomies and that strategies of movement are as important as the underlying morphologies. For example, in Taï Forest, similarly sized species using broadly equivalent forest strata during feeding can move in strikingly different ways as a function of cryptic (*Cercopithecus campbelli*) vs. conspicuous foraging strategies (*C. diana*) (McGraw, 1998a, 2000).

Certainly the best test to assess the flexibility of guenon positional behavior would be to study a single species at two different sites (Doran and Hunt, 1994; Garber and Pruetz, 1995). This has not yet been done for guenons, though Gebo and Chapman (1995b) and McGraw (1996) examined behavior of the guenon communities in structurally different portions of the Kibale and Taï Forests, respectively. They arrived at different conclusions: Kibale guenon locomotion varied with forest type while that of the Taï guenons did not. Comparison of two members of the same species group may shed additional light on the issue. *Cercopithecus petaurista* and *C. ascanius*, two similarly sized members of the *cephus* superspecies (Lernould, 1988; Oates, 1988), differed significantly in their locomotion between the Taï and Kibale Forests despite them having very similar morphologies (Gebo and Sargis, 1994; McGraw, 2000). Possible explanations for the variation include differences in methodology, forest structure, diet and other unknown factors. These results hint at the possibility that closely related and morphologically similar guenons are quite flexible in their movement strategies.

Compared to apes and platyrrhines (Dagosto and Gebo, 1998), quantitative positional data are rare for guenons. This is perhaps due to the perception that they do not warrant additional study: if you know the positional tendencies of one *Cercopithecus* species, you know them all. This is a curious attitude given the level of diversity in other aspects of guenon behavior and the time and effort spent documenting and explaining it (Gautier-Hion *et al.*, 1988). While we know comparatively little about guenon locomotion and posture, existing data reveal significant diversity in arboreal behavior. Viewing the group as an anatomically uniform radiation that differs primarily in where arboreal locomotion occurs overlooks important variation in frequencies of ground movement, quadrupedism and climbing. By exploring this diversity we may gain additional insight into the natural history of this successful primate radiation. The unanswered questions raised in this paper should lead to additional research on guenon positional behavior.

Summary

I assess the positional diversity characterizing the guenons and examine the correlation between locomotion and posture with other aspects of guenon biology including diet, body size and limb length. I discuss data from three field studies on six species representing most guenon species groups. Additional data on ground dwelling were culled from the literature. Results indicate that terrestriality varies widely among guenons. Current data indicate that *Cercopithecus solatus* may be the most ground-dwelling member of this group though additional data are needed for confirmation. Most guenons are predominately quadrupedal, most leap approximately 10% of the time and climbing is the most variable locomotor behavior. Climbing and leaping are not correlated with body size; intermembral index is a much better predictor of leaping. Certain elements of positional behavior are associated with guenon diets. Climbing is negatively correlated with fruit intake while species that eat more insects sit less and employ more transitional postures. Additional field studies are needed to understand more fully the diversity of positional behavior within the guenon radiation, and thus to allow more stringent tests of the relationships among locomotion, posture, diet, body size and limb length.

ACKNOWLEDGMENTS

I thank Mary Glenn and Marina Cords for inviting me to participate in the guenon symposium in Adelaide and to contribute to this volume. Thanks to them, Russell Tuttle and Joe Holomuzski for helpful comments on an earlier version of this paper. Many thanks to the skilled assistants of the Taï Monkey Project and the project's founder, Ronald Noë. For permission to work in the Ivory Coast, I thank the Ministere d'Enseinment Superior et Recherche Scientifique, the Ministere d'Agriculture et Resources Animales de Cote d'Ivoire and the directorate of the Inistitute d'Ecologie Tropicale. The continued support and logistical assistance provided by the Centre Suisse de Recherche Scientifique and its director, Dr. Olivier Girardin are gratefully acknowledged. Financial support was generously provided by the Ohio State University, Conservation International, American Society of Primatology, Primate Conservation Incorporated and Wildlife Conservation International.

References

Ashton, E. H., and Oxnard, C. E. 1964a. Locomotor patterns in primates. *Proc. Zool. Soc. London* **142**:1–28.

Ashton, E. H., and Oxnard, C. E. 1964b. Functional adaptations in the primate shoulder girdle. *Proc. Zool. Soc. London* **142**:49–66.

Ashton, E. H., Oxnard, C. E., and Spence, T. E. 1965. Scapular shape and primate classification. *Proc. Zool. Soc. London* **145**:125–142.

Booth, A. H. 1955. Speciation in the Mona monkeys. *J. Mammal.* **36**:434–449.

Booth, A. H. 1956. The cercopithecidae of the Gold and Ivory Coasts: geographic and systematic observations. *Ann. Mag. Nat. Hist.* **12**:476–480.

Booth, A. H. 1958. The zoogeography of West African primates: a review. *Bull. Inst. Fr. Afr. Noire-A* **20**:587–622.

Bourlière, F., Hunkeler, C., and Bertrand, M. 1970. Ecology and behavior of Lowe's guenon (*Cercopithecus campbelli lowei*) in the Ivory Coast. In: J. R. Napier, and P. H. Napier (eds.), *Old World Monkeys: Evolution, Systematics and Behavior*, pp. 297–350. Academic Press, New York.

Brain, C. K. 1965. Observations on the behavior of vervet monkeys, *Cercopithecus aethiops. Zool. Africana* **1**:13–27.

Brugière, D., Gautier, J.-P., and Lahm, S. 1998. Additional data on the distribution of *Cercopithecus (lhoesti) solatus. Folia Primatol.* **69**:331–336.

Chism, J., and Rowell, T. E. 1988. The natural history of patas monkeys. In: A. Gautier-Hion, F. Bourlière, J.-P. Gautier, and J. Kingdon (eds.), *A Primate Radiation: Evolutionary Biology of the African Guenons*, pp. 412–438. Cambridge University Press, Cambridge.

Cords, M. 1984. Mating patterns and social structure in redtail monkeys (*Cercopithecus ascanius*). *Z. Tierpsychol.* **64**:313–329.

Cords, M. 1986. Interspecific and intraspecific variation in diet of two forest guenons, *Cercopithecus ascanius* and *C. mitis. J. Anim. Ecol.* **55**:811–827.

Cords, M. 1987. Mixed-species association of Cercopithecus monkeys in the Kakamega Forest, Kenya. *Univ. Calif. Pub. Zool.* **117**:1–109.

Dagosto, M., and Gebo, D. L. 1998. Methodological issues in studying positional behavior: meeting Ripley's challenge. In: E. Stasser, J. Fleagle, A. Rosenberger, and H. McHenry (eds.), *Primate Locomotion: Recent Advances*, pp. 5–30. Plenum Press, New York.

Doran, D. M., and Hunt K. D. 1994. Comparative locomotor behavior of chimpanzees and bonobos: species and habitat differences. In: R. W. Wrangham, W. C. McGrew, F. B. de Waal, and P. G. Heltne (eds.), *Chimpanzee Cultures*, pp. 93–108. Harvard University Press, Cambridge.

Fedigan, L., and Fedigan, L. M. 1988. *Cercopithecus aethiops*: a review of field studies. In: A. Gautier-Hion, F. Bourlière, J.-P. Gautier, and J. Kingdon (eds.), *A Primate Radiation: Evolutionary Biology of the African Guenons*, pp. 389–411. Cambridge University Press, Cambridge.

Fleagle, J. G. 1984. Primate locomotion and diet. In: D. Chivers, B. A. Wood, and A. Bilsborough (eds.), *Food Acquisition and Processing in Primates*, pp. 105–118. Plenum Press, New York.

Fleagle, J. G. 1999. *Primate Adaptation and Evolution*. Academic Press, New York.

Fleagle, J. G., and Mittermeier, R. A. 1980. Locomotor behavior, body size and comparative ecology of seven Surinam monkeys. *Am. J. Phys. Anthro.* **52**:301–314.

Fleagle, J. G., Stern, J. T., Jungers, W. L., Susman, R. L., Vangor, A. K., and Wells, J. P. 1981. Climbing: a biomechanical link with brachiation and bipedalism. In: M. H. Day (ed.), *Vertebrate Locomotion. Symp. Zool. Soc. London* **48**:359–375.

Galat, G., and Galat-Luong, A. 1985. La communautá de primates diurnes de la forêt de Taï, Côte d'Ivoire. *Terre Vie* **30**:3–30.

Garber, P. A., and Pruetz, J. D. 1995. Positional behavior in moustached tamarin monkeys: effects of habitat on locomotor variability and locomotor stability. *J. Hum. Evol.* **28**:411–426.

Gartlan, J. S., and Brain, C. K. 1968. Ecology and social variability in *Cercopithecus aethiops* and *C. mitis*. In: P. Jay (ed.), *Primates: Studies in Adaptation and Variability*, pp. 253–292. Holt, Rhinehart and Winston, New York.

Gartlan, J. S., and Struhsaker, T. T. 1972. Polyspecific associations and niche separation of rainforest anthropoids in Cameroon, West Africa. *J. Zool.* **168**:221–266.

Gautier, J.-P. 1988. Interspecific affinities among guenons as deduced from vocalizations. In: A. Gautier-Hion, F. Bourlière, J.-P. Gautier, and J. Kingdon (eds.), *A Primate Radiation: Evolutionary Biology of the African Guenons*, pp. 194–226. Cambridge University Press, Cambridge.

Gautier, J.-P., Moysan, F., Feistner, A. T. C., Loireau, J. N., and Cooper, R. W. 1992. The distribution of *Cercopithecus (l'hoesti) solatus*, an endemic guenon of Gabon. *Terre Vie* **47**:367–381.

Gautier, J.-P., Vercauteren Drubbel, R., and Deleporte, P. 2002. Phylogeny of the *Cercopithecus lhoesti* group revisited: combining multiple character sets. In: M. E. Glenn, and M. Cords (eds.), *The Guenons: Diversity and Adaptation in African Monkeys*, pp. 37–48. Kluwer Academic Publishers, New York.

Gautier-Hion, A. 1971. L'écologie du talapoin du Gabon (*Miopithecus talapoin*). *Terre Vie* **25**:427–490.

Gautier-Hion, A. 1978. Food niches and coexistence in sympatric primates in Gabon. In: D. J. Chivers and J. Herbert (eds.), *Recent Advances in Primatology, Volume II*, pp. 270–286. Academic Press, New York.

Gautier-Hion, A. 1988a. The diet and dietary habits of forest guenons. In: A. Gautier-Hion, F. Bourlière, J.-P. Gautier, and J. Kingdon (eds.), *A Primate Radiation: Evolutionary Biology of the African Guenons*, pp. 257–283. Cambridge University Press, Cambridge.

Gautier-Hion, A. 1988b. Polyspecific associations among forest guenons: ecological, behavioral and evolutionary aspects. In: A. Gautier-Hion, F. Bourlière, J.-P. Gautier, and J. Kingdon (eds.), *A Primate Radiation: Evolutionary Biology of the African Guenons*, pp. 452–476. Cambridge University Press, Cambridge.

Gautier-Hion, A., and Gautier, J.-P. 1974. Les associations polyspécifiques des Cercopitheques du plateau de M'passa, Gabon. *Folia Primatol.* **26**:165–184.

Gautier-Hion, A., and Gautier, J.-P. 1978. Le singe de Brazza: une stratágie originale. *Z. Tierpsychol.* **46**:84–104.

Gautier-Hion, A., Bourlière, F., Gautier, J.-P., and Kingdon, J. (eds.). 1988. *A Primate Radiation: Evolutionary Biology of the African Guenons*. Cambridge University Press, Cambridge.

Gebo, D. L. 1996. Climbing, brachiation, and terrestrial quadrupedalism: historical precursors of hominid bipedalism. *Am. J. Phys. Anthropol.* **101**:55–92.

Gebo, D. L., and Sargis, E. J. 1994. Terrestrial adaptations in the postcranial skeletons of guenons. *Am. J. Phys. Anthropol.* **93**:341–371.

Gebo, D. L., and Chapman, C. A. 1995a. Positional behavior in five sympatric Old World monkeys. *Am. J. Phys. Anthropol.* **97**:49–76.

Gebo, D. L., and Chapman, C. A. 1995b. Habitat, annual, and seasonal effects on positional behavior in red colobus monkeys. *Am. J. Phys. Anthropol.* **96**:73–82.

Gebo, D. L., and Chapman, C. A. 2000. Locomotor behavior in Ugandan monkeys. In: P. F. Whitehead, and C. J. Jolly (eds.), *Old World Monkeys*, pp. 480–495. Cambridge University Press, New York.

Gonzalez-Kirchner, J. P. 1996. Notes on the habitat use by the Russet-eared guenon (*Cercopithecus erythrotis* Waterhouse 1838) on Bioko Island, Equatorial Guinea. *Trop. Zool.* **9**:297–304.

Haddow, A. J. 1952. Field and laboratory studies of an African monkey, *Cercopithecus ascanius schmidti*. *Proc. Zool. Soc. London* **122**:297–394.

Hall, K. R. L. 1965. Behavior and ecology of the wild patas monkey, *Erythrocebus patas*, in Uganda. *J. Zool. London* **148**:15–87.

Hall, K. R. L., and Gartlan, S. J. 1965. Ecology and behavior of the vervet monkey, *Cercopithecus aethiops*, Lolui Island, Lake Victoria. *Proc. Zool. Soc. London* **145**:37–56.

Harrison, M. J. S. 1988. A new species of guenon (genus *Cercopithecus*) from Gabon. *J. Zool. London* **215**:561–575.

Hunt, K. D., Cant, J. G. H., Gebo, D. L., Rose, M. D., Walker, S. E., and Youlatos, D. 1996. Standardized descriptions of primate locomotor and postural modes. *Primates* **37**:363–387.

Isbell, L. A., Pruetz, J. D., Lewis, M., and Young, T. P. 1998. Locomotor activity differences between sympatric patas monkeys (*Erythrocebus patas*) and vervet monkeys (*Cercopithecus aethiops*): implications for the evolution of long hindlimb length in *Homo*. *Am. J. Phys. Anthropol.* **105**:199–207.

Jolly, C. J. 1965. *Origins and Specialisations of the Long-faced Cercopithecoidea*. Ph.D. Thesis. University of London, London.

Jungers, W. L. 1984. Aspects of size and scaling in primate biology with special reference to the locomotor skeleton. *Yrbk. Phys. Anthropol.* **27**:73–97.

Jungers, W. L., and Burr, D. B. 1994. Body size, long bone geometry and locomotion in quadrupedal monkeys. *Z. Morph. Anthropol.* **80**:89–97.

Jungers, W. L., Burr, D. B., and Cole, M. S. 1998. Body size and scaling of long bone geometry, bone strength, and positional behavior in cercopithecoid primates. In: E. Strasser, J. Fleagle, A. Rosenberger, and H. McHenry (eds.), *Primate Locomotion: Recent Advances*, pp. 309–330. Plenum Press, New York.

Kaplin, B. A. 2002. Terrestriality and the maintenance of the disjunct geographical distribution in the *lhoesti* group. In: M. E. Glenn, and M. Cords (eds.), *The Guenons: Diversity and Adaptation in African Monkeys*, pp. 49–59. Kluwer Academic Publishers, New York.

Kaplin, B. A., and Moermond, T. C. 2000. Foraging ecology of the mountain monkey (*Cercopithecus l'hoesti*): implications for its evolutionary history and use of disturbed forest. *Am. J. Primatol.* **50**:227–246.

Kavanagh, M. 1980. Invasion of the forest by an African savannah monkey: behavioral adaptations. *Behavior* **73**:238–260.

Kingdon, J. 1988. Comparative morphology of hands and feet in the genus *Cercopithecus*. In: A. Gautier-Hion, F. Bourlière, J.-P. Gautier, and J. Kingdon (eds.), *A Primate Radiation: Evolutionary Biology of the African Guenons*, pp. 184–193. Cambridge University Press, Cambridge.

Lambert, J. E. 2002. Resource switching and species coexistence in guenons: a community analysis of dietary flexibility. In: M. E. Glenn, and M. Cords (eds.), *The Guenons: Diversity and Adaptation in African Monkeys*, pp. 309–323. Kluwer Academic Publishers, New York.

Larson, S. G. 1998. Unique aspects of quadrupedal locomotion in nonhuman primates. In: E. Strasser, J. Fleagle, A. Rosenberger, and H. McHenry (eds.), *Primate Locomotion: Recent Advances*, pp. 157–174. Plenum Press, New York.

Larson, S. G., and Stern, J. T. 1989. The role of supraspinatus in the quadrupedal locomotion of vervets (*Cercopithecus aethiops*): implications for interpretations of humeral morphology. *Am. J. Phys. Anthropol.* **79**:369–377.

Lawes, M. J. 1990. The distribution of the samango monkey (*Cercopithecus mitis erythrarchus* Peters 1852 and *Cercopithecus mitis labiatus* I. Geoffroy, 1843) and forest history in southern Africa. *J. Biogeogr.* **17**:669–680.

Leakey, M. 1988. Fossil evidence for the evolution of guenons. In: A. Gautier-Hion, F. Bourlière, J.-P. Gautier, and J. Kingdon (eds.), *A Primate Radiation: Evolutionary Biology of the African Guenons*, pp. 7–12. Cambridge University Press, Cambridge.

Lernould, J. M. 1988. Classification and geographical distribution of guenons: a review. In: A. Gautier-Hion, F. Bourlière, J.-P. Gautier, and J. Kingdon (eds.), *A Primate Radiation: Evolutionary Biology of the African Guenons*, pp. 54–78. Cambridge University Press, Cambridge.

Manaster, B. J. 1979. Locomotor adaptations within the *Cercopithecus* genus: a multivariate approach. *Am. J. Phys. Anthropol.* **50**:169–182.

McCrossin, M. L., Benefit, B. R., Gitau, S. N., Palmer, A. K., and Blue, K. T. 1998. Fossil evidence for the origins of terrestriality among Old World higher primates. In: E. Strasser, J. Fleagle, A. Rosenberger, and H. McHenry (eds.), *Primate Locomotion: Recent Advances*, pp. 353–396. Plenum Press, New York.

McGraw, W. S. 1994. Census, habitat preference and polyspecific associations of six monkeys in the Lomako Forest, Zaire. *Am. J. Primatol.* **34**:295–307.

McGraw, W. S. 1996. Cercopithecid locomotion, support use, and support availability in the Taï Forest, Ivory Coast. *Am. J. Phys. Anthropol.* **100**:507–522.

McGraw, W. S. 1998a. Comparative locomotion and habitat use of six monkeys in the Taï Forest, Ivory Coast. *Am. J. Phys. Anthropol.* **105**:493–510.

McGraw, W. S. 1998b. Locomotion, support use, maintenance activities, and habitat structure: the case of the Taï Forest cercopithecids. In: E. Strasser, J. Fleagle, A. Rosenberger, and H. McHenry (eds.), *Primate Locomotion: Recent Advances*, pp. 79–94. Plenum Press, New York.

McGraw, W. S. 1998c. Posture and support use of Old World monkeys (Cercopithecidae): the influence of foraging strategies, activity patterns, and the spatial distribution of preferred food items. *Am. J. Primatol.* **46**:229–250.

McGraw, W. S. 2000. Positional behavior of *Cercopithecus petaurista*. *Int. J. Primatol.* **21**:157–182.

Meldrum, D. J. 1989. *Terrestrial Adaptations in the Feet of African Cercopithecines*. Ph.D. Thesis. State University of New York, Stony Brook.

Meldrum, D. J. 1991. Kinematics of the cercopithecine foot on arboreal and terrestrial substrates with implications for the interpretation of hominid terrestrial adaptations. *Am. J. Phys. Anthropol* **84**:273–289.

Moffett, M. W. 2000. What's "up"? A critical look at the basic terms of canopy biology. *Biotropica* **32**:569–596.

Napier, J. 1962. Monkeys and their habits. *New Scient.* **295**:88–92.

Oates, J. F. 1985. The Nigerian guenon, *Cercopithecus erythrogaster*: ecological, behavioral, systematic and historical observations. *Folia Primatol.* **45**:25–43.

Oates, J. F. 1988. The distribution of *Cercopithecus* monkeys in West African forest. In: A. Gautier-Hion, F. Bourlière, J.-P. Gautier, and J. Kingdon (eds.), *A Primate Radiation: Evolutionary Biology of the African Guenons*, pp. 79–103. Cambridge University Press, Cambridge.

Peignot, P., Fontaine, B., and Wickings, E. J. 1999. Habitat exploitation, diet and some data on reproductive behavior in a semi-free ranging colony of *Cercopithecus l'hoesti solatus*, a guenon species recently discovered in Gabon. *Folia Primatol.* **70**:29–36.

Pickford, M., and Senut, B. 1988. Habitat and locomotion in Miocene cercopitheoids. In: A. Gautier-Hion, F. Bourlière, J.-P. Gautier, and J. Kingdon (eds.), *A Primate Radiation: Evolutionary Biology of the African Guenons*, pp. 35–53. Cambridge University Press, Cambridge.

Quris, R. 1976. Données compatives sur la sociécologie de huit espèces de Cercopithecidae vivant dans une meme zone de forêt primitive périodiquement inondée (N-E Gabon). *Terre Vie* **30**:193–209.

Rollinson, J., and Martin, R. D. 1981. Comparative aspects of primate locomotion, with special reference to arboreal cercopithecines. *Symp. Zool. Soc. London* **48**:377–427.

Rose, M. D. 1974. Postural adaptations in New and Old World Monkeys. In: F. A. Jenkins (ed.), *Primate Locomotion*, pp. 201–222. Academic Press, New York.

Rose, M. D. 1978. Feeding and associated positional behavior of black and white colobus monkeys (*Colobus guereza*). In: G. Montgomery (ed.), *The Ecology of Arboreal Folivores*, pp. 253–262. Smithsonian Institute Press, Washington DC.

Rose, M. D. 1979. Positional behavior of natural populations: some quantitative results of a field study of *Colobus guereza* and *Cercopithecus aethiops*. In: M. E. Morbeck, H. Preuschoft, and G. Gomberg (eds.), *Environment, Behavior and Morphology: Dynamic Interactions in Primates*, pp. 75–93. Gustav Fischer, New York.

Rowe, N. 1996. *The Pictorial Guide to the Living Primates*. Pogonias Press, East Hampton, NY.

Schmitt, D. O. 1998. Forelimb mechanics during arboreal and terrestrial quadrupedalism in Old World monkeys. In: E. Strasser, J. Fleagle, A. Rosenberger, and H. McHenry (eds.), *Primate Locomotion: Recent Advances*, pp. 175–200. Plenum Press, New York.

Schultz, A. H. 1970. The comparative uniformity of the Cercopithecoidea. In: J. R. Napier, and P. H. Napier (eds.), *Old World Monkeys*, pp. 39–52. Academic Press, New York.

Smith, R. J., and Jungers, W. L. 1997. Body mass in comparative primatology. *J. Hum. Evol.* **32**:523–559.

Strasser, E. 1987. Pedal evidence for the origin of and diversification of cercopithecid clades. *J. Hum. Evol.* **17**:225–246.

Strasser, E. 1992. Hindlimb proportions, allometry, and biomechanics in Old World monkeys (Primates, Cercopithecidae). *Am. J. Phys. Anthropol.* **87**:187–213.

Strasser, E. 1994. Relative development of the hallux and pedal digit formulae in Cercopithecidae. *J. Hum. Evol.* **26**:413–440.

Struhsaker, T. T. 1967. Behavior of vervet monkeys. *Univ. Calif. Pub. Zool.* **82**:1–64.

Struhsaker, T. T. 1969. Correlates of ecology and social organization among African cercopithecines. *Folia Primatol.* **11**:80–118.

Struhsaker, T. T. 1978. Food habits of five monkey species in the Kibale Forest, Uganda. In: D. J. Chivers, and J. Herbert (eds.), *Recent Advances in Primatology, Volume II*, Academic Press, pp. 225–248. New York.

Struhsaker, T. T. 1980. Comparison of the ecology of red colobus and redtail monkeys in the Kibale Forest, Uganda. *Afr. J. Ecol.* **18**:33–51.

Struhsaker, T. T. 1981. Polyspecific associations among tropical rainforest primates. *Z. Tierpsychol.* **57**:268–304.

Struhsaker, T. T., and Leland, L. 1979. Socioecology of five sympatric monkey species in the Kibale forest, Uganda. *Adv. Stud. Behav.* **9**:159–228.

Tappen, N. C. 1960. Problems of distribution and adaptation of the African monkeys. *Curr. Anthrop.* **1**:91–120.

Thomas, S. C. 1991. Population densities and patterns of habitat use among anthropoid primates of the Ituri Forest, Zaire. *Biotropica* **23**:68–83.

Tosi, A. J., Buzzard, P. J, Morales, J. C., and Melnick, D. J. 2002. Y-chromosomal window onto the history of terrestrial adaptation in the Cercopithecini. In: M. E. Glenn, and M. Cords (eds.), *The Guenons: Diversity and Adaptation in African Monkeys*, pp. 15–26. Kluwer Academic Publishers, New York.

Tuttle, R. H. 1969. Terrestrial trends in the hands of the anthropoidea: a preliminary report. *Proc. 2nd Int. Con. Primatol., Vol. 2*, pp. 192–200. Karger, Basel.

Tuttle, R. H. 1975. Parallelism, brachiation and hominoid phylogeny. In: W. P. Luckett and F. S. Szalay (eds.), *Phylogeny of the Primates*, pp. 447–480. Plenum Press, New York.

Vilensky, J. A., and Gankiewicz, E. 1990. Effects of growth and speed on hindlimb joint angular displacement patterns in vervet monkeys (*Cercopithecus aethiops*). *Am. J. Phys. Anthropol.* **81**:441–449.

Wachter, B., Schabel, M., and Noe, R. 1997. Diet overlap and polyspecific associations of red colobus and diana monkeys in the Taï National Park. *Ethology* **103**:514–526.

Whitehead, P. F., and Larson, S. G. 1994. Shoulder motion during quadrupedal walking in *Cercopithecus aethiops*: integration of cineradiographic and electromyographic data. *J. Hum. Evol.* **26**:525–544.

Unique Behavior of the Mona Monkey (*Cercopithecus mona*): All-Male Groups and Copulation Calls

10

MARY E. GLENN, REIKO MATSUDA,
and KEITH J. BENSEN

Introduction

Central to understanding any species or group of species is the manner in which they reproduce, and by extension, the manner and frequencies with which males and females interact with one another and with members of their own sex. Group formation and interactions among group members provide insight into individual reproductive strategies. Unique social groupings within a genus are significant because they expand the spectrum of behavioral possibilities and, more importantly, provide greater context for behavioral patterns of other species within the genus. The same is true of unique

MARY E. GLENN • Department of Anthropology, Humboldt State University, Arcata, CA 95521, USA.   REIKO MATSUDA • Department of Anthropology, Graduate School and University Center of the City University of New York, New York, NY 10016, USA.   KEITH J. BENSEN • Windward Islands Research and Education Foundation, 11 East Main Street, Bayshore, New York, NY 11706, USA.

The Guenons: Diversity and Adaptation in African Monkeys, edited by Glenn and Cords. Kluwer Academic/Plenum Publishers, New York, 2002.

vocalizations, especially ones associated with mating, such as calls that attract mates, deter competitors, or accompany copulation.

Male relationships with other males away from females and mixed-sex groups have rarely been reported for forest guenons. It has been assumed that males that have emigrated from their natal groups but have not yet become established within a mixed-sex group remain solitary (Gartlan, 1973). What these surplus males do and with whom they associate have been difficult to observe, and these difficulties have been attributed mainly to males' highly inconspicuous behavior (Rowell, 1988).

Beyond their role in helping researchers unravel mating behavior and male and female reproductive strategies, vocalizations may also reflect evolutionary relationships. Genetic factors play a predominant role in the development of vocal repertoire. Jürgens (1979, 1990) demonstrated that the motor patterns underlying vocalizations do not have to be learned by imitation. Gautier and Gautier (1977), Newman and Symmes (1982) and Brockelman and Schilling (1984) have also demonstrated via hybridization studies in monkeys that vocal behavior is inherited and not learned. Accordingly, guenon phylogenies have been proposed using vocalization repertoires (Struhsaker, 1970; Gautier, 1988). Gautier *et al.* (2002) incorporated vocalizations in analyses aimed at constructing multi-character phylogenies. Vital to conducting such analyses is knowledge of the complete vocal repertoire of each species.

Originally, most forest guenons were thought to adhere to a limited range of group compositions and vocalizations. Although ongoing research on less studied forest guenons has been expanding our appreciation of the behavioral variation within the group, we did not expect any surprises from our studies on *Cercopithecus mona* until we found that they form all-male groups and emit copulation calls.

Our research on *Cercopithecus mona* has spanned both the specific range on mainland Africa (Glenn *et al.*, unpub. report a; Matsuda, in prep.) and introduced ranges on islands in the Gulf of Guinea and the Caribbean (Glenn, 1996, 1997, 1998; Glenn and Bensen, 1998; Glenn *et al.*, unpub. report b) (Fig. 1). Our efforts concentrated on the description of mona monkey behavior and ecology. Here we provide detailed descriptions of all-male groups and copulation vocalizations at four study sites and discuss the significance of both behaviors in relation to other *Cercopithecus* species and guenons as a whole.

Methods

The Grand Etang Forest Reserve (12°6′ N, 61°42′ W) in Grenada receives an average of 4060 mm of rain per year (Grenada Government and Organization of American States, 1988) and consists of primary and secondary

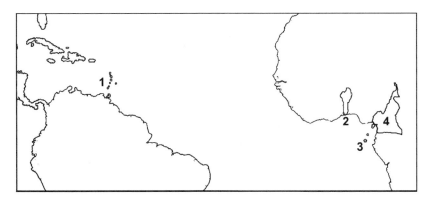

Fig. 1. Map of the four study sites: (1) Grand Etang Forest Reserve, Grenada, (2) Lama Forest Reserve, Benin, (3) São Tomé and Príncipe, and (4) Bimbia Bonadikombo Community Forest, Cameroon.

lowland tropical rain forest and montane tropical rain forest. Mona monkeys were introduced to Grenada *ca.* 200–300 years ago, and are the only non-human primates on the island. Glenn (1996) studied them for 28 months between September 1992 and April 1995. The population density averaged 42.1 individuals/km^2 (SD = 35.7) (Glenn, 1998).

The Lama Forest Reserve (6°55′–7°00′ N, 2°04′–2°12′ E) in southern Republic of Benin is a dry seasonal forest ($\bar{x}$ = 1100 mm rain/year) with two distinct wet and dry seasons. The vegetation is a mosaic of primary and secondary forest and regenerating farm bush. Aside from *Cercopithecus mona*, other nonhuman primates within the reserve are *C. aethiops tantalus*, *C. erythrogaster erythrogaster*, *Colobus vellerosus*, *Procolobus verus*, and *Galago senegalensis*. Matsuda conducted studies for 21 months between August 1995 and June 1997. The mona monkey population density averaged 56 individuals/km^2 (SD = 40.4) (Matsuda, unpub. data).

Glenn *et al.* (unpub. report a) conducted a study in the central mountains of the islands of São Tomé (0°10′ N, 6°40′ E) and Príncipe (1°40′ N, 7°25′ E) from June to October 1998. Survey areas on both São Tomé and Príncipe consisted of mosaics of primary tropical rain forest, montane tropical rain forest, secondary tropical rain forest, and abandoned tree crop farmlands. Average annual rainfall varied from 3000 mm to >8000 mm between the various survey sites on both islands. Mona monkeys, the only nonhuman primate existing on either island, were probably introduced sometime between 160 and 500 years ago. Glenn *et al.* (unpub. report a) estimated the mona population density averaged 18.8 individuals/km^2 (SD = 20.12) on São Tomé and 21.4 individuals/km^2 (SD = 21.39) on Príncipe.

Glenn *et al.* (unpub. report b) also conducted a study from April through June 1999 in the Bimbia Bonadikombo Community Forest (4°1′ N, 8°6′ E) on the coast of southwestern Cameroon. The area receives >11,000 mm of rain per year and comprises a highly fragmented mosaic of primary and secondary lowland tropical rain forest, patches of cleared agricultural land, and estuarine mangrove forest. Beside *Cercopithecus mona*, other diurnal nonhuman primates include *C. nictitans* and *C. erythrotis*. Mona monkey population density was not recorded.

At all four field sites, we collected group composition and behavioral data using *ad libitum* sampling and recorded vocalizations during regular trail surveys. We gathered population and behavioral data in Benin, Grenada, and São Tomé and Príncipe during census studies using modified line transect methods. At the Grenada site, Glenn (1997) also collected data during periods when monkeys frequented the forested areas surrounding the field research station. We defined a group as a cluster of individuals that engaged in feeding, grooming or travel together.

Results

The duration of most monkey encounters at each site was short because none of the study populations was habituated, except for males that frequented the field station in Grenada. In Grenada, visual encounters with unhabituated monkeys lasted an average of 17.1 minutes (range 1–310, SD = 28.8, $n = 308$). At Lama, although many of the observation periods were short, Matsuda often repeatedly observed the same groups of individuals, especially four easily identifiable males (three adults and one subadult), over the course of several hours and on different days. On São Tomé and Príncipe, all encounters were <5 minutes ($n = 59$), while at Bimbia Bonadikombo encounters averaged 13 minutes (range 1–55, SD = 13.6, $n = 51$).

All-Male Groups

Glenn (1997) encountered habituated all-male groups 1164 times in Grenada in the forest immediately surrounding the field station where they were usually within observational range for many hours at a time. All-male groups with identifiable individuals occurred throughout the year (Fig. 2). None of the individuals occurred in a mixed-sex group at any time during their tenure in an all-male group. Each of the 18 different individuals (eight adults, eight subadults, and two juveniles) belonged at one time or another to one or more of 33 different all-male groups observed over three years. These all-male groups contained between two and four individuals ($\bar{x} = 2.4$, SD = 0.6). Several male pairs formed long-term, stable associations; one pair was together for more than two years, while three other pairs contained the same members for

Fig. 2. All-male mona monkey pair at the Grand Etang Forest Reserve, Grenada. This pair remained together for more than two years. Photograph by Heather Bruce.

more than one year. Pairs remained together for an average of 7.7 months (SD = 7.1, min = 1, max = 25, $n = 12$). This average was lower for male trios ($\bar{x} = 2.4$ months, SD = 1.1, min = 0.5, max = 5, $n = 18$) and foursomes ($\bar{x} = 2.5$ months, SD = 1.3, min = 1.5, max = 4, $n = 3$). Within the 33 groups, 13 combinations of age classes occurred of which subadult/subadult pairs ($n = 5$) and adult/subadult pairs ($n = 4$) were the most frequent. Solitary juvenile, subadult and adult male monkeys did occur ($n = 9$); however, most of them had also belonged to an all-male group at some point.

Glenn (1997) encountered unhabituated and unidentifiable all-male groups away from the field station in Grenada on 26 occasions. These groups consisted of two to three individuals, most of which (77%) were pairs of subadult monkeys ($n = 8$). Glenn and Bensen saw unhabituated solitary males eight times in Grenada.

At Lama, Matsuda observed all-male groups 58 times (range 1–10/month, $\bar{x}$/month = 4.8, SD = 2.4) between July 1996 and June 1997. Group size varied from two to five individuals, among which adult pairs were most frequent ($n = 14$).

Matsuda had two sightings of all-male groups containing a juvenile and also two sightings of groups that included a *Cercopithecus erythrogaster* male. Matsuda also observed solitary males on 11 occasions. Sightings of all-male groups occurred throughout the year at Lama and there was no relationship between the number of sightings per month and rainfall. Groups with identifiable individuals often changed size within a day and between days. During intergroup encounters, it was often difficult to determine whether males observed together belonged to an all-male group or a mixed-sex group because many mixed-sex groups at Lama contained more than one adult male.

Glenn *et al.* (unpub. report a and b) did not observe all-male groups or solitary males in São Tomé or Príncipe or at Bimbia Bonadikombo. Monkey hunters from São Tomé, however, reported that they had frequently seen all-male groups of mona monkeys.

Members of all-male groups interacted with one another often at each of the study sites and usually stayed in close proximity to one another, even while moving through the forest. In Grenada, grooming was the most common affiliative behavior followed by resting together, playing, and homosexual mounting. No member of an all-male group in Grenada was observed emitting adult male double boom loud calls (Struhsaker, 1969; Glenn, 1996). Interactions between all-male groups and mixed-sex groups were observed only twice in Grenada; on both occasions, members of the all-male group were silent and vigilant.

At Lama, feeding, traveling, grooming, and sitting together were frequent behaviors in all-male groups. There was no mounting behavior between males. Encounters between all-male and mixed-sex groups containing either one male or more than one male were more commonly observed at Lama than in Grenada. At Lama, male loud calls, chases, and other aggressive behavior between males of all-male groups and males of mixed-sex groups were frequent during intergroup encounters. Vocalizations within all-male groups were infrequent and restricted to soft contact calls.

Copulation Calls

The copulation call series is made up of two call types, warbles and grunts (Fig. 3), each given by a different individual in the mating pair. We were not able to determine which sex made which call type. The general nature of the call series does not appear to vary across study sites. Warbles are characterized by a continuous whining, which rises and falls erratically in pitch. Grunts are choppy, low-toned calls that are rapidly repeated. Warbles and grunts are given simultaneously, alternately, or singly during call series. Both calls are loud and can be heard >200 m away. We heard copulation vocalizations only from monkeys in mixed-sex groups and only when an adult male mounted a female. We were not able to determine if the calls were associated with

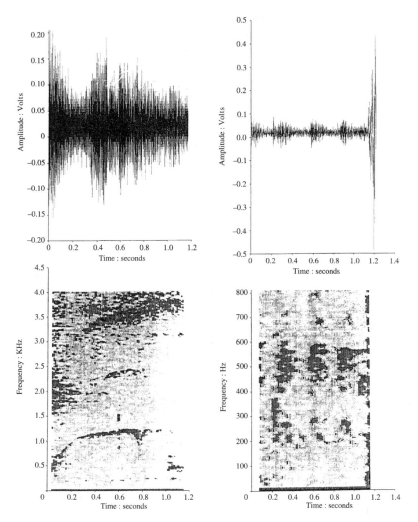

Fig. 3. Sound spectrographs of warble call (on left), and three grunt calls (on right) from a mona monkey copulation call series from the Grenada field site.

ejaculations or given during all adult male/female mounts, regardless of ejaculation. Glenn (1996) noted that local hunters in Grenada specifically associated the calls only with copulatory behavior. Males mounting males in all-male groups did not vocalize.

Copulation calls occurred 104 times throughout the year in Grenada. Half of the call series consisted of simultaneous warbles and grunts. In 45% of all copulation calls, only warbles were emitted, while in 5% of them only grunts were emitted. The mean duration of all copulation call series was 7.9 seconds

(SD = 3.6). In 26% of all copulation calls, adult male loud calls occurred simultaneously (given by individuals outside of the mating pair).

At Lama, 54% of 56 copulation calls were simultaneous warbles and grunts, 12% warbles only, and 34% grunts only. Adult male loud calls often occurred immediately before and immediately after copulation calls. The distribution of calls across the study period showed a distinct seasonal pattern, clustering over a period of six months from the end of the major wet season to near the end of the major dry season.

No copulation calls occurred during any surveys on São Tomé or Príncipe, and only one copulation call series occurred at Bimbia Bonadikombo during the study.

Discussion

Descriptions of *Cercopithecus mona* group composition and vocal repertoire were first documented by Struhsaker (1969) and Gartlan and Struhsaker (1972) from primate surveys in southwestern Cameroon. Neither study concentrated solely on mona monkeys and neither author mentions all-male groups, though Struhsaker (1969, 1970) documented their copulation calls. Howard (1977) studied unhabituated mona monkeys in southwestern Cameroon as part of a niche separation study that included *Cercopithecus erythrotis* and *C. nictitans*. He did not observe all-male groups of *Cercopithecus mona*, but heard many copulation calls.

All-Male Groups

Apparently, no other *Cercopithecus* species regularly forms as closely bonded all-male groups as those of *C. mona*. Documented all-male associations exist for only two other *Cercopithecus* species. Bourlière *et al.* (1970) provided an account of two subadult male *Cercopithecus campbelli lowei* at Adiopodoumé in Ivory Coast that were in regular physical contact with one another after they had emigrated from a mixed-sex group. Apart from this account, no other all-male association has been documented for this species. Blue monkey males (*Cercopithecus mitis stuhlmanni*) not associated with mixed-sex groups have been observed feeding in close proximity to one another (Tsingalia and Rowell, 1984) and "loosely associating" with one another (Cords *et al.*, 1986). In addition, Cords (pers. com.) has seen several blue monkey males moving in small, relatively coordinated groups away from larger mixed-sex groups and grooming one another on rare occasions. These all-male groups, however, last for only 2–3 months.

Patas monkeys (*Erythrocebus patas*) form all-male groups (Hall, 1965) that appear to be most like those of mona monkeys. Unlike the loose male associations described for blue monkeys, patas monkey all-male groups are discrete social units in which individual members interact regularly and form close bonds (Gartlan, 1975). The non-agonistic interactions of members of all-male patas groups are similar to those we have observed in mona monkey all-male groups. Most interactions between mona monkeys in all-male groups were affiliative and appeared to strengthen the bonds between individual males. Gartlan (1973) theorized that the brightly colored scrota of patas monkeys attract them to one another much like they attract females. He also hypothesized that *Cercopithecus* species would not form all-male groups because of their relatively small and less colorful scrota as well as their arboreal lifestyle. Our observations of mona monkey all-male groups do not support these hypotheses.

It is not clear why all-male groups are so rare among guenons. Resident males belonging to mixed-sex groups of forest-dwelling guenons are intolerant of other males, and often chase and threaten male intruders (Galat and Galat-Luong, 1978; Struhsaker and Leland, 1979; Cords, 1984; Tsingalia and Rowell, 1984; Butynski, 1990). Cords (1987) hypothesized that aggression among male guenons developed as a result of competition for access to females and that it inhibits male–male social bonding. Aggression between resident and intruder males, however, should not prevent extragroup males from forming bonds. Extragroup males may benefit from associations with other males by possibly enhancing food detection or increasing predator awareness (Terborgh, 1983; Pulliam and Caraco, 1984; van Schaik and Horstermann, 1994). In addition, coalitions of extragroup males might have a better chance than a solitary male to infiltrate a mixed-sex group and copulate with receptive females. Support for this hypothesis comes from Lama (Matsuda, in prep.) and the Mungo Reserve in Cameroon (Howard, 1977) where extragroup males invaded mixed-sex groups that often contained >1 male. At neither site, however, was there confirmation that these extragroup males were members of all-male groups acting in concert. Male–male interactions may also benefit extragroup males by serving as a kind of social salve, allowing males that have emigrated from their natal mixed-sex groups to continue to interact with other monkeys. Similar reasoning was used by Fleury and Gautier Hion (1997) to explain the regular close associations observed between single *Cercopithecus pogonias* males and *Colobus satanas* groups.

Copulation Calls

Like all-male groups, copulation calls in *Cercopithecus mona* seem to be widespread throughout their original and introduced ranges. Howard's (1977) study of mona monkeys in southwestern Cameroon provides additional information about the distinctiveness and regularity of the copulation call and its social context. He presented evidence that the behavior is specific to *Cercopithecus mona*, and not attributable to habitat variables. From February 1972 to

January 1973, Howard heard 2102 copulation calls. Mona monkeys emitted them during 52 of 72 copulations he observed. Like us, he noted that copulation calls were given only by adult heterosexual pairs in mixed-sex groups. As in Grenada, but unlike Lama, Howard heard calls throughout the year, though there was a significant increase in copulatory activity at his site during July to December. Howard also regularly heard mona male loud calls concurrently with copulation calls within the same mixed-sex group, which is similar to our observations.

Based on vocal repertoire, Gautier (1988) proposed a phylogenetic dendrogram for guenons within which *Cercopithecus mona* is shown to have one of the more derived guenon vocal repertoires. His arrangements generally agreed with the karyological analyses of Dutrillaux *et al.* (1982) and the protein electrophoretic analyses of Ruvolo (1988), which placed *Cercopithecus mona* as a more derived species. *Cercopithecus solatus*, the only other species of *Cercopithecus* known to make copulation calls, might be ancestral to *C. mona*. *Erythrocebus*, *Allenopithecus*, and *Miopithecus* spp. also emit copulation calls. Members of these genera also are shown to be ancestral to *Cercopithecus mona* in all three of the proposed phylogenies. Gautier (1988) considers the presence of a copulation call to be an ancestral trait because it occurs in many other species of the Cercopithecidae thought to be ancestral to guenons. Although his analysis did not rely solely on a specific call type, the mona copulation call creates a significant incongruity. It is conceivable that a species could possess a mixture of derived and ancestral calls if different call types are subject to different selective pressures (Gautier, 1988). The presence of a copulation call in the repertoire of *Cercopithecus mona* and in no other member of the *mona* superspecies (or even in other more derived arboreal *Cercopithecus* spp.) makes the issue difficult to resolve.

It is possible that the copulation vocalization in mona monkeys developed as an evolutionary isolating mechanism. The striking similarity among all the members of the *mona*-superspecies is notable. The only significant pelage difference between *Cercopithecus campbelli* and *C. mona* is the absence/presence of white rump spots, while the only significant differences between *C. pogonias* and *C. mona* are the presence/absence of a cranial crest, yellow belly, and white rump spots. In addition, their vocal repertoires are similar as is their habitat use (Struhsaker, 1970). Perhaps *Cercopithecus campbelli* and *C. pogonias* individuals would react inappropriately to a sudden strange call by a *C. mona* during copulation, resulting in unsuccessful mating.

Summary

We conducted systematic behavioral observations and censuses of mona monkeys (*Cercopithecus mona*) between 1992 and 1999 at four different study sites in both the monas' original African range and on islands to which populations were introduced. We observed all-male groups and heard copulation calls at three sites. All-male groups contained between two and five individuals consisting of adults, subadults, and/or juveniles. Individual tenure within habituated all-male groups varied from a few days to more than two years. We rarely observed agonistic behavior within all-male groups but affiliative behavior was common. Copulation calls consisted of two call types, warbles and grunts, given either simultaneously or alone. We only heard adult heterosexual pairs of mixed-sex groups giving these calls while copulating. The nature of the calls did not vary between study sites. Before our studies, the presence of relatively stable all-male groups was unknown in the genus *Cercopithecus* and only one other species besides *C. mona, C. solatus,* was known to emit vocalizations while copulating.

ACKNOWLEDGMENTS

Funding for the research conducted by M. E. G. and K. J. B. was provided by the National Geographic Society, ECOFAC, Windward Islands Research and Education Foundation, The Rockefeller University, Yerkes Regional Primate Research Center, Wildlife Conservation Society, and the Center for Environmental Research and Conservation at Columbia University. M. E. G. and K. J. B. are grateful to the Governments of Grenada, São Tomé and Príncipe, and Cameroon, the Grenada National Parks, the São Tomé and Príncipe ECOFAC staff, the Cameroon Ministry of Environment and Forests, and the Mount Cameroon Project for allowing us to conduct this research. Special thanks go to Heather Bruce, Kevin Jensen, and Patricia Bekhuis for their help in the field in Grenada, São Tomé and Príncipe, and Cameroon, respectively.

Research in the Lama Forest of Republic of Benin was funded by the National Science Foundation, Primate Conservation Inc., Leopold Schepp Foundation, and Wenner-Gren Foundation for Anthropological Research. R. M. is grateful to the Office National du Bois (ONAB) and to the Mission Forestière de Allemande (MIFOR-GTZ) for permission to use various facilities in Lama and in Cotonou and for help with logistics. R. M.'s study would not have been possible without the help of Akakpo, Justin, Roger, Aline, and many other people in Koto Village and its periphery. R. M. is particularly indebted to Mr and Mrs Roland Hilbert (Conseiller Forestier, MIFOR) for their generous moral support and friendship throughout the study. R. M. would also like to thank Peter Neuenshwander of the International Institute of Tropical Agriculture (IITA) in Abomey–Calavi for logistic and moral support, and guidance.

Finally, we would like to thank Marina Cords, Thelma Rowell, and Russell Tuttle for providing helpful comments for this chapter and Christopher Lloyd for producing the map in Figure 1.

References

Bourlière, F., Hunkeler, C., and Bertrand, M. 1970. Ecology and behavior of Lowe's guenon (*Cercopithecus campbelli lowei*) in the Ivory Coast. In: J. R. Napier, and P. H. Napier (eds.), *Old World Monkeys: Evolution, Systematics and Behaviour*, pp. 297–350. Academic Press, New York.

Brockelman, W., and Schilling, D. 1984. Inheritance of stereotyped gibbon calls. *Nature* (London) **312**:634–636.

Butynski, T. M. 1990. Comparative ecology of blue monkeys (*Cercopithecus mitis*) in high- and low-density subpopulations. *Ecol. Monogr.* **60**(1): 1–26.

Cords, M. 1984. Mating patterns and social structure in redtail monkeys (*Cercopithecus ascanius*). *Z. Tierpsychol.* **64**:313–329.

Cords, M. 1987. Forest guenons and patas monkeys: male–male competition in one-male groups. In: B. B. Smuts, D. L. Cheney, R. M. Seyfarth, R. W. Wrangham, and T. T. Struhsaker (eds.), *Primate Societies*, pp. 98–111. University of Chicago Press, Chicago.

Cords, M., Mitchell, B. J., Tsingalia, H. M., and Rowell, T. E., 1986. Promiscuous mating among blue monkeys in the Kakamega Forest, Kenya. *Ecology*. **72**:214–226.

Dutrillaux, B., Muleris, M., and Couturier, J. 1988. Chromosomal evolution of Cercopithecinae. In: A. Gautier-Hion, F. Bourlière, J.-P. Gautier, and J. Kingdon (eds.), *A Primate Radiation: Evolutionary Biology of the African Guenons*, pp. 150–159. University Press, Cambridge, England.

Fleury, M.-C., and Gautier-Hion, A. 1997. Better to live with allogenerics than to live alone? The case of single male *Cercopithecus pogonias* in troops of *Colobus satanas*. *Int. J. Primatol.* **18**: 967–974.

Galat, G., and Galat-Luong, A. 1978. Diet of green monkeys in Senegal. In: D. J. Chivers, and J. Herbert (eds.), *Recent Advances in Primatology, Vol. 1 (Behaviour)*, pp. 257–258. Academic Press, New York.

Gartlan, J. S. 1973. Influences of phylogeny and ecology on variations in the group organization of primates. In: E. W. Menzel (ed.), *Precultural Primate Behavior, Symp. IVth Int. Congr. Primatol., Vol. 1*, pp. 88–101. Karger, Basel.

Gartlan, J. S. 1975. Adaptive aspects of social structure in Erythrocebus patas. In: S. Kondo, M. Kawai, A. Ehara, and S. Kawamura (eds.), *Proceedings from the Symposia of the Fifth Congress of the International Primatological Society*, pp. 161–171. Japan Science Press, Tokyo.

Gartlan, J. S., and Struhsaker, T. T. 1972. Polyspecific associations and niche separation of rainforest anthropoids in Cameroon, West Africa. *J. Zool., Lond.* **168**:221–266.

Gautier, J.-P. 1988. Interspecific affinities among guenons as deduced from vocalizations. In: A. Gautier-Hion, F. Bourlière, J.-P. Gautier, and J. Kingdon (eds.), *A Primate Radiation: Evolutionary Biology of the African Guenons*, pp. 194–226. University Press, Cambridge, England.

Gautier, J.-P., and Gautier, A. 1977. Communication in Old World monkeys. In: T. A. Seebok (ed.), *How Animals Communicate*, pp. 890–964. Indiana University Press, Bloomington, Indiana.

Gautier, J.-P., Vercauteren Drubbel, R., and Deleporte, P. 2002. Phylogeny of the *Cercopithecus lhoesti* group revisited: Combining multiple character sets. In: M. E. Glenn, and M. Cords (eds.), *The Guenons: Diversity and Adaptation in African Monkeys*, pp. 37–48. Kluwer Academic Publishers, New York.

Glenn, M. E. 1996. *The Natural History and Ecology of the Mona Monkey* (Cercopithecus mona *Schreber 1774) on the Island of Grenada, West Indies*, Ph.D. Dissertation. Northwestern University, Evanston, Illinois, USA.

Glenn, M. E. 1997. Group size and group composition of the mona monkey (*Cercopithecus mona*) on the island of Grenada, West Indies. *Am. J. Primatol.* **43**(2): 167–173.

Glenn, M. E. 1998. Population density of *Cercopithecus mona* on the Caribbean island of Grenada. *Folia Primatol.* **69**:167–171.

Glenn, M. E., and Bensen, K. J. 1998. Capture techniques and morphological measurements of the mona monkey (*Cercopithecus mona*) on the island of Grenada, West Indies. *Am. J. Phys. Anthropol.* **105**:481–491.

Glenn, M. E., Bensen, K. J., and Bekhuis, P. (unpub. report, a). Surveys of Mona Monkeys in the Bimbia Bonadikombo Community Forest Reserve, Cameroon. Research Report to the Government of Cameroon, October 2000, 20 pp.

Glenn, M. E., Bensen, K. J., and Jensen, K. F. (unpub. report, b). Comparative Studies of Island and Mainland Dwelling Populations of Mona Monkeys: São Tomé and Príncipe, Research Report to AGRECO and ECOFAC Africa, May 1999, 37 pp.

Grenada Government and Organization of American States (1988). *Plan and Policy for a System of National Parks and Protected Areas*, Dept. Reg. Dev., OAS, Washington, D.C.

Hall, K. R. L. 1965. Behaviour and ecology of the wild patas monkey, *Erythrocebus patas* in Uganda. *J. Zool., Lond.* **148**:15–87.

Howard, R. 1977. *Niche Separation among Three Sympatric Species of* Cercopithecus *Monkeys*, Ph.D. Dissertation, The University of Texas at Austin.

Jürgens, U. 1979. Vocalizations as an emotional indication. A neuroethological study on the squirrel monkeys. *Behaviour* **69**:88–117.

Jürgens, U. 1990. Vocal communication in primates. In: R. P. Kesner, and D. S. Olton (eds.), *Neurobiology of Comparative Cognition*, pp. 51–76. Lawrence Erlbaum Associates, Hillsdale, New Jersey.

Matsuda, R. (in prep.). *Behavior and Ecology of the Mona Monkey in the Lama Forest of Republic of Benin, West Africa*. Ph.D. Dissertation, Graduate School and University Center of the City University of New York, New York.

Newman, J. D., and Symmes, D. 1982. Inheritance and experience in the acquisition of primate acoustic behavior. In: C. T. Snowdon, C. H. Brown, and M. R. Petersen (eds.), *Primate Communication*, pp. 259–278. Cambridge University Press, Cambridge, Massachusetts.

Pulliam, H. R., and Caraco, T. 1984. Living in groups: Is there an optimal group size? In: J. R. Krebs, and N. B. Davies (eds.), *Behavioural Ecology: An Evolutionary Approach*, pp. 122–147. Sinauer Associates, Sunderland, Massachusetts.

Rowell, T. E. 1988. The social system of guenons, compared with baboons, macaques and mangabeys. In: A. Gautier-Hion, F. Bourlière, J.-P. Gautier, and J. Kingdon (eds.), *A Primate Radiation: Evolutionary Biology of the African Guenons*, pp. 439–451. University Press, Cambridge, England.

Ruvolo, M. 1988. Genetic evolution in the African guenons. In: A. Gautier-Hion, F. Bourlière, J.-P. Gautier, and J. Kingdon (eds.), *A Primate Radiation: Evolutionary Biology of the African Guenons*, pp. 128–139. University Press, Cambridge, England.

Struhsaker, T. T. 1969. Correlates of ecology and social organization among African cercopithecines. *Folia Primatol.* **11**:80–118.

Struhsaker, T. T. 1970. Phylogenic implications of some vocalizations of *Cercopithecus* monkeys. In: J. R. Napier, and P. H. Napier (eds.), *Old World Monkeys: Evolution, Systematics and Behaviour*, pp. 365–411. Academic Press, London.

Struhsaker, T. T., and Leland, L. 1979. Socioecology of five sympatric monkey species in the Kibale Forest, Uganda. In: J. S. Rosenblatt, R. A. Hinde, C. Beer, and M.-C. Busnel (eds.), *Advances in the Study of Behaviour, Vol. 9*, pp. 159–228. Academic Press, London.

Terborgh, J. 1983. *Five New World Primates: A Study in Comparative Ecology*, Princeton University Press, Princeton, New Jersey.

Tsingalia, H. M., and Rowell, T. E. 1984. The behaviour of adult male blue monkeys. *Z. Tierpsychol.* **64**:253–268.

van Schaik, C. P., and Horstermann, M. 1994. Predation risk and the number of adult males in a primate group: a comparative test. *Behav. Ecol. Sociobiol.* **35**:261–272.

Group Fission in Red-tailed Monkeys (*Cercopithecus ascanius*) in Kibale National Park, Uganda

11

TAMMY L. WINDFELDER and JEREMIAH
S. LWANGA

Introduction

Benefits attributed to group-living in general and larger group size in particular include increased foraging efficiency (Cody, 1971; Rudran, 1978; Wrangham, 1980; Garber, 1988) and enhanced protection from predators (Hamilton, 1971; Pulliam, 1973; Curio, 1978; Hoogland, 1979b; Bertram, 1980). As group size continues to increase, costs due to increased feeding competition (Chapman and Chapman, 2000), aggression, and increased conspicuousness to predators (Hoogland, 1979a) begin to outweigh advantages associated with further group size increases. Observed group sizes and composition likely reflect the trade-off between costs and benefits (Pulliam and Caraco, 1984).

Group fission is one response to the pressures exerted by group size increases beyond a certain maximum supportable size. Group fission occurs

TAMMY L. WINDFELDER • Department of Anthropology, University of Notre Dame, Notre Dame, IN 46556, USA.   JEREMIAH S. LWANGA • Makerere University Biological Field Station, P.O. Box 409, Fort Portal, Uganda.
The Guenons: Diversity and Adaptation in African Monkeys, edited by Glenn and Cords. Kluwer Academic/Plenum Publishers, New York, 2002.

in various primate species, including two forest guenons: red-tailed monkeys, *Cercopithecus ascanius* (Struhsaker and Leland, 1988) and blue monkeys, *C. mitis* (Cords and Rowell, 1986). Both red-tailed monkeys and blue monkeys show exclusive use of parts of their home ranges and defend territorial boundaries (Struhsaker and Leland, 1979; Cords, 1987).

For Schmidt's red-tailed monkeys (*Cercopithecus ascanius schmidti*) at the Ngogo study site in Kibale National Park, Uganda, the maximum supportable group size appears to be *ca.* 50 individuals. This is roughly 50% larger than the *Cercopithecus ascanius* average group size (30–35 individuals) at this site (Struhsaker, 1988; Windfelder *et al.*, unpub. data). On two occasions, and possibly a third, group fission occurred when a red-tailed monkey group reached 50 individuals. We describe the events surrounding a case of group fission in red-tailed monkeys in 1998. The group also was involved in a fission in 1980 (Struhsaker and Leland, 1988). Thus, we have established a frequency of group fission for this particular group of one per 18 years, which is significantly higher than previous estimates. We use data collected before and after the group split to examine the ecological and behavioral consequences of group fission. In addition to changes in home range use, we investigate patterns of reproduction, travel, intragroup aggression, and intergroup conflicts in our assessment of group fission in red-tailed monkeys.

Methods

Study Site

The Ngogo study site in Kibale National Park, Uganda is primarily medium-altitude evergreen rain forest interspersed with small grasslands (Ghiglieri, 1984; Butynski, 1990; Struhsaker, 1997).

Data Collection

Investigators have studied red-tailed monkeys at Ngogo since 1975 (Struhsaker, 1977, 1988; Struhsaker and Leland, 1988; Struhsaker *et al.*, 1988; Struhsaker and Pope, 1991; Mitani *et al.*, 2000). Subjects of this study were members of group Ss and daughter groups Ss1 and Ss2. We observed members of Ss for 2 months before fission (January 1998–March 1998), and we studied members of the resulting daughter groups for 29 months after fission (March 1998–August 2000). To determine group compositions, we conducted censuses opportunistically for all red-tailed monkey groups. Additional data on the two daughter groups, especially Ss2, are available from a separate study of polyspecific associations between red-tailed monkeys and mangabeys

(Windfelder *et al.*, in prep.). We followed focal groups for five days each month. We collected systematic behavioral data via group scan samples (Fragaszy *et al.*, 1992) taken at 30-minute intervals. Scans lasted for five minutes, during which the observer noted the behavior of all visible individuals.

At the end of each scan, we marked the location of the center of the group on a map of the study area. Using these maps, we measured the distance traveled during 30-minute intervals throughout the day. Daily travel distance equals the sum of the measurements between 0730–1800. We used only data from days in which we followed subjects for ≥ 10 hours to investigate changes in average daily travel distance before and after group fission.

We recorded all instances of aggression *ad libitum* (Altmann, 1974). We examined territorial conflicts between groups separately from aggressive interactions between members of the same group.

Results

The Process of Group Fission

Ss increased to approximately 50 individuals including 23 adult females and four to five new infants before the fission. In January 1998, only one adult male (Male #1) was with Ss, but in February 1998 a second adult male also began to associate with Ss.

When systematic observations began in January 1998, the members of Ss were already beginning to form transient subgroups, with small groups temporarily splitting from the main group for several hours at a time. They varied in composition, typically containing three to 11 adult females along with several subadults and juveniles. Male #1 sometimes left with one of the subgroups, while at other times he remained with the main group. The frequency and duration of forming subgroups increased over a period of several weeks.

In March 1998, the fission became more obvious with distinct subgroups spending several days apart. Although the split of group Ss was an ongoing, gradual process, we identified mid-March as the time of the fission because at that time the two subgroups were more stable in composition and remained separated for several days at a time. Furthermore, when the groups encountered one another in March, their meetings usually were associated with aggressive behaviors typical of territorial conflicts between neighboring groups, but daughter groups were more likely to tolerate the presence of one another temporarily when gray-cheeked mangabeys (*Lophocebus albigena*) were present.

In April 1998, individuals of the original Ss group were split fairly evenly between the two new daughter groups, Ss1 and Ss2. Compositions of the two new daughter groups for the two years following group fission are in Table I.

Table I. Changes in Group Composition for Two Years following the Fission of Group Ss

	Groups											
	Ss1				Ss2				RT15			
Year	M	F	I	Total	M	F	I	Total	M	F	I	Total
April 1998	1	12 (9–17)	2–3	31–36	1	11	2	25				
Late 1998	1	11	4–5	43	2	11–12	4	25–27				
1999	1–2	11	10	43–54	2	12	0	27				
2000	2–3	11	5	~50	2–3	12	6	32–33	1	6	3	17

Numbers reflect the number of individuals in each category. M = adult males; F = adult females; I = infants.

Ss1 initially had 31–36 individuals, including one adult male (Male #1), ⩾11 adult females and two to three infants. Ss2 had approximately 25 individuals, including one adult male, 11 adult females and two infants. There were a few subsequent transfers of individuals from Ss2 to Ss1. As a result of those transfers, new births, and the appearance of several other new individuals, Ss1 was notably larger than Ss2 by late 1998 (Table I).

There was an increase in the number of males associated with Ss1. Male #1, the resident male at the time of group fission, remained the sole adult male in Ss1 for more than six months after fission. On 10 November 1998, we discovered that he had a snare on his left wrist. He eventually left the group and remained solitary before finally disappearing in 1999. A subadult male took over the position of resident male for a short time until a fully adult male took the resident male position between mid-January and mid-February 1999. Later in 1999 another subadult/adult male joined the group.

Two adult males were associated with Ss2, the resident male and a second male that also spent time with a different red-tailed monkey group (BTP).

Rate of Group Fission

The members of Ss were last involved in group fission in 1980 when group S split into Sn and Ss (Struhsaker and Leland, 1988). Our observations of the fission of Ss in 1998 establish a known period of 18 years between fission events for this group.

The observed rate of group fission (one per 18 years) exceeds population-wide estimates for the rate of group fission in red-tailed monkeys. In 2000, 18 red-tailed monkey groups were present in the same area where only 12 groups existed in 1980. The appearance of six new groups over a 20-year period is consistent with a rate of group fission much slower than one per 18 group-years. Population-level estimates of group fission derived from the work at Ngogo over three years (1998–2000) also suggest a slower rate of group fission: one per 34 group-years.

Reproductive Rates

In 1998, during and immediately after the fission, a few infants were born in each daughter group. In 1999, the year after the fission, birth rates were significantly different between the two subgroups: 10 infants were born in Ss1 whereas no infant appeared in Ss2 (Table I). These differences do not reflect differences in the number of females with young infants, who may have been less likely to conceive: three of 11 Ss1 females and four of 12 Ss2 females had young infants in 1998, and yet all three Ss1 mothers and no Ss2 mothers gave birth again in 1999. In 2000, two years after the fission, birth rates in the two groups were roughly equivalent with five infants born in Ss1 and six infants born in Ss2 (Table I).

Ecological Consequences

Home range

The original home range of Ss was *ca.* 0.63 km^2 [Fig. 1(a)]. After the fission, Ss1 continued to use almost all of the original range of Ss while Ss2 shifted to the southwestern part of the original home range, expanding further west and southwest [Fig. 1(b)]. The size of Ss2's home range following the fission was *ca.* 0.47 km^2. The approximate size of Ss1's home range was 0.68 km^2.

The area where Ss2 established its home range, west and south of the original home range of Ss, exploited more of the Southwest Grassland and regenerating forest along its edges. Other red-tailed monkey group(s) that had previously used this area were displaced as Ss2 took over the northern part of

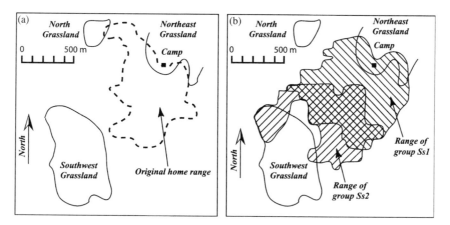

Fig. 1. (a) Home range of Ss before group fission. (b) Home ranges of daughter groups Ss1 and Ss2 after group fission.

Table II. Rates of Aggression

	Ss			Ss1			Ss2		
	D	C	F	D	C	F	D	C	F
Pre-fission Jan–March 1998	0 (59.5)	0 (59.5)	0.05 (59.5)						
Fission March 1998				0 (28)	0 (28)	0 (28)	0.10 (40)	0.05 (40)	0.05 (40)
Post-fission I April–July 1998				0.15 (54)	0.10 (54)	0.07 (54)	0.03 (62)	0.09 (62)	0.13 (62)
Post-fission II Nov 1998				0.07 (52.5)	0.48 (52.5)	0.19 (52.5)			

D = displacements/hour (total hours observed); C = chases/hour (total hours observed); F = fights/hour (total hours observed).

the grassland. Although the original range of Ss also included grassland and regenerating forest, Ss2 started to exploit the grassland habitat more with its new home range.

Daily travel distances

Before the completion of group fission, Ss traveled *ca.* 1 km/day ($\bar{x} = 1.08$ km/day, $n = 5$ days). One day in March, shortly after the completion of the fission, Ss1 traveled a slightly longer 1.23 km. Ss2 traveled distances of 0.77 km and 1.09 km on two days in March after the fission was complete. In April 1998, Ss2 tended to exhibit longer daily travel distances ($\bar{x} = 1.41$ km/day, $n = 4$ days). Daily travel distances for Ss1 remained relatively low throughout the remainder of 1998 ($\bar{x} = 0.91$ km/day, $n = 28$ days), and early 1999 (January–April 1999: $\bar{x} = 1.05$ km/day, $n = 20$ days).

Intragroup aggression

During 59.5 contact hours before group fission, no displacement, no chase, and only a few fights (rate $= 0.05$ fights/h) occurred among the members of Ss (Table II). In March 1998 (during and immediately after the split), no aggression occurred among the members of Ss1, while low rates of displacements, chases, and fights occurred in Ss2 (Table II). Rates of aggression increased after group fission (Table II). Over 50 contact hours with Ss1 in November 1998 revealed continued high rates of intragroup aggression well after the fission was completed.

Intergroup conflicts

We observed no territorial encounter during 59.5 contact hours before fission, but in late March 1998 we recorded a total of 11 territorial encounters over five days. Ten of them were between Ss1 and Ss2. During this time, they localized their activity in a relatively small area along the boundary between the two new territories. The two new daughter groups exchanged aggressive growls, chirps, and chatter during the confrontations, and often members of one daughter group displaced or chased members of the other one. The encounters escalated to physical aggression on several occasions. Females were the primary aggressors during the encounters. Males participated only occasionally by exchanging hack or pop vocalizations with the male of the other group, and very rarely engaging in physical aggression.

Following group fission, the rate of territorial encounters (April–October 1998) was slightly lower. Fights between Ss1 and Ss2 were less frequent, but there were more conflicts with other neighboring groups.

Another group fission?

A small group of 17 individuals (RT15) appeared in 2000, comprising one adult male, six adult females, seven subadult and juvenile individuals, and three infants. The group used an area exclusively within the home range of Ss1, but was clearly subordinate to Ss1. Ss1 displaced this new group whenever they were in the same area.

Discussion

The Process of Group Fission

In 1998, we witnessed group fission in a group of red-tailed monkeys (group Ss) in Kibale National Park, Uganda. The group fission we observed involved >50 individuals. Similarly, Struhsaker and Leland (1988) report an earlier fission in a group of 45–50 red-tailed monkeys at the same study site. In the Kakamega Forest, Kenya, a group of blue monkeys (*Cercopithecus mitis*) split after reaching 46 individuals (Cords and Rowell, 1986). Together these observations suggest an upper limit to group size in these species of approximately 50 individuals.

Group fission was a gradual process occurring over a period of months, as has been reported for other guenon groups (Cords and Rowell, 1986; Struhsaker and Leland, 1988). The subgroups formed by group Ss were initially transient and relatively rare, but increased in frequency and duration with time.

Social relationships between individuals also changed gradually during the process of group fission. Initially, there was no aggression associated with individuals in subgroups leaving and then rejoining the original group. Two months later, however, when individuals from the two subgroups met, encounters often were associated with aggressive behaviors typical of territorial conflict.

The new daughter groups were initially similar in size. In other cases of group fission in guenons, however, group sizes were more dissimilar, with smaller daughter groups containing less than half the number of individuals in larger daughter groups (Cords and Rowell, 1986; Struhsaker and Leland, 1988). Our study groups reached a similar size disparity within a year after fission. Several individuals moved from Ss2 to Ss1 during that year, but the higher reproductive rates of Ss1 females relative to Ss2 females was the primary factor leading to the size disparity.

Struhsaker and Leland (1988) report the rapid turnover of at least seven adult males in the larger daughter group following the fission of a group of red-tailed monkeys. We also observed a temporary increase in the number of males associated with the larger daughter group within a year of group fission, but this was more likely related to the disappearance of the resident male than an effect of group fission. Male #1, the resident male at the time of group fission, remained the sole adult male in Ss1 for ⩾6 months after fission. He left the group after incurring a serious snare injury, and it was only after his disappearance that additional males joined Ss1 resulting in an increase in the number of males associated with the larger daughter group.

Rate of Group Fission

The observed interval between group fission events for the individuals in group Ss was 18 years. The observed frequency of group fission (one per 18 years) for group Ss is double that of population-wide estimates (one per 34 group-years) as well as previous estimates of the rate of group fission for this population (one per 36.3 group-years: Struhsaker and Leland, 1988).

Why is the frequency of group fission for Ss apparently higher than the rates for other groups and the Ngogo population as a whole? There is no obvious difference in food availability in the Ss home range vs. other sites. Perhaps Ss suffered less predation by virtue of its proximity to camp (Fig. 1). It is possible that guenons closer to human encampments are less likely to be targeted by their main predators, crowned hawk eagles (*Stephanoaetus coronatus*) and chimpanzees (*Pan troglodytes*) if the predators avoid humans. The red-tailed monkeys are certainly not exempt from predation as a result of their proximity to camp, however. Chimpanzees have eaten red-tailed monkeys within the area, and eagle attack rates are similar to those for other groups (Windfelder *et al.*, unpub. data). Thus, it is unclear why Ss should experience such high reproductive success and subsequent growth relative to other groups.

Reproductive Rates

In our study, reproductive rates declined markedly in the smaller daughter group as no surviving offspring were seen the year following group fission. These data differ significantly from other studies. Struhsaker and Leland (1988) reported higher birth rates for both daughter groups in the two years post-fission. Similarly, Cords and Rowell (1986) found that births per adult female blue monkey increased in both splinter groups during the first year after division. As discussed by Struhsaker and Leland (1988), data on group fission in Japanese macaques (*Macaca fuscata*) and olive baboons (*Papio anubis*) follow the same pattern, providing additional support for the hypothesis that group fission leads to increased reproductive output. The data from the smaller daughter group that we studied stand in contrast to this general pattern.

One possibility for the unexpectedly low birth rate we observed in the smaller daughter group immediately following fission is that the females were stressed as a result of expanding their range into new habitat and engaging in frequent territorial encounters. Changes in habitat type seem an unlikely explanation for the decrease in reproductive rate immediately after fission, because this rate did not remain low in the second year in the new range.

Ecological Consequences of Group Fission

After the fission, the larger daughter group continued to use the original home range while the smaller daughter group shifted its home range to the southwestern part of the original home range and expanded further west and southwest. The area where Ss2 established its territory exploited more grassland and regenerating forest habitat. The long-term consequences of this shift, if any, have yet to be determined.

The approximate size of the smaller daughter group's home range following the fission was smaller than that of the original home range. The approximate size of the larger daughter group's home range, on the other hand, was slightly larger than the original home range reflecting increased excursions to the west into the other daughter group's new home range, usually during association with gray-cheeked mangabeys (*Lophocebus albigena*). Red-tailed monkeys are responsible for initiating and maintaining polyspecific associations with gray-cheeked mangabeys at Kibale. When mangabeys were present, both daughter groups were more likely to tolerate the temporary presence of the other daughter group in their newly established territories.

The smaller daughter group (Ss2) tended to travel further per day after group fission, while travel patterns of Ss1 did not significantly change after fission. The increased daily travel distances of Ss2 were likely related to the displacement of this group and the necessity of establishing a new home range. Whereas group fission is expected to reduce travel costs associated with large group size, many studies have found no relationship between group size and day range in primates (Struhsaker and Leland, 1987; Struhsaker & Leland, 1988; Butynski, 1990; Bronikowski and Altmann, 1996). In a study of red colobus monkeys (*Procolobus badius*), travel costs associated with larger group size were evident only after food availability was taken into account (Gillespie and Chapman, 2001). The changes in daily travel distances we observed in the smaller, displaced daughter group may have been strongly affected by differences in food availability as they exploited new habitat with more grassland and regenerating forest and less primary forest. More information is necessary to evaluate the influence of habitat change on this group's travel patterns.

For both daughter groups, rates of displacements, chases and fights increased after the fission was completed. The increases in intragroup aggression may have been related to a period of instability as group composition was solidifying; however, similarly high rates of intragroup aggression well after the fission was completed shed doubt on this hypothesis.

The rate of intergroup conflicts was especially high in March 1998, during group fission. Most conflicts at that time occurred between the two daughter groups, Ss1 and Ss2. The high level of territorial conflict was apparently the result of the two daughter groups fighting to establish a border between two new territories. The restricted range use in March 1998, localized in a relatively small area along the boundary between the two new territories, is consistent with this interpretation. Following group fission, there were more territorial conflicts with other neighboring groups, but the number of territorial conflicts between the two daughter groups decreased resulting in lower overall territorial encounter rates.

Due to the continued high rate of increase exhibited by Ss1, they again approached 50 individuals by May 2000. If the maximum supportable group size were 50, another fission would be expected. A small group of 17 individuals (RT15) appeared in 2000. They probably split off from Ss1, but there was no individual with distinct markings, thus rendering it impossible to verify this hypothesis. RT15 uses an area exclusively within the range of Ss1 and their group size suggests that they may have split from Ss1, but we were unable to confirm a decrease in Ss1 group size after the appearance of RT15. If indeed RT15 split from Ss1, they are exhibiting an extremely high rate of growth and fission.

Summary

We describe group fission in Schmidt's red-tailed monkeys (*Cercopithecus ascanius schmidti*) at the Ngogo study site, Kibale National Park, Uganda. The process of group fission lasted several months, beginning when the original group (Ss) reached just over 50 individuals. For both daughter groups, rates of both intragroup and intergroup aggression increased following fission. Rates of intergroup aggression were especially high immediately following fission as the two daughter groups fought frequently while establishing a new territorial border. The larger daughter group continued to use the original home range of Ss whereas the smaller daughter group shifted its home range. The displaced daughter group increased daily travel distances following group fission as they established a new territory, and no infants were born the year after fission. The members of Ss were last involved in a group fission 18 years ago, suggesting a faster rate of group fission compared to population-level estimates (one per 35 group-years). The especially high frequency of group fission relative to the rest of the population is deserving of further study.

ACKNOWLEDGMENTS

We thank Mary Glenn and Marina Cords for the invitation to participate in the IPS symposium "The Genus *Cercopithecus*: An Update." We gratefully acknowledge the helpful discussions and invaluable advice received from Thomas T. Struhsaker. We thank John Mitani for encouraging our study of red-tailed monkeys and mangabeys at Ngogo. Permission to conduct research in the Kibale National Park was granted by the Uganda National Council of Science and Technology, Ugandan Wildlife Authority, and Makerere University Biological Field Station. John Kasenene, Gilbert Isabirye-Basuta, and the staff of the Makerere University Biological Field Station provided logistical assistance. We are grateful to John Mitani, David Watts, and Charles Businge for support in the field. Marina Cords, Mary Glenn, Thomas T. Struhsaker, Russell H. Tuttle, and an anonymous reviewer provided helpful suggestions for improving this manuscript. We thank Ian Kuijt for preparing Figure 1. This study was made possible by grant support from the National Science Foundation and the National Geographic Society.

References

Altmann, J. 1974. Observational study of behaviour: sampling methods. *Behaviour* **49**:227–267.

Bertram, B. C. R. 1980. Vigilance and group size in ostriches. *Anim. Behav.* **28**:278–286.

Bronikowski, A. M., and Altmann, J. 1996. Foraging in a variable environment: weather patterns and the behavioral ecology of baboons. *Behav. Ecol. Sociobiol.* **39**:11–25.

Butynski, T. M. 1990. Comparative ecology of blue monkeys (*Cercopithecus mitis*) in high- and low-density subpopulations. *Ecol. Monogr.* **60**:1–26.

Chapman, C. A., and Chapman, L. J. 2000. Interdemic variation in mixed-species association patterns: common diurnal primates of Kibale National Park, Uganda. *Behav. Ecol. Sociobiol.* **47**:129–139.

Cody, M. L. 1971. Finch flocks in the Mohave Desert. *Theor. Popul. Biol.* **2**:142–158.

Cords, M. 1987. Forest guenons and patas monkeys: male–male competition in one-male groups. In: B. B. Smuts, D. L. Cheney, R. M. Seyfarth, R. W. Wrangham, and T. T. Struhsaker (eds.), *Primate Societies*, pp. 98–111. University of Chicago Press, Chicago.

Cords, M., and Rowell, T. E. 1986. Group fission in blue monkeys of the Kakamega Forest, Kenya. *Folia Primatol.* **46**:70–82.

Curio, E. 1978. The adaptive significance of avian mobbing: I. Teleonomic hypotheses and predictions. *Z. Tierpsychol.* **48**:175–183.

Fragaszy, D. M., Boinski, S., and Whipple, J. 1992. Behavioral sampling in the field: comparison of individual and group sampling methods. *Am. J. Primatol.* **26**:259–275.

Garber, P. A. 1988. Diet, foraging patterns, and resource defense in a mixed species troop of *Saguinus mystax* and *Saguinus fuscicollis* in Amazonian Perú. *Behaviour* **105**:18–34.

Ghiglieri, M. 1984. *The Chimpanzees of Kibale Forest*. Columbia University Press, New York.

Gillespie, T. R., and Chapman, C. A. 2001. Determinants of group size in the red colobus monkey (*Procolobus badius*): an evaluation of the generality of the ecological-constraints model. *Behav. Ecol. Sociobiol.* **50**:329–338.

Hamilton, W. D. 1971. Geometry for the selfish herd. *J. Theor. Biol.* **7**:295–311.

Hoogland, J. L. 1979a. Aggression, ectoparasitism and other possible costs of prairie dog (Sciuridae: *Cynomys* spp.) coloniality. *Behaviour* **69**:1–35.

Hoogland, J. L. 1979b. The effect of colony size on individual alertness of prairie dogs (Sciuridae: *Cynomys* spp.). *Anim. Behav.* **27**:394–407.

Mitani, J., Struhsaker, T., and Lwanga, J. 2000. Primate community dynamics in old growth forest over 23.5 years at Ngogo, Kibale National Park, Uganda: Implications for conservation and census methods. *Int. J. Primatol.* **21**:269–286.

Pulliam, H. R. 1973. On the advantage of flocking. *J. Theor. Biol.* **38**:419–422.

Pulliam, H. R., and Caraco, T. 1984. Living in groups: Is there an optimal group size? In: J. R. Krebs, and N. B. Davies (eds.), *Behavioural Ecology: An Evolutionary Approach, 2nd ed*, pp. 122–147. Blackwell Scientific, Oxford.

Rudran, R. 1978. Socioecology of the blue monkeys (*Cercopithecus mitis stuhlmanni*) of the Kibale Forest, Uganda. *Smith. Contri. Zool.* **49**:1–88.

Struhsaker, T. T. 1977. Infanticide and social organization in the redtail monkey (*Cercopithecus ascanius schmidti*) in the Kibale Forest, Uganda. *Z. Tierpsychol.* **45**:75–84.

Struhsaker, T. T. 1988. Male tenure, multi-male influxes, and reproductive success in redtail monkeys (*Cercopithecus ascanius*). In: A. Gautier-Hion, F. Bourlière, J.-P. Gautier, and J. Kingdon (eds.), *A Primate Radiation: Evolutionary Biology of the African Guenons*, pp. 340–363. Cambridge University Press, Cambridge, UK.

Struhsaker, T. T. 1997. *Ecology of an African Rain Forest: Logging in Kibale and the Conflict Between Conservation and Exploitation*. University Press of Florida, Gainesville, FL.

Struhsaker, T. T., and Leland, L. 1979. Socioecology of five sympatric monkey species in the Kibale Forest, Uganda. *Adv. Stud. Behav.* **9**:159–228.

Struhsaker, T. T., and Leland, L. 1987. Colobines: infanticide by adult males. In: B. B. Smuts, D. L. Cheney, R. M. Seyfarth, R. W. Wrangham, and T. T. Struhsaker (eds.), *Primate Societies*, pp. 83–97. University of Chicago Press, Chicago.

Struhsaker, T. T., and Leland, L. 1988. Group fission in redtail monkeys (*Cercopithecus ascanius*) in the Kibale Forest, Uganda. In: A. Gautier-Hion, F. Bourlière, J.-P. Gautier, and J. Kingdon (eds.), *A Primate Radiation: Evolutionary Biology of the African Guenons*, pp. 364–388. Cambridge University Press, Cambridge, UK.

Struhsaker, T. T., and Pope, T. R. 1991. Mating system and reproductive success: a comparison of two African forest monkeys (*Colobus badius* and *Cercopithecus ascanius*). *Behaviour* **117**:182–205.

Struhsaker, T. T., Butynski, T. M., and Lwanga, J. 1988. Hybridization between redtail monkeys (*Cercopithecus ascanius schmidti*) and blue (*C. mitis stuhlmanni*) monkeys in the Kibale Forest, Uganda. In: A. Gautier-Hion, F. Bourlière, J.-P. Gautier, and J. Kingdon, (eds.), *A Primate Radiation: Evolutionary Biology of the African Guenons*, pp. 477–497. Cambridge University Press, Cambridge, UK.

Wrangham, R. W. 1980. An ecological model of female-bonded primate groups. *Behaviour* **75**: 262–300.

Interindividual Proximity and Surveillance of Associates in Comparative Perspective

12

ADRIAN TREVES and PASCAL BAGUMA

Introduction

The variability and rich diversity of social relationships, group dynamics, and group compositions (hereafter, social organization) in the Order Primates has prompted many efforts to classify the social systems of primates and analyze their determinants (Sterck *et al.*, 1997; Koenig *et al.*, 1998). Attention has been directed to within-group behavior such as patterns of interindividual proximity and visual monitoring of associates (conspecific members of the same group) as useful indicators of social organization (Kummer, 1968; Richard, 1974; Chance, 1976; Keverne *et al.*, 1978; Borries, 1993). For example, individuals that feed and rest in close proximity are usually kin, allies, or potential mates, whereas close proximity is rarely maintained for long between antagonists (Sekulic, 1983; Silk, 1991). Similarly, high levels of visual monitoring of associates reflect the risk of aggression or competition within

ADRIAN TREVES • Center for Applied Biodiversity Science, Conservation International, 65 Greenway Terrace, Princeton, NJ 08540, USA. PASCAL BAGUMA • Makerere University Biological Field Station, P.O. Box 409, Fort Portal, Uganda.
The Guenons: Diversity and Adaptation in African Monkeys, edited by Glenn and Cords. Kluwer Academic/Plenum Publishers, New York, 2002.

groups (Keverne *et al.*, 1978; Caine and Marra, 1988; Watts, 1998; Alberts, 1994; Treves, 1999a, 2000; Treves and Pizzagalli, 2002).

Despite the role of interindividual proximity and visual monitoring of associates in group dynamics, recent hypotheses about the evolution of primate social organization focus on dispersal, grooming, and dyadic conflict over resources, which are rare, conspicuous behaviors (Wrangham, 1980; van Schaik, 1989, 1996; Mitchell *et al.*, 1991; van Hooff and van Schaik, 1991; Cheney, 1992). Both rare, conspicuous social behaviors and more common, subtle interactions (e.g., interindividual proximity, social monitoring, approach/avoid) have fitness consequences. Individuals that lose contests may be injured, reproduce more slowly or die (Hamilton, 1985; Pusey *et al.*, 1997). Likewise, individuals without close associates spend more time engaged in costly surveillance of their surroundings than their more affiliative associates (Treves, 2000; Treves *et al.*, 2001). Moreover, spatial isolation may increase the risk of predation (Boinski *et al.*, 2000). How should one weigh the relative contributions of subtle, frequent, low-cost social interactions against conspicuous, rare, high-cost interactions? Do the two types of evidence produce similar views of social organization?

We examined the behavior of female monkeys in three species to determine whether common, subtle social behaviors (interindividual proximity and within-group vigilance) are predicted by a classificatory system based on rarer, more conspicuous patterns of social behavior. We compared Schmidt's red-tailed monkeys (*Cercopithecus ascanius schmidtii*) to red colobus monkeys (*Procolobus badius tephrosceles*) and black howler monkeys (*Alouatta pigra*). These three species are ideal for this aim, because they differ greatly in features of social organization that are central to current theory about the ecology of female primate social relationships (Mitchell *et al.*, 1991; van Schaik, 1996; Sterck *et al.*, 1997). Red-tailed monkeys live in stable groups from which only mature males typically disperse. Female red-tailed monkeys engage in allomothering and are frequently seen grooming each other. Aggression between females or female and young is very rare and not severe (Struhsaker, 1977, 1978, 1980, 1988; Struhsaker and Leland, 1979, 1988; Jones and Bush, 1988; Treves, 1998, 1999a). Red-tailed monkey females can therefore be classified as resident and egalitarian, and hence we expected high levels of female–female proximity, low levels of female–male proximity, and low levels of within-group vigilance. By contrast, red colobus monkeys females disperse as juveniles or even adults. Allomothering is rare or absent in red colobus monkeys. Aggression between females and between females and juveniles is common over food, preferred resting sites, and during sexual competition (Struhsaker, 1975, 1978, 1980; Struhsaker and Oates, 1975; Struhsaker and Leland, 1979, 1985; Marsh, 1979a,b; Isbell, 1983; Treves, 1998, 1999a; unpub. data). Red colobus females can be classified as dispersing and despotic; hence we expected low levels of female–female proximity, high levels of female–male proximity and high levels of within-group vigilance. Black howler monkey females disperse as juveniles but female–female relations within their small groups are almost invariably calm and cohesive. In >4000 hours of observation, we saw <10 instances of female–female aggression, none of which led to physical contact or injury. Allomothering is not uncommon (Horwich, 1983; Horwich and Gebhard, 1983, 1986; Treves *et al.*, 2001, unpub. data). Black howler monkey females can be classified as dispersing and egalitarian, and hence we expect them to show moderate levels of female–female proximity, high levels of female–male proximity, and low levels of within-group vigilance. Female–male proximity is expected to reflect the need for male protection. For example, red colobus monkeys and black howler monkey females should maintain higher levels of proximity to adult males because, immigration often requires male tolerance and protection (Marsh, 1979b, Sekulic, 1983; Glander, 1992). By contrast, guenon groups remain stable despite the total absence of males (Cords and Rowell, 1986; Jones and Bush, 1988; Struhsaker, 1988).

Methods

We measured vigilance and interindividual proximity of red-tailed monkeys at Kanyawara in Kibale National Park for 16 months from 1994–1995 (Treves, 1997, 1998, 1999a,b, 2002). Kanyawara is described in Struhsaker (1997). We studied four groups containing 19, 22, 23 and 29 individuals (Treves, 1998a). We studied red colobus monkeys at the same site over the same period in five groups consisting of 20, 32, 58, 66 and 76 individuals (Treves, 1998, 1999a,b, 2002). We established individual identification for only a small portion of the females in these studies as priority was given to sampling many groups. Treves and his assistants measured the same behaviors in black howler monkeys between 1997 and 2001 at Lamanai Reserve in Belize (Treves, 2001; Treves *et al.*, 2001). Data from March–August 1999 are presented from five intensively studied groups consisting of three, five, five, six and ten individuals, respectively. Individual identification was complete for these five groups.

We recorded vigilance (visual search beyond the immediate vicinity of the focal animal) and interindividual proximity in continuous focal animal samples (Treves, 1998, 1999a,b, 2000; Treves *et al.*, 2001). Here, we analyzed only the presence or absence of ⩾1 adult female within 2 m (ADULT FEMALE), the presence or absence of ⩾1 adult male within 2 m (ADULT MALE), and the presence or absence of ⩾1 glances towards associates over the course of the focal animal sample (WITHIN-GROUP VIGILANCE). The sample duration for black howler monkeys (2 minutes) was twice as long as that used for the other species (1 minute); hence we halved the former's measure of WITHIN-GROUP VIGILANCE. Our measure did not distinguish multiple glances within one sample from a single glance, but a subsequent study revealed that only 10% ($n = 589$) of samples contained more than one glance at an associate among black howler monkeys and only 3% contained more than two glances to associates (Treves, unpub. data). Nevertheless, we did not measure the duration of single bouts of within-group vigilance, so error remains in the use of a frequency measure to index investment in within-group vigilance if species differ in average glance duration.

Here, we analyzed only the behavior of adult females that contributed $\geqslant 10$ samples. This has left 2768 samples (973 of red-tailed monkeys, 1069 of red colobus monkeys and 726 of black howler monkeys). Fewer samples contained information about neighbors because neighbors $\geqslant 2$ m could not be always classified by age–sex class in dense vegetation. Although some individuals were sampled more often than others because of the vicissitudes of observing animals in the canopy (Treves, 1998), individual differences in vigilance have proven to be minimal within age–sex classes (Treves et al., 2001).

To compare the three species of monkeys, we analyzed average differences directly (raw scores) using pairwise Mann-Whitney U tests. Individual samples were used to generate pooled averages for species. As this procedure may inflate differences between species if there is a non-independence between samples, we accept only $p < 0.01$ for tests of raw scores. We also compared species using averages corrected for the availability of associates in each group. For example, the number of females in proximity to a focal female may reflect the number of females in her group. Thus each group has an expected number of females within two m of a focal female, calculated as follows:

$$\text{expected} = (\text{population mean number of all associates within 2 m})$$
$$\times (\text{number of adult females in group} - 1)/(\text{group size} - 1) \quad (1)$$

We calculated the expected value for ADULT MALE, similarly, but the expected value for WITHIN-GROUP VIGILANCE in each group differed as follows:

$$\text{expected} = (\text{population mean proportion of samples with} \geqslant 1 \text{ glance}$$
$$\text{to an associate}) \times (\text{group size} - 1)/(\text{population mean}$$
$$\text{group size}) \quad (2)$$

The expected values for ADULT FEMALE, ADULT MALE, and WITHIN-GROUP VIGILANCE for each group (Eqs. 1 and 2) were then subtracted from the raw score for that same group to yield a deviation from the expected. Large negative deviations reflect lower than expected proximity or vigilance, while positive values indicate higher than expected proximity or vigilance. Deviations for each group were tested between species using a Kruskal-Wallis H test (df $= 2$, corrected for ties with criterion for significance set at $p \leqslant 0.05$), and within species, using a Wilcoxon signed-ranks test ($n = $ number of groups).

Both raw and per capita scores were analyzed because we had no *a priori* reason to discard one or the other. For example, a group of 50 may have more associates within view if it is more compact than a group with 10 individuals, but the group of 50 may disperse over a wider area and therefore have a similar number of individuals within view.

Results

Table I reveals heterogeneity between species in raw scores for ADULT FEMALE. Species with female dispersal (red colobus and black howler monkeys) should show less cohesion among females, than those with female residence (red-tailed monkeys). This prediction (i) was not borne out as female red-tailed monkeys were significantly less cohesive than the other females (Mann-Whitney U test: ADULT FEMALE Red-tailed monkeys vs. red colobus: $Z = 13.6$, $p < 0.0001$; vs. black howler monkeys: $Z = 2.7$, $p = 0.0071$). This result held whether the focal females were foraging or resting. This is consistent with data from the same site collected using different methods (Struhsaker, 1980), but fails to support prediction (i).

Species with female dispersal (red colobus and black howler monkeys) should have higher proximity between adult females and adult males than those with female residence (red-tailed monkeys). This prediction (ii) received support as both female-dispersal species showed higher proximity to adult

Table I. Interindividual Proximity and Within-group Vigilance (WGV) in Female Monkeys from Three Species

Species	Female dispersal	Competitive regime		Interindividual proximity[a]		
				ADULT FEMALE (%)	ADULT MALE (%)	WGV (%)
Red-tailed monkeys	rare	egalitarian	mean	25.6	5.1	67.6
			SE	1.6	0.7	1.5
Red colobus monkeys	common	despotic	mean	56.1	17.2	67.7
			SE	1.7	1.3	1.4
Black howler monkeys	common	egalitarian	mean	32.0	9.4	17.2
			SE	1.8	1.1	0.9

[a] Raw scores: not corrected for the availability of associates.
ADULT FEMALE: % of samples in which one or more adult females was within 2 m of the focal female;
ADULT MALE: % of samples in which one or more adult males was within 2 m of the focal female;
WITHIN-GROUP VIGILANCE: % of samples containing $\geqslant 1$ glance to conspecific members of the same group.

males than did the female-resident red-tailed monkeys (Table I: red-tailed monkeys vs. red colobus monkeys: $Z = -8.1$, $p < 0.0001$; black howler monkeys: $Z = 3.3$, $p = 0.0009$). This result held whether females were foraging or resting.

However, direct comparison of raw scores may be deceptive if large groups were more compact than small groups because associates would then have closer proximity by chance. Hence we repeated the analyses using the deviations from the expected for each group (Eq. 1 and Table II). The three species did not differ in deviations for ADULT FEMALE (Kruskal-Wallis $H = 4.2$, $p = 0.12$), again contrary to prediction (i). The three species did differ in deviations for ADULT MALE ($H = 9.0$, $p = 0.011$), because all five black howler monkey groups showed lower than expected proximity between adult males and focal females (Wilcoxon signed-ranks for ADULT MALE: $Z = 2.0$, $p = 0.043$). These results do not support prediction (ii).

Species with egalitarian relationships (red-tailed and black howler monkeys) should monitor associates less frequently than those with more despotic interactions (red colobus monkeys, prediction (iii)). Red-tailed monkeys and red colobus monkeys did not differ in raw scores for WITHIN-GROUP VIGILANCE ($Z = -0.05$, $p = 0.96$). However, red-tailed monkeys

Table II. Observed and Expected Interindividual Proximity and Within-group Vigilance Split by Group

Species	Group	N	ADULT FEMALE		ADULT MALE		WITHIN-GROUP VIGILANCE	
			Observed (%)	Expected (%)[a]	Observed (%)	Expected (%)[a]	Observed (%)	Expected (%)[a]
Red-tailed monkeys								
	B	118	42	37	4	5	83	61
	C	237	25	38	5	3	68	52
	F	447	25	41	5	2	67	81
	U	171	26	39	5	3	72	68
Red colobus monkeys								
	BC	308	69	64	16	12	67	77
	C	134	65	58	15	12	62	101
	P	336	49	52	17	18	68	42
	SC	170	58	53	20	21	74	26
	66	121	63	42	15	13	47	87
Black howler monkeys								
	A	112	26	27	19	27	28	25
	O	106	0	13	13	40	41	50
	Q	173	36	22	3	22	40	31
	S	29	55	30	0	30	28	22
	T	93	50	48	5	48	31	14

[a] see Eqs. (1) and (2) for calculation of expected values.

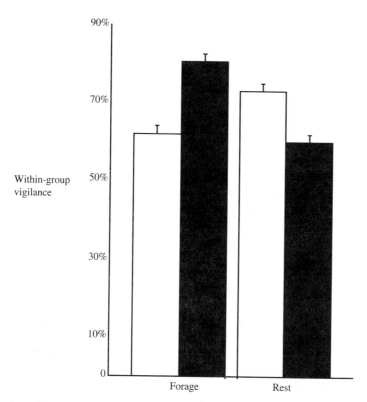

Fig. 1. Surveillance of associates in sympatric populations of red-tailed monkeys (open bars) and red colobus monkeys (dark bars). WITHIN-GROUP VIGILANCE = the proportion of focal samples containing ⩾1 glance at a conspecific group member. Data are shown for focal females engaged in foraging (left pair of bars) and resting (right pair of bars). Error bars denote 1 SE.

were significantly more likely to glance at associates while resting ($Z = 4.6$, $p < 0.0001$), and red colobus monkeys were significantly more likely to glance at associates while foraging ($Z = -6.0$, $p < 0.0001$). Pooling these two activities eliminated this interspecific difference (Fig. 1).

By contrast, female red-tailed monkeys and black howler monkeys did not differ in WITHIN-GROUP VIGILANCE while foraging ($Z = -0.8$, $p = 0.42$), but did differ while resting ($Z = 13.7$, $p < 0.0001$). Although high levels of within-group vigilance among resting female red-tailed monkeys remain unexplained, these findings support prediction (iii), because the despotic red colobus monkeys had the highest levels and the egalitarian species were similar during foraging.

Correcting for availability of associates (Eq. 2) revealed no interspecific differences in deviation for WITHIN-GROUP VIGILANCE ($H = 0.22$,

$p = 0.89$). None of the groups showed a consistent pattern of deviations within species (Table II), contradicting prediction (iii).

Discussion

Social systems with female dispersal should be associated with lower levels of familiarity and relatedness among females than social systems with female residence. Thus, we predicted that female–female spatial proximity would be higher in the female-resident red-tailed monkeys than in the female-dispersing red colobus monkeys or black howler monkeys. This prediction was not borne out as red-tailed monkey females foraged and rested further from each other than the other two species. The demands of foraging for scattered prey may account for the large spread seen among the insectivorous–frugivorous red-tailed monkey females, but the persistence of relatively large interindividual spacing during resting periods is puzzling.

Similarly, we predicted that the red colobus monkey females and black howler monkey females would maintain higher levels of proximity to adult males because immigration often requires male tolerance and protection (Marsh, 1979b, Sekulic, 1983; Glander, 1992). By contrast, guenon females seem to remain cohesive despite the absence of males (Cords and Rowell, 1986; Jones and Bush, 1988; Struhsaker, 1988). Our prediction received equivocal support. The raw data (Table I) supported the prediction because red colobus monkey females and black howler monkey females showed two to three times higher average values for proximity to adult males than did red-tailed monkeys. However, correcting for the availability of adult males eliminated the difference between red-tailed monkeys and red colobus monkeys (Table II), while indicating that black howler monkey females and adult males avoided close proximity. Perhaps our measure of interindividual spacing (associates within 2 m) was too restrictive if females are satisfied with being in the same tree as other males or females. On the other hand, corrected scores assume that females select any adult male or female for proximity, whereas some dyads consistently show greater cohesion than others. This would argue for the use of raw scores averaged across individuals. Further work is needed to determine whether interindividual proximity reflects patterns of female dispersal and female competitive regimes in primate groups.

Also, we predicted that within-group vigilance—a behavior strongly associated with competition and threats within the group (Treves, 2000)—would be lower in the egalitarian species (red-tailed monkeys and black howler monkeys) than in the despotic red colobus monkeys. This prediction also received equivocal support. The despotic red colobus monkey females were far more likely to glance at associates during foraging than were the egalitarian females in either of the other species. Correcting for availability of associates eliminated these interspecific differences. However, previous work on within-group vigilance suggests that the raw scores will be most informative. Within-group vigilance is directed to associates that are most salient to the vigilant animal rather than being directed to all associates (Watts, 1998; Treves, 2000). Moreover, within-group vigilance, as we have measured it here is unresponsive to dramatic changes in group size (Treves, 1999a). If further study of within-group vigilance is concordant with our findings, we suggest that measures of surveillance of associates support classifications of social organization based on contest competition.

If glances to associates are strongly associated with the perceived threat of within-group competition, as indicated by many studies (Treves, 2000), then female red colobus monkeys and black howler monkeys perceived greater risk from associates during foraging than did the red-tailed monkey females. This is consistent with increased female–female competition over food when females are unrelated, and it is also consistent with interference due to higher proximity of red colobus monkeys and black howler monkey females during foraging. If we can explain why these females foraged in close proximity, we may understand why they glanced at associates so frequently. The type of food eaten may play a role, if, for example, palatable leaves or fruits selected by red colobus monkeys and black howler monkeys were found in smaller patches than those selected by red-tailed monkeys. Alternately, the vagaries of insect pursuit may separate red-tailed monkey females from each other, while simultaneously promoting scanning of foliage and branches over monitoring of associates.

We find that female–female proximity did not match the classificatory schemes based on rare occurrences of female dispersal or female–female contest competition (Sterck *et al.*, 1997). By contrast, female–male proximity and within-group vigilance appeared to match such classificatory schemes. Therefore, we propose that subtle, common behaviors like female–male proximity and within-group vigilance can be used to classify primate social systems. Indeed, these data may be more efficient than the rarer behaviors, as they can be collected rapidly, and they provide a quantitative measure of female–female relationships rather than a simple dichotomy. More efficient measures that do not require the observation of rare events may accelerate the development of analytical models to explain the diversity and evolution of primate social systems.

Summary

Theory about the evolution of primate social organization typically focuses on rare, conspicuous behaviors such as female dispersal, grooming patterns or dyadic contests over resources which are felt to shape female relationships within groups. We evaluated if other, more subtle and more common measures of female behavior agree with the classificatory schemes derived from the former measures. Red-tailed monkeys (*Cercopithecus ascanius*), classified as egalitarian and female-resident on the basis of dispersal, grooming, and contest, were compared to red colobus monkeys (*Procolobus badius tephrosceles*) classified as despotic and female-dispersing, and to black howler monkeys (*Alouatta pigra*) classified as egalitarian and female-dispersing. Whether a species was classified as female-dispersal or female-resident did not predict patterns of female–female proximity but did predict patterns of female–male proximity. Despotic vs. egalitarian competitive regimes among females predicted patterns of within-group vigilance. Measurement of female–male proximity and within-group vigilance can therefore provide an alternative, more rapid and continuous measure of female social relationships within primate groups.

ACKNOWLEDGMENTS

A.T. was supported by National Geographic at both sites and the University of Wisconsin-Madison Graduate School at Lamanai. We are indebted to Makerere University Biological Field Station and field assistance from A. Drescher, N. Ingrisano, K. Jones, and F. Mugurusi. M. Cords, M. Glenn, R. Tuttle and one anonymous reviewer offered helpful comments.

References

Alberts, S. C. 1994. Vigilance in young baboons: Effects of habitat, age, sex and maternal rank on glance rate. *Anim. Behav.* **47**:749–755.

Boinski, S., Treves, A., and Chapman, C. 2000. A critical evaluation of the influence of predators on primates: Effects on group movement. In: S. Boinski, and P. Garber (eds.), *On the Move: How and Why Animals Travel in Groups*, pp. 43–72. University of Chicago Press, Chicago.

Borries, C. 1993. Ecology of female social relationships: Hanuman langurs (*Presbytis entellus*) and the van Schaik model. *Folia Primatol.* **61**:21–30.

Caine, N. G., and Marra, S. L. 1988. Vigilance and social organization in two species of primates. *Anim. Behav.* **36**:897–904.

Chance, M. R. A. 1976. Attention structure as the basis of primate rank orders. In: M. R. A. Chance, and R. R. Larsen (eds.), *The Social Structure of Attention*, pp. 11–28. Wiley and Sons, London.

Cheney, D. L. 1992. Intragroup cohesion and intergroup hostility: The relation between grooming distributions and intergroup competition among female primates. *Behav. Ecol.* **3**:334–345.

Cords, M., and Rowell, T. E. 1986. Group fission in blue monkeys of the Kakamega Forest, Kenya. *Folia Primatol.* **46**:70–82.

Glander, K. E. 1992. Dispersal patterns in Costa Rican mantled howling monkeys. *Int. J. Primatol.* **13**(4):415–436.

Hamilton, W. J. I. 1985. Demographic consequences of a food and water shortage to desert chacma baboons, *Papio ursinus. Int. J. Primatol.* **6**:451–466.

van Hooff, J. A. R. A. M., and van Schaik, C. P. 1991. Cooperation in competition: The ecology of primate bonds. In: A. H. Harcourt, and F. B. M. deWaal (eds.), *Us Against Them: Coalitions and Alliances in Humans and Other Animals,* pp. 110–123. Oxford University Press, Oxford.

Horwich, R. H. 1983. Breeding behavior in the black howler monkey (*Alouatta pigra*) of Belize. *Primates* **24**:222–230.

Horwich, R. H., and Gebhard, K. 1983. Roaring rhythms in black howler monkeys (*Alouatta pigra*) of Belize. *Primates* **24**:290–296.

Horwich, R. H., and Gebhard, K. 1986. Relation of allomothering to infant age in howlers, *Alouatta pigra*, with reference to Old World monkeys. In: D. M. Taub, and F. A. King (eds.), *Current Perspectives in Primate Social Dynamics,* pp. 66–88. van Nostrand Reinhold Co., New York.

Isbell, L. A. 1983. Daily ranging behavior of red colobus monkeys (*Colobus badius tephrosceles*) in Kibale Forest, Uganda. *Folia Primatol.* **41**:34–48.

Jones, W. T., and Bush, B. B. 1988. Movement and reproductive behavior of solitary male red-tailed guenons (*Cercopithecus ascanius*). *Am. J. Primatol.* **14**:203–222.

Keverne, E. B., Leonard, R. A., Scruton, D. M., and Young, S. K. 1978. Visual monitoring in social groups of talapoin monkeys (*Miopithecus talapoin*). *Anim. Behav.* **26**:933–944.

Koenig, A., Beise, Chalise, M. K., and Ganzhorn, J. U. 1998. When females should contest for food-testing hypotheses about resource density, distribution, size, and quality with Hanuman langurs (*Presbytis entellus*). *Behav. Ecol. Sociobiol.* **42**:225–237.

Kummer, H. 1968. *Social Organization of Hamadryas Baboons.* University of Chicago Press, Chicago.

Marsh, C. W. 1979a. Comparative aspects of social organization in the Tana River red colobus, *Colobus badius rufomitratus. Z. Tierpsychol.* **51**:337–362.

Marsh, C. W. 1979b. Female transference and mate choice among Tana River red colobus. *Nature* **281**:568–569.

Mitchell, C. L., Boinski, S., and van Schaik, C. P. 1991. Competitive regimes and female bonding in two species of squirrel monkeys (*Saimiri oerstedi* and *S. sciureus*). *Behav. Ecol. Sociobiol.* **28**:55–60.

Pusey, A. E., Williams, J., and Goodall, J. 1997. The influence of dominance rank on the reproductive success of female chimpanzees. *Science* **277**:828–831.

Richard, A. 1974. Intra-specific variation in the social organization and ecology of *Propithecus verreauxi. Folia Primatol.* **22**:178–192.

van Schaik, C. P. 1989. The ecology of social relationships amongst female primates. In: V. Standen, and R. A. Foley (eds.), *Comparative Socioecology: The Behavioural Ecology of Humans and Other Mammals,* pp. 195–218, Blackwell Scientific, Oxford.

van Schaik, C. P. 1996. Social evolution in Primates: The role of ecological factors and male behaviour. *Proc. British Acad.* **88**:9–31.

Sekulic, R. 1983. Spatial relationships between recent mothers and other troop members in red howler monkeys (*Alouatta seniculus*). *Primates* **24**:475–485.

Silk, J. B. 1991. Mother-infant relations in bonnet macaques: Sources of variation in proximity. *Int. J. Primatol.* **12**:21–39.

Sterck, E. H. M., Watts, D. P., and van Schaik, C. P. 1997. The evolution of female social relationships in nonhuman primates. *Behav. Ecol. Sociobiol.* **41**:291–309.

Struhsaker, T. T. 1975. *The Red Colobus Monkey.* University of Chicago Press, Chicago.

Struhsaker, T. T. 1977. Infanticide and social organization in the red-tailed monkey (*Cercopithecus ascanius schmidtii*) in the Kibale Forest, Uganda. *Z. Tierpsychol.* **45**:75–84.

Struhsaker, T. T. 1978. Food habits of five monkey species in the Kibale Forest, Uganda. In: D. J. Chivers, and J. Herbert (eds.), *Recent Advances in Primatology,* pp. 225–248. Academic Press, London.

Struhsaker, T. T. 1980. Comparison of the behaviour and ecology of red colobus and red-tailed monkeys in the Kibale Forest, Uganda. *Afr. J. Ecol.* **18**:33–51.

Struhsaker, T. T. 1988. Male tenure, multi-male influxes, and reproductive success in red-tailed monkeys (*Cercopithecus ascanius*). In: A. Gautier-Hion, F. Bourlière, J.-P. Gautier, and J. Kingdon (eds.), *A Primate Radiation: Evolutionary Biology of the African Guenon*, pp. 340–363. Cambridge University Press, Cambridge.

Struhsaker, T. T. 1997. *Kibale: A Case Study of Logging in a Tropical Rainforest*. University of Florida Press, Gainesville, FL.

Struhsaker, T. T., and Leland, L. 1979. Socioecology of five sympatric monkey species in the Kibale Forest, Uganda. In: J. S. Rosenblatt, R. A. Hinde, C. Beer, and M. C. Busnel (eds.), *Advances in the Study of Behavior*, pp. 159–228. Academic Press, New York City.

Struhsaker, T. T., and Leland, L. 1985. Infanticide in a patrilineal society of red colobus monkeys. *Z. Tierpsychol.* **69**:89–132.

Struhsaker, T. T., and Leland, L. 1988. Group fission in red-tailed monkeys (*Cercopithecus ascanius*) in the Kibale Forest, Uganda. In: A. Gautier-Hion, F. Bourlière, J.-P. Gautier, and J. Kingdon (eds.), *A Primate Radiation: Evolutionary Biology of the African Guenons*, pp. 364–388. Cambridge University Press, Cambridge.

Struhsaker, T. T., and Oates, J. F. 1975. Comparison of the behavior and ecology of red colobus and black-and-white colobus in Uganda: a summary. In: R. H. Tuttle (ed.), *Socioecology and Psychology of Primates*, pp. 103–123. Mouton, The Hague.

Treves, A. 1997. Vigilance and use of micro-habitat in solitary rainforest mammals. *Mammalia* **61**:511–525.

Treves, A. 1998. The influence of group size and near neighbors on vigilance in two species of arboreal primates. *Behav.* **135**:453–482.

Treves, A. 1999a. Within-group vigilance in red colobus and red-tailed monkeys. *Am. J. Primatol.* **48**:113–126.

Treves, A. 1999b. Has predation shaped the social systems of arboreal primates? *Int. J. Primatol.* **20**:35–53.

Treves, A. 2000. Theory and method in studies of vigilance and aggregation. *Anim. Behav.* **60**:711–722.

Treves, A. 2001. Reproductive consequences of variation in the composition of howler monkey social groups. *Behav. Ecol. Sociobiol.* **50**:61–71.

Treves, A. 2002. Predicting predation risk for foraging, arboreal monkeys. In: L. Miller (ed.), *Eat or Be Eaten: Predator Sensitive Foraging in Nonhuman Primates*, pp. 222–241. Cambridge University Press, Cambridge.

Treves, A., and Pizzagalli, D. 2002. Vigilance and perception of social stimuli: Views from ethology and social neuroscience. In: M. Bekoff, C. Allen, and G. Burghardt (eds.), *The Cognitive Animal: Empirical and Theoretical Perspectives on Animal Cognition*, pp. 463–469. MIT Press, Cambridge, MA.

Treves, A., Drescher, A., and Ingrisano, N. 2001. Vigilance and aggregation in black howler monkeys (*Alouatta pigra*). *Behav. Ecol. Sociobiol.* **50**:90–95.

Watts, D. P. 1998. A preliminary study of selective visual attention in female mountain gorillas (*Gorilla gorilla beringei*). *Primates* **39**:71–78.

Wrangham, R. W. 1980. An ecological model of female-bonded primate groups. *Behav.* **75**:262–300.

Why Vervet Monkeys (*Cercopithecus aethiops*) Live in Multimale Groups

13

LYNNE A. ISBELL, DOROTHY L. CHENEY, and ROBERT M. SEYFARTH

Introduction

Explanations of patterns of male residence in primate groups have long been sought by behavioral ecologists. The permanent co-existence of multiple males with groups of females is unusual in mammals, but primates include a large number of such species. In cercopithecine primates alone, 17 of 44 species (39%) live in multimale groups year-round (Smuts *et al.*, 1987). Suggested determinants of male residence patterns in primates include phylogenetic constraints (Struhsaker, 1969), female defensibility (numbers of females, or temporal and spatial distribution of females) (Clutton-Brock and Harvey, 1977; Wrangham, 1979, 1980; van Schaik and van Hooff, 1983; Terborgh, 1983; Andelman, 1986; Ridley, 1986; Dunbar, 1988; Altmann, 1990; Janson, 1992; Mitani *et al.*, 1996; Nunn, 1999), predation (Struhsaker, 1969; Crook, 1972; Henzi, 1988; van Schaik and van Noordwijk, 1989;

LYNNE A. ISBELL • Department of Anthropology, University of California, Davis, CA 95616, USA. DOROTHY L. CHENEY • Department of Biology, University of Pennsylvania, Philadelphia, PA 19104, USA. ROBERT M. SEYFARTH • Department of Psychology, University of Pennsylvania, Philadelphia, PA 19104, USA.

The Guenons: Diversity and Adaptation in African Monkeys, edited by Glenn and Cords. Kluwer Academic/Plenum Publishers, New York, 2002.

Baldellou and Henzi, 1992; van Schaik and Hörstermann, 1994), cooperative defense of food (Isbell *et al.*, 1991) or females (Mitani *et al.*, 1996), and trade-offs between time spent in feeding and defense of females (Terborgh and Janson, 1986). None of these factors appears to account for the co-existence of multiple males in female groups of vervet monkeys (*Cercopithecus aethiops*). Vervet monkeys are exceptional not only among guenons, the majority of which have only one resident male in their groups year-round, but also among all primates, and alternative explanations for their multimale social organization must be sought. Here we briefly discuss why current hypotheses are not adequate for vervet monkeys and propose a new alternative hypothesis.

The Inadequacy of Current Explanations

Phylogenetic Constraints

Cercopithecines show a phylogenetic pattern of male residence and female philopatry. In all species, females typically remain in their natal groups throughout life, whereas males typically disperse around sexual maturity. Savannah baboons (*Papio* spp.) and macaques (*Macaca* spp.) live in multimale groups throughout the year, whereas most guenons (*Cercopithecus* spp.) live in single-male groups outside the breeding season. Patas monkeys (*Erythrocebus patas*), which are so similar to other guenons that they are sometimes included in the genus *Cercopithecus* (Gautier-Hion *et al.*, 1988; Groves, 2001), display the typical guenon single-male social system (Chism and Rowell, 1988). Although phylogeny may play a role in behavioral differences between the Cercopithecini and Papionini in general, it cannot explain the multimale social organization of vervet monkeys, which are more closely related to other guenons than to baboons and macaques.

Defensibility of Females

Numerous studies indicate that male numbers in groups of primates are influenced by some quality of females, which could be the number of females in the group, the length of the breeding season, the degree of estrus synchrony, or the spatial spread of the females (Andelman, 1986; Ridley, 1986; Altmann, 1990; van Hooff and van Schaik, 1992; Mitani *et al.*, 1996; Nunn, 1999). In guenons such as patas monkeys, Schmidt's redtailed monkeys (*Cercopithecus ascanius schmidti*), and Stuhlmann's blue monkeys (*C. mitis stuhlmanni*), multiple males sometimes reside temporarily with larger groups of females during the breeding season (Cords, 1984, 2000; Cords *et al.*, 1986; Harding and Olson, 1986; Carlson and Isbell, 2001). Multimale influxes have never been reported outside of the breeding season, which suggests that exclusion of other males is easier when there is no possibility of reproducing. In patas monkeys, the resident male actively excludes extragroup males throughout the year, sometimes fatally wounding them (Isbell, unpub. data). Why the resident male should bother to exclude other males outside the breeding season is puzzling, given that females cannot conceive then.

Vervet monkeys are similar to other guenons in numbers of females per group, length of the breeding season, numbers of females mating per day within groups, and group spreads (Table I). Vervet monkeys meet all the requirements for successful exclusion of extra males, and yet they are nearly always multimale, even outside the breeding season. Factors other than female defensibility must favor multiple males (Henzi, 1988), because vervet monkeys differ from other guenons most during the non-breeding season.

Predation

Vervet monkeys live in savannah-woodlands, which are more open than forests, and carnivores are common in the savannah-woodlands. It has long been assumed that primates living in the more open habitats suffer higher predation rates than primates in more forested habitats (Altmann, 1974; Dunbar, 1988; Isbell, 1994; Olupot and Waser, 2001). Vervet monkeys can suffer high levels of predation (Cheney *et al.*, 1988; Isbell, 1990). Predation on vervet monkeys in Amboseli National Park, Kenya, the population for which predation is best documented, is mirrored by similarly high predation in vervet monkeys on Segera Ranch on the Laikipia Plateau, Kenya. Segera is a semi-arid savannah-woodland with an intact community of potential predators of primates, including leopards (*Panthera pardus*), lions (*P. leo*), black-backed jackals (*Canis mesomelas*), and martial eagles (*Polemaetus bellicosus*) (Isbell, 1998). Using the criteria of Cheney *et al.* (1988) and modified by Isbell (1990) for Amboseli vervet monkeys, Isbell and Enstam (2002) found, for example, that a minimum of 10 of 18 (56%) adult female vervet monkeys on Segera have died of suspected or confirmed predation.

In a population of vervet monkeys in South Africa, Baldellou and Henzi (1992) tested the hypothesis that predation could account for the vervet multimale social organization by determining whether males provide an advantage to females against predators. They found that although males were more vigilant than females, they were not better than females at detecting predators. In addition, the highest-ranking male was more vigilant and more active against predators than all other males. This led them to suggest that vervet monkey groups have multiple males because supernumerary males attach themselves to groups to minimize their own risk of predation.

If this hypothesis is correct, we would expect patas monkeys, which live in more open habitats and are more terrestrial than vervet monkeys, to live in permanent multimale, multifemale groups also, but they do not. Like vervet

Table I. Traditional Factors that Affect Female Defensibility in Vervet Monkeys vs. Other Guenons

Species, location	No. females in group	No. males in group on same day	Length of birth season	No. females mating per day	Group spread	Sources
Vervet monkeys, Amboseli	range: 1–8 (n = 6 groups)	1–6 yr-round	3 months	mean: 1.8 range: 1–3	mean: 55 m ± 0.97 SE	Isbell, L. A., unpub. data
Vervet monkeys, Segera	2–4 (small group) 3–9 (large group)	1–3 (small group) 0–10 (large group) year-round	3–4 months		mean: 128.0 m ± 10.2 SE	Isbell and Enstam (2002)
Patas Monkeys, Segera	6–15 (one group)	1 outside breeding season; 0–6, breeding season	3–4 months	range: 0–7	mean: 151.8 m ± 6.8 SE	Enstam et al., (2002); Isbell and Enstam (2002); Carlson and Isbell (2001)
Red-tailed Monkeys, Kibale	range: 5–16	1 outside breeding season; 1–6, breeding season	most, if not all, months		median: 51–55 m	Struhsaker and Leland (1979); Butynski (1988); Struhsaker (1988)
Red-tailed Monkeys, Kakamega	10	1 outside breeding season; 1–4, breeding season	6 months	maximum: 3	mean: 56 m range: 20–85 m	Cords (1984, 1987a,b)
Blue Monkeys, Kibale	mean: 4.5–10	mean: 1	most months		median: 46–50 m	Struhsaker and Leland (1979); Butynski (1988)
Blue Monkeys, Kakamega	range: 6–18	mean: 5.9; range: 2–11, breeding season	most months	mean: 3.8 range: 1–8	mean: 109 m range: 60–190 m	Cords (1986, 1987b)

monkeys, Segera patas monkeys also suffer high mortality, much of which is from predation. Of 34 adult female patas monkeys that died during the study, 16 (47%) died of suspected or confirmed predation. We were unable to determine the cause of disappearance for 12 adult females (35%) because we were often unable to find the group in their large home range within our three day window of opportunity for assigning disappearances to predation (Isbell and Young, in prep.). We may have underestimated predation.

Cords (2000) suggested that predation could still explain the differences between vervet monkeys and patas monkeys if male patas monkeys have different strategies for dealing with predators that enable them to avoid predation better than male vervet monkeys do. However, Enstam and Isbell (in press) found that when vervet monkeys and patas monkeys were in the same habitat, they responded similarly to alarm calls, i.e., in areas with short trees, they both fled on the ground and did not attempt to hide from predators.

Home Range Defense

In vervet monkeys, male survival can depend on access to food resources (Wrangham, 1981) and male reproductive success appears to be influenced more by longevity than by dominance status (Cheney *et al.*, 1988). An examination of the dynamics of group disintegration and fusion in Amboseli revealed that adult males and adult females behaved similarly to maintain groups at the minimum group size of two adults regardless of the sex of the adults (Isbell *et al.*, 1991). Vervet monkeys are one of very few multimale, territorial species (Wrangham, 1980) and it has been suggested that male vervet monkeys benefit directly from home range defense (Isbell *et al.*, 1991). Cords (2000) has raised a counter-argument, however, that weakens the home range defense hypothesis: vervet monkeys exhibit substantial variation in the frequency or intensity of home range defense but this variation is not mirrored by the vervet monkey grouping pattern. In populations wherein home range defense is infrequent, vervet monkeys still live in multimale groups (Kavanagh, 1981; Cheney, 1987).

Cooperative Defense of Females

Mitani *et al.* (1996) suggested that cooperative associations of males to defend females could help explain extra males in primate groups, i.e., groups that had too few females to account for the number of males in them. However, this does not explain the vervet multimale social system. Male vervet monkeys do not form coalitions within groups to defend access to females, and, of all males, the alpha male is most frequently involved in aggressive intergroup encounters (Cheney, 1981).

Time Constraints

Terborgh and Janson (1986) suggested that multimale groups could result if males are forced to make trade-offs between obtaining their food and defending females. Noting that frugivorous primates tend to live in multimale groups whereas folivorous primates tend to live in single-male groups, they suggested that this difference might exist because frugivorous species require more time for feeding or foraging, which reduces the time available for defending exclusive access to groups of females. However, this does not explain why the vervet social system differs from that of most other guenons. Though vervet monkeys are more frugivorous than folivorous, so are most guenon species that live in single-male groups (Cords, 1987a). More importantly, vervet, red-tailed, and blue monkeys all spend *ca.* 30% of time feeding (Struhsaker and Leland, 1979; Isbell and Young, 1993), and vervet monkeys spend less time foraging than patas monkeys do (Isbell *et al.*, 1998).

An Alternative: The Limited Dispersal Hypothesis

We suggest here an alternative hypothesis to account for the vervet multimale social system. The limited dispersal hypothesis proposes that two elements, configuration of the habitat and costs of dispersal, determine the number of groups available for a dispersing male to join. If the configuration of the habitat results in a small number of adjacent groups and the costs of dispersal are high, dispersing males limit their movements to adjacent groups and do not transfer often in their lifetimes. Immigrant males refrain from committing infanticide in the groups they join because they are sufficiently related genetically to members of their new groups that committing infanticide would decrease their inclusive fitness. Multimale groups form because the minimal risk of infanticide also favors tolerance of immigrant males by group members. However, as the configuration of the habitat allows groups to share borders with a greater number of adjacent groups and the costs of dispersal decrease, it becomes increasingly likely that dispersing males will join groups without relatives. Such males will be more likely to attempt infanticide and members of groups will be less tolerant of them, even to the point of excluding them. The vervet social system should be facultatively multimale because habitat configuration and costs of dispersal are dependent on local environmental conditions.

Vervet monkeys typically live within narrow belts of vegetation along rivers (Wolfheim, 1983). Adjacent to them are drier, more sparsely treed habitats, which are unsuitable for vervet monkeys. A vervet monkey group is usually bordered by two groups, one at either end of its linear home range. There are exceptions, of course, the best known being Amboseli, where vervet monkeys lived along swamps rather than rivers (Cheney *et al.*, 1988). The less restricted configuration of the habitat in Amboseli enabled up to five groups to border the home ranges of each group (Cheney and Seyfarth, 1983). Unprovisioned vervet monkeys nearly always live near water, which limits their distribution in savannah-woodlands.

Dispersers often face higher risks of mortality than philopatric animals (Shields, 1982; Isbell *et al.*,1990, 1993; Alberts and Altmann, 1995; Olupot and Waser, 2001). Dispersal costs are difficult to estimate, however, because individuals that disappear may have emigrated to distant groups or died. That dispersal can be costly to male vervet monkeys is suggested by greater mortality of female and immature vervet monkeys when they moved into unfamiliar areas (Isbell *et al.*, 1990, 1993). Because vervet groups have little home range overlap, dispersal typically involves moving into unfamiliar areas.

Segera Vervet Monkeys

The vervet monkeys on Segera are typical of ones elsewhere in Africa. No Segera vervet monkey group lives in a home range without access to a river. Home ranges are small (10–40 ha), linear, and contiguous, and when population density is sufficiently high, they occupy all habitat along the rivers. Outside the home ranges, there is no suitable habitat. Since the long-term study began in 1992, we saw a vervet monkey only once *ca.* 2 km from a river. Each vervet monkey group is bordered by two adjacent groups, one on either end of its linear home range.

Few Segera males survive long enough in their natal groups to disperse, but those that do typically emigrate to either of two adjacent groups. Of 27 males either born into two Segera study groups or not yet fully grown at the beginning of the study, 21 (78%) either died ($n = 14$) or disappeared ($n = 7$) (Fig. 1). Because four that disappeared were one to six months old and too young to transfer, they presumably died. The other three were five to six years old and could have transferred to non-adjacent groups, though we did not see them in those groups. Six males (22%) survived to disperse, and stayed an average of 16 months in the new groups. Five of them (83%) transferred to adjacent groups. Two were maternal brothers that transferred into the same group one year apart. Thus, even with high mortality, closely related males can disperse into the same group.

We also monitored the movements of 27 non-natal males. Four died before they could transfer again (Fig. 1). Fifteen (56%) are unexplained disappearances. The high percentage of disappearances of non-natal males relative to natal males suggests that some of them could have transferred to non-adjacent groups, though we did not see them there. Eight non-natal males (30%) transferred again, and all of them moved to adjacent groups. Including movements into the study groups, one transferred ⩾3 times,

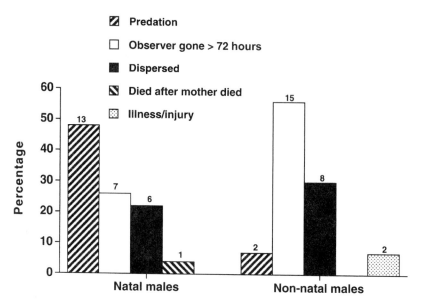

Fig. 1. Fates of natal and non-natal male vervet monkeys on Segera Ranch, Laikipia, Kenya, over a nine-year period. Suspected and confirmed predation were included in the category of predation.

with ≥2 of the transfers in the same direction. Four males transferred at least twice. The average length of tenure for non-natal males was 20 months ($n = 28$ transfer events).

As Cheney and Seyfarth (1983) have pointed out for Amboseli vervet monkeys, the effect of such non-random dispersal is that males are genetically more closely related to members of the groups that they join than if dispersal were random. We further suggest that because male vervet monkeys typically have fewer dispersal options than other species, they are also genetically more closely related to members of their new groups than are dispersing males of other species.

Not only might maternal brothers disperse into the same group but also their fathers might have dispersed into the same group before them, and reproduced. Consider the following scenario: a male transfers into Group 1 during the breeding season. He fathers a son in that group, stays for 20 months, and then transfers to adjacent Group 2. In Group 2, the male fathers a daughter and a son. By the time his daughter is reproductively mature (at four years; Cheney *et al.*, 1988), his son from Group 1 is five to six years old and ready to transfer. If his son transfers to Group 2 (a high probability since there are only two groups available), he will have two paternal sibs in the new group, one of which is a potential mate.

Limited dispersal should favor mutual tolerance by increasing the average degree of relatedness between immigrants and their new groupmates.

Immigrant males would be expected to be more tolerant of the offspring of other males, including unweaned offspring. Immigrants would not commit infanticide because doing so would decrease their inclusive fitness. Resident males and females would be expected to be more tolerant of non-group males, even to the extent that they are allowed into the groups, because these males are related and will not attempt to commit infanticide. Indeed, no suspected or confirmed infanticides by immigrant males have occurred on Segera over nine years and 58 births and 20 immigrations.

Dispersal to adjacent home ranges by both natal and non-natal males and dispersal of siblings to the same group on Segera are remarkably consistent with those from Amboseli (Cheney and Seyfarth, 1983). In Amboseli, where swamps enabled more vervet monkey groups to share home range borders than is possible along rivers, males from the same natal group tended to restrict their movements to a subset of the groups available to them. Similarly, Henzi and Lucas (1980) reported dispersal to adjacent groups and joint dispersal by males in a provisioned population in South Africa. Limiting dispersal to adjacent groups may help males mitigate the high costs of dispersal (Pusey and Packer, 1987; Isbell and Van Vuren, 1996). Although males in other species also disperse to adjacent groups, sometimes with siblings (Pusey and Packer, 1987), vervet monkeys appear to differ from other primates in having a more limited number of adjacent groups into which dispersing males can go.

In Amboseli, where the number of adjacent groups was higher for each group than at Segera, the estimated rate of infanticide was also higher. Infanticide by immigrant males was suspected in the deaths of three of 148 infants (2%) during 66 immigration events over 11 years. The estimated infanticide rate, weighted by the number of births and immigration events, was 0.03 infanticides per 100 immigration events (three infanticides/148 births)/66 immigration events ×100). We suggest that the absence of infanticide of vervet monkeys on Segera results from their living along a river in a habitat, where dispersal costs are high, which limits dispersal options to one of two groups, whereas Amboseli vervet monkeys' higher infanticide rate is a result of their living around swamps, which provided more options for dispersal into adjacent groups. Nonetheless, the rate of infanticide in Amboseli was still quite low, which may be a consequence of the tendency of natal groupmates to disperse to the same groups.

Testing the Limited Dispersal Hypothesis

The limited dispersal hypothesis predicts that the vervet multimale social system is facultative and dependent upon the costs of dispersal and the configuration of the habitat. If limited dispersal minimizes the risk of infanticide so that exclusion of other males is possible but not necessary, then less limited dispersal should increase the risk of infanticide and, at some point, make exclusion necessary. This prediction is supported by the behavior of vervet monkeys in Barbados.

Barbados vervet monkeys are relatively evenly distributed, with opportunities for gene flow throughout the island (Horrocks, 1984). In addition, because their only predators are humans and domestic dogs and they feed extensively on cultivated food (Horrocks and Hunte, 1986) the costs of dispersal appear to be relatively low. This suggests that dispersal options are greater for Barbados vervet monkeys than for vervet monkeys on Segera and in Amboseli. Barbados vervet monkeys usually live in single-male groups (Horrocks and Hunte, 1986). As in other species with single-male groups, extragroup males and aggressive takeovers occur in the population (Horrocks and Baulu, 1988). Horrocks (1986) suspected infanticide in the deaths of two of 22 (9%) infants during two immigration events. Infanticide appears to be a serious cause of infant mortality (Horrocks and Baulu, 1988). The estimated infanticide rate among Barbados vervet monkeys is 4.5 infanticides per 100 immigration events. Detailed information on dispersal distance, time spent as an extragroup male, likelihood of becoming a resident male, and tenure with female groups would help to clarify the effect of dispersal on behavioral differences between males living in single-male and multimale populations of vervet monkeys.

Single-male groups also occur in Naivasha, Kenya. Turner *et al.* (2001) collected genetic material from them as well as from populations with multimale groups. Accordingly, it should be possible to test the limited dispersal hypothesis via genetic analyses. It would also be useful to compare vervet monkeys in single-male and multimale populations with other guenons. Like vervet monkeys that live in single-male groups, other guenons in single-male groups are predicted to have a lower degree of relatedness between immigrant males and the groups they join than vervet monkeys that live in multimale groups.

Dispersal Options and Infanticide in Other Guenons

Unlike most vervet monkeys, forest guenons and patas monkeys live in home ranges surrounded by additional suitable habitat that could be used by multiple groups. Male forest guenons and patas monkeys are thus likely to have more groups into which they might disperse. We expect immigrant male forest guenons and patas monkeys to be much less genetically related to the members of their new groups than vervet monkeys living in multi-male groups. Consequently, we also expect immigrant male forest guenons and patas monkeys to attempt infanticide more often than immigrant male vervet monkeys in multimale groups.

Infanticide has been strongly suspected or directly observed in Stuhlmann's blue monkeys, Schmidt's red-tailed monkeys, Lowe's monkeys (*Cercopithecus campbelli lowei*), and patas monkeys (Struhsaker, 1977; Galat-Luong and Galat, 1979; Butynski, 1982; Enstam *et al.*, 2002). It appears to be rare in patas monkeys because males that are present for the birth season were also usually present during the previous breeding season and thus are potential fathers. In a case in which a male became resident after the breeding season and stayed for the next birth season, he attacked an infant, which we found dead five days later. Bite wounds on its body, female behavior toward the male, and the male's behavior before and after the infant's death indicated that he caused the death. If we view infanticide as an extreme form of male–male competition rather than a male strategy against females, it becomes much easier to reconcile the occurrence of infanticide in species that were not originally expected to have infanticide, i.e., multimale species such as baboons (*Papio ursinus*: Palombit *et al.*, 2000) and seasonally breeding species such as patas monkeys. We suggest that for guenons living in single-male groups, the resident male stays with the group outside the breeding season not to defend access to the females but instead to protect his putative infant offspring from infanticidal males.

Summary

Although the presence of either one male or multiple males in primate groups appears in general to be a function of the number of adult females in those groups, vervet monkeys are an exception to this pattern. Given the small number of adult females per group, one would expect vervet monkey groups to have only one adult male, but instead they typically have multiple adult males. Several other possible determinants of multimale social organizations have been proposed, including compressed temporal distribution of estrous females, large group spread, heavy predation, phylogenetic history, and feeding constraints. We discussed and dismissed each of these for vervet monkeys, and provided an alternative hypothesis. The limited dispersal hypothesis proposes that habitat configuration and costs of dispersal favor multimale groups of vervet monkeys by limiting dispersal options for males. Limited dispersal options increase the genetic relatedness between immigrants and members of their new groups, which selects against infanticidal behavior by immigrant males. Exclusion of immigrant males thus becomes unnecessary, resulting in multimale groups. We provided nine years of demographic data on male vervet monkeys of Segera Ranch, Kenya, as an example of a vervet monkey population with limited dispersal options. Comparison of infanticide rates of Segera vervet monkeys with those of vervet monkey populations having greater dispersal options provided support for the hypothesis. Multimale groups of vervet monkeys appear to be facultative responses to local environmental conditions that affect dispersal options. A similar comparison of multimale vervet monkeys with single-male forest guenons and patas monkeys led us to suggest that in these species the resident male remains with females outside the breeding season not to defend females but to defend his putative offspring from infanticidal males.

ACKNOWLEDGMENTS

Many thanks to M. Cords and M. Glenn for inviting L.A.I. to participate in the symposium at the XVIIIth Congress of the International Primatological Society on which this paper is based and to K. Bensen, M. Cords, M. Glenn, T. Rowell, and R. Tuttle for their helpful reviews. We are grateful to the Office of the President, Republic of Kenya, for permitting us to conduct research in Kenya; J. Else, C. Bambra, and J. Mwenda, for facilitating local sponsorship through the Institute of Primate Research; the Fonville family, especially R. Fonville, and J. Gleason and J. Ruggieri for allowing L.A.I. to conduct research on Segera; J. Wreford-Smith, G. Prettijohn, and P. Valentine for logistical assistance on Mpala and Segera Ranches; P. Lee, S. Andelman, and M. Hauser for their contributions to the long-term data base in Amboseli; J. Pruetz, A. Carlson, K. Enstam, and S. and V. Cummins for similar contributions on Segera; B. Musyoka Nzuma for field assistance in Amboseli and Segera; C. Molel, R. Mohammed, and F. Ramram for field assistance on Segera; D. Johnson of the Cisco Ducks for a valuable discussion while warming up for softball; and T. Young for assistance in innumerable ways. The research was supported by grants to L.A.I. from the National Science Foundation, the Wenner-Gren Foundation, the L.S.B. Leakey Foundation, Rutgers University Research Council, the UC Davis Bridge and Faculty Research Grant Programs, and the California Regional Primate Research Center (through NIH grant #RR00169) and to DLC and RMS from the NSF, National Institutes of Health, and the University of Pennsylvania.

References

Alberts, S. C., and Altmann, J. 1995. Balancing costs and opportunities: dispersal in male baboons. *Am. Nat.* **145**:179–306.

Altmann, J. 1990. Primate males go where the females go. *Anim. Behav.* **39**:193–195.

Altmann, S.A. 1974. Baboons, space, time, and energy. *Am. Zool.* **14**:221–248.

Andelman, S. J. 1986. Ecological and social determinants of Cercopithecine mating patterns. In: D. I. Rubenstein, and R. W. Wrangham (eds.), *Ecological Aspects of Social Evolution*, pp. 201–216. Princeton, Princeton University Press.

Baldellou, M., and Henzi, S. P. 1992. Vigilance, predator detection and the presence of supernumerary males in vervet monkey troops. *Anim. Behav.* **43**:451–461.

Butynski, T. M. 1982. Harem-male replacement and infanticide in the blue monkey (*Cercopithecus mitis stuhlmanni*). *Am. J. Primatol.* **3**:1–22.

Butynski, T. M. 1988. Guenon birth seasons and correlates with rainfall and food. In: A. Gautier-Hion, F. Bourlière, J.-P. Gautier, and J. Kingdon (eds.), *A Primate Radiation: Evolutionary Biology of the African Guenons*, pp. 284–322. Cambridge University Press, New York.

Carlson, A. A., and Isbell, L. A. 2001. Causes and consequences of single-male and multimale mating in free-ranging patas monkeys (*Erythrocebus patas*). *Anim. Behav.* **62**:1047–1058.

Cheney, D. L. 1981. Intergroup encounters among free-ranging vervet monkeys. *Folia Primatol.* **35**:124–146.

Cheney, D. L. 1987. Interactions and relationships between groups. In: B. B. Smuts, D. L. Cheney, R. M. Seyfarth, R. W. Wrangham, and T. T. Struhsaker (eds.), *Primate Societies*, pp. 267–281. University of Chicago Press, Chicago.

Cheney, D. L., and Seyfarth, R. M., 1983. Nonrandom dispersal in free-ranging vervet monkeys: social and genetic consequences. *Am. Nat.* **122**:392–412.

Cheney, D. L., Seyfarth, R. M., Andelman, S. J., and Lee, P. C. 1988. Reproductive success in vervet monkeys. In: T. H. Clutton-Brock (ed.), *Reproductive Success: Studies of Individual Variation in Contrasting Breeding Systems*, pp. 384–402. University of Chicago Press, Chicago.

Chism, J., and Rowell, T. E. 1988. The natural history of patas monkeys. In: A. Gautier-Hion, F. Bourlière, J.-P. Gautier, and J. Kingdon (eds.), *A Primate Radiation: Evolutionary Biology of the African Guenons*, pp. 412–438. Cambridge University Press, Cambridge.

Clutton-Brock, T. H., and Harvey, P. H. 1977. Primate ecology and social organization. *J. Zool., Lond.* **183**:1–39.

Cords, M. 1984. Mating patterns and social structure in redtail monkeys (*Cercopithecus ascanius*). *Z. Tierpsychol.* **64**:313–329.

Cords, M. 1987a. Forest guenons and patas monkeys: male-male competition in one-male groups. In: B. B. Smuts, D. L. Cheney, R. M. Seyfarth, R. W. Wrangham, and T. T. Struhsaker (eds.), *Primate Societies*, pp. 98–111. University of Chicago Press, Chicago.

Cords, M. 1987b. *Mixed-Species Associations of* Cercopithecus *Monkeys in the Kakamega Forest, Kenya*, University of California Press, Berkeley.

Cords, M. 2000. The number of males in guenon groups. In: P. M. Kappeler (ed.), *Primate Males: Causes and Consequences of Variation in Group Composition*, pp. 84–96. Cambridge University Press, New York.

Cords, M., Mitchell, B. J., Tsingalia, H. M., and Rowell, T. E., 1986. Promiscuous mating among blue monkeys in the Kakamega Forest, Kenya. *Ethology* **72**:214–226.

Crook, J. H. 1972. Sexual selection, dimorphism, and social organization in the primates. In: B. G. Campbell (ed.), *Sexual Selection and the Descent of Man*, 1871–1971, pp. 231–281. Aldine, Chicago.

Dunbar, R. I. M. 1988. *Primate Social Systems*, Ithaca, New York, Cornell University Press.

Enstam, K. L., and Isbell, L. A. (in press). Comparison of responses to alarm calls by patas (*Erythrocebus patas*) and vervet (*Cercopithecus aethiops*) monkeys in relation to habitat structure. *Am. J. Phys. Anthropol.*

Enstam, K. L., Isbell, L. A., and de Maar, T. W. 2002. Male demography, female mating behavior, and infanticide in wild patas monkeys (*Erythrocebus patas*). *Int. J. Primatol.* **23**:85–104.

Galat-Luong, A., and Galat, G., 1979. Conséquences comportementales des perturbations sociales repetées sur une troupe de Mones de Lowe *Cercopithecus campbelli lowei* de Côte d'Ivoire. *Terre et Vie* **33**:4–57.

Gautier-Hion, A., Bourlière, F., Gautier, J.-P., and Kingdon, J. (eds.), 1988. *A Primate Radiation: Evolutionary Biology of the African Guenons*, Cambridge University Press, Cambridge.

Groves, C. 2001. *Primate Taxonomy*, Smithsonian Institution Press, Washington, D.C.

Harding, R. S. O., and Olson, D. K. 1986. Patterns of mating among male patas monkeys (*Erythrocebus patas*) in Kenya. *Am. J. Primatol.* **11**:343–358.

Henzi, S. P. 1988. Many males do not a multimale troop make. *Folia Primatol.* **51**:165–168.

Henzi, S. P., and Lucas, J. W. 1980. Observations on the inter-troop movement of adult vervet monkeys (*Cercopithecus aethiops*). *Folia Primatol.* **33**:220–235.

Horrocks, J. A. 1984. *Aspects of the Behavioural Ecology of* Cercopithecus aethiops sabaeus *in Barbados, West Indies*, Ph.D. dissertation, University of the West Indies, Barbados.

Horrocks, J. A. 1986. Life-history characteristics of a wild population of vervet monkeys (*Cercopithecus aethiops sabaeus*) in Barbados, West Indies. *Int. J. Primatol.* **7**:31–47.

Horrocks, J. A., and Baulu, J. 1988. Effects of trapping on the vervet (*Cercopithecus aethiops sabaeus*) in Barbados. *Am. J. Primatol.* **15**:223–233.

Horrocks, J. A., and Hunte, W. 1986. Sentinel behaviour in vervet monkeys: who sees whom first? *Anim. Behav.* **34**:1566–1567.

Isbell, L. A. 1990. Sudden short-term increase in mortality of vervet monkeys (*Cercopithecus aethiops*) due to leopard predation in Amboseli National Park, Kenya. *Am. J. Primatol.* **21**:41–52.

Isbell, L. A. 1994. Predation on primates: ecological causes and evolutionary consequences. *Evol. Anthropol.* **3**:61–71.

Isbell, L. A. 1998. Diet for a small primate: insectivory and gummivory in the (large) patas monkey (*Erythrocebus patas pyrrhonotus*). *Am. J. Primatol.* **45**:381–398.

Isbell, L. A., Cheney, D. L., and Seyfarth, R. M. 1990. Costs and benefits of home range shifts among vervet monkeys (*Cercopithecus aethiops*) in Amboseli National Park, Kenya. *Behav. Ecol. Sociobiol.* **27**:351–358.

Isbell, L. A., Cheney, D. L., and Seyfarth, R. M. 1991. Group fusions and minimum group sizes in vervet monkeys (*Cercopithecus aethiops*). *Am. J. Primatol.* **25**:57–65.

Isbell, L. A., Cheney, D. L., and Seyfarth, R. M. 1993. Are immigrant vervet monkeys (*Cercopithecus aethiops*) at greater risk of mortality than residents? *Anim. Behav.* **45**:729–734.

Isbell, L. A., and Enstam, K. L. 2002. Predator (in)sensitive foraging in sympatric female vervet monkeys (*Cercopithecus aethiops*) and patas monkeys (*Erythrocebus patas*): a test of ecological models of group dispersion. In: L. E. Miller (ed.), *Eat or be Eaten: Predator Sensitive Foraging in Nonhuman Primates*, pp. 154–168. Cambridge University Press, New York.

Isbell, L. A., Pruetz, J. D., Lewis, M., and Young, T. P. 1998. Locomotor activity differences between sympatric vervet monkeys (*Cercopithecus aethiops*) and patas monkeys (*Erythrocebus patas*): implications for the evolution of long hindlimb length in *Homo. Am. J. Phys. Anthropol.* **105**:199–207.

Isbell, L. A., and Van Vuren, D. 1996. Differential costs of locational and social dispersal and their consequences for female group-living primates. *Behaviour* **133**:1–36.

Isbell, L. A., and Young, T. P. 1993. Social and ecological influences on activity budgets of vervet monkeys, and their implications for group living. *Behav. Ecol. Sociobiol.* **32**:377–385.

Janson, C. H. 1992. Evolutionary ecology of primate social structure. In: E. A. Smith, and B. Winterhalder (eds.), *Evolutionary Ecology and Human Behavior*, pp. 95–130. Aldine de Gruyter, New York.

Kavanagh, M. 1981. Variable territoriality among Tantalus monkeys in Cameroon. *Folia Primatol.* **36**:76–98.

Mitani, J. C., Gros-Louis, J., and Manson, J. H. 1996. Number of males in primate groups: comparative tests of competing hypotheses. *Am. J. Primatol.* **38**:315–332.

Nunn, C. L. 1999. The number of males in primate social groups: a comparative test of the socioecological model. *Behav. Ecol. Sociobiol.* **46**:1–13.

Olupot, W., and Waser, P. M. 2001. Activity patterns, habitat use and mortality risks of mangabey males living outside social groups. *Anim. Behav.* **61**:1127–1235.

Palombit, R. A., Cheney, D. L., Fischer, J., Johnson, S., Rendall, D., Seyfarth, R. M., and Silk, J. B. 2000. Male infanticide and defense of infants in chacma baboons. In: van C. P. Schaik, and C. H. Janson (eds.), *Infanticide by Males and its Implications*, pp. 123–151. Cambridge University Press, Cambridge.

Pusey, A. E., and Packer, C. 1987. Dispersal and philopatry. In: B. B. Smuts, D. L. Cheney, R. M. Seyfarth, R. W. Wrangham, and T. T. Struhsaker (eds.), *Primate Societies*, pp. 250–266. University of Chicago Press, Chicago.

Ridley, M. 1986. The number of males in a primate troop. *Anim. Behav.* **34**:1848–1858.

Shields, W. M. 1982. *Philopatry, Inbreeding, and the Evolution of Sex*, State University of New York Press, Albany.

Smuts, B. B., Cheney, D. L., Seyfarth, R. M., Wrangham, R. W., and Struhsaker, T. T. (eds.), 1987. *Primate Societies*, University of Chicago Press, Chicago.

Struhsaker, T. T. 1969. Correlates of ecology and social organization among African Cercopithecines. *Folia Primatol.* **11**:80–118.

Struhsaker, T. T. 1977. Infanticide and social organization in the redtail monkey (*Cercopithecus ascanius schmidti*) in the Kibale Forest, Uganda. *Z. Tierpsychol.* **45**:75–84.

Struhsaker, T. T. 1988. Male tenure, multi-male influxes, and reproductive success in redtail monkeys (*Cercopithecus ascanius*). In: A. Gautier-Hion, F. Bourlière, J.-P. Gautier, and J. Kingdon (eds.), *A Primate Radiation: Evolutionary Biology of the African Guenons*, pp. 340–363. Cambridge University Press, Cambridge.

Struhsaker, T. T., and Leland, L. 1979. Socioecology of five sympatric monkey species in the Kibale Forest, Uganda. *Adv. Study Behav.* **9**:159–228.

Terborgh, J. 1983. *Five New World Primates*, Princeton University Press, Princeton.

Terborgh, J., and Janson, C. H. 1986. The socioecology of primate groups. *Ann. Rev. Ecol. Syst.* **17**:111–135.

Turner, T. R. Whitten, P. L., and Gray, J. P. 2001. An ecological approach to vervet monkey life history. Paper presented at the *XVIIIth Cong. Int. Primatol. Soc.*, Adelaide, Australia.

van Hooff, J. A. R. A. M., and van Schaik, C. P. 1992. Cooperation in competition: the ecology of primate bonds. In: A. Harcourt, and F. B. M. de Waal (eds.), *Coalitions and Alliances in Humans and Other Animals*, pp. 357–389. Oxford University Press, New York.

van Schaik, C. P., and van Hooff, J. A. R. A. M. 1983. On the ultimate causes of primate social systems. *Behaviour* **85**:91–117.

van Schaik, C. P., and Hörstermann, M. 1994. Predation risk and the number of adult males in a primate group: a comparative test. *Behav. Ecol. Sociobiol.* **35**:261–272.

van Schaik, C. P., and van Noordwijk, M. A. 1989. The special role of male *Cebus* monkeys in predation avoidance and its effect on group composition. *Behav. Ecol. Sociobiol.* **24**:265–276.

Wolfheim, J. H. 1983. *Primates of the World*, University of Washington Press, Seattle, Washington.

Wrangham, R. W. 1979. On the evolution of ape social systems. *Soc. Sci. Info.* **18**:335–368.

Wrangham, R. W. 1980. An ecological model of female-bonded primate groups. *Behaviour* **75**:262–300.

Wrangham, R. W. 1981. Drinking competition in vervet monkeys. *Anim. Behav.* **29**:904–910.

When are there Influxes in Blue Monkey Groups?

14

MARINA CORDS

Introduction

An enduring question in behavioral studies of group-living primates is what determines the number of males in a group (Kappeler, 2000). Variation in male number occurs on several scales: there may be persistent differences among species, among populations or groups of single species, and even within single groups over time. The variables that explain variation on these different scales are not necessarily the same (Henzi, 1988).

I discuss factors that explain variation within single groups over time in forest-dwelling blue monkeys (*Cercopithecus mitis*). Like most other forest guenons, as well as closely related patas monkeys (*Erythrocebus patas*), blue monkey groups include only one adult male most of the time, especially outside the breeding season. During the breeding season, however, the number of males in a group of blue monkeys, and in some other guenon species, is more variable (Cords, 1987a, 1988, 2000; González-Martinez, 1998; Kaplin *et al.*, 1998; Macleod, 2000). Studies of blue monkeys, redtailed monkeys (*Cercopithecus ascanius*) and patas monkeys have revealed how the one-male group persists during some breeding seasons: the male that has been with the females previously continues to accompany them, and is the only male continuously present. Other males may make occasional, brief visits, but they do not remain

MARINA CORDS • Department of Ecology, Evolution and Environmental Biology, Columbia University, New York, NY 10027, USA.

The Guenons: Diversity and Adaptation in African Monkeys, edited by Glenn and Cords. Kluwer Academic/Plenum Publishers, New York, 2002.

in the group. In blue and patas monkeys, most male visitors are known to be non-resident in any heterosexual group, but a few are residents from neighboring groups, which typically make their visits at territorial boundaries or during intergroup encounters. At the other extreme are breeding seasons in which the one-male group structure breaks down completely and several adult males come into the group, often for longer periods, as part of a multimale influx (Cords, 1988). Prior residents may be ousted during an influx, or they may weather influxes and persist as the sole resident when breeding subsides.

While intermediate cases exist, it seems legitimate to dichotomize breeding periods into influx and non-influx years. For the Kakamega blue monkey population, there is no overlap in the values of distinguishing criteria. Specifically, during influx years there is a more conspicuous and continuous presence of multiple males (50–94% of days in six influx seasons vs. 4–28% of days in ten non-influx seasons), and both the average number of males per day (1.6–3.8 in an influx, 1.0–1.3 in a non-influx year) and the maximum number of males per day (11 in an influx, four in a non-influx year) are higher than in non-influx years (Cords, 2000, with additional data from 1998–2001). The duration of the typical male visit also differs, lasting days or weeks in influx years, but only a few hours in non-influx years.

My goal is to explain variation in the occurrence of multimale influxes. Using long-term data from 23 years of monitoring breeding seasons in one blue monkey population and detailed records of the behavior of individually identified males and females in two to four study groups over seven years, I examine correlates of influx breeding seasons. I predicted that influxes should be most likely when a resident male's ability to exclude other males is low. Indeed, blue monkey males behave as if they want to keep other males away from their females, being highly vigilant—seemingly toward rivals, rather than predators—and likely to chase intruding rivals if detected. However, intruders appear to be attracted to groups that offer reproductive opportunities, and they often copulate with females that may prefer them as mates. When many females are mating, it should be especially difficult for residents to exclude other males, and more likely that multiple males are present.

Female blue monkeys are not necessarily fertile (ovulating, non-pregnant) when they mate (Pazol *et al.*, 2002). While it might benefit males to distinguish fertile from infertile females, there is no evidence that they do so. Female blue monkeys do not have external signs of fertility, and olfactory signals also seem poorly developed (males rarely sniff females). Therefore I assume that female mating behavior, i.e., estrus, is the best indicator of reproductive opportunities available to males.

Methods

I studied blue monkeys (*Cercopithecus mitis stuhlmanni*) inhabiting the Kakamega Forest, western Kenya (Cords, 1987b). This rain forest fragment, whose main block is approximately 86 km^2 (Brooks *et al.*, 1999) at about 1650 m, receives an average annual rainfall of >2000 mm. Two rainy and two dry seasons per year can usually be distinguished, but there is much interannual variation.

Kakamega blue monkeys occur at a density of *ca.* 170–220 individuals/km^2 (Cords, 1987b; Fashing and Cords, 2000), with the higher figures more representative of the periods in the last eight years when reproductive behavior was most intensively sampled. This is a high population density for the species.

Several groups have been monitored to various degrees since 1979 and provide general information on the occurrence of multimale influxes (Fig. 1). The two main study groups, T and G, have been followed on a nearly daily basis during most of the breeding season since 1979 and 1994, respectively. Both have fissioned within this period, T in 1984 and G in 1999. Monitoring of Te, one of the daughter groups of the original T group, lasted only until 1989, after which a change in ranging patterns prevented continued contact. Tw, the other daughter group, has been monitored through the present. Both daughter groups of G (Gn and Gs) have been monitored since they separated from each other in 1999.

The most detailed records on male presence and sexual behavior are available from Tw, G, Gn and Gs groups from 1995 to 2001, when each was followed daily through the breeding season by a team of two to three observers. During this period of intensive monitoring, we aimed to keep

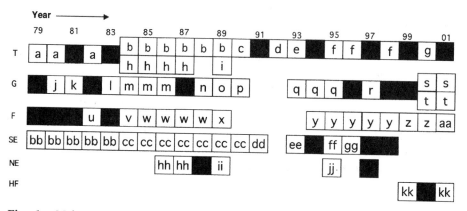

Fig. 1. Males present in study groups of blue monkeys during the 1979–2001 breeding seasons. Each row represents one group, two of which fissioned (T in early 1984, G in late 1999). Each box represents a particular breeding season for a given group in a given year. Letters are used to designate the identity of individual resident males in non-influx years; black-shaded boxes represent multimale influx years. If no box is present for a particular year, the group was not monitored thoroughly enough to determine the number of males present.

continuous contact with sexually active adults, and regularly patrolled the group's periphery searching for intruding males. In earlier years with less intensive observations, we also recorded which females mated and which had infants the following breeding season. Before 1998, some conceptions may have been missed if an infant was born and died during gaps in observations, which were limited to the breeding season.

Information on the frequency of multimale influxes is also available for four additional groups in the population, all of which are neighbors of the main study groups. They were followed intermittently and opportunistically to determine how many and which males were present during the breeding season.

All adult males and females in the T and G groups were individually recognized using natural characteristics. In the neighboring groups, only the adult males were individually recognized. Adult males typically have more distinctive characteristics than females and juveniles, including scars, stiff or missing digits, and distinctive pyow vocalizations. In T and G groups, female parity was estimated based on nipple elongation early in the study, and was known from longitudinal records later on.

Some of the analyses required a distinction between sexually active (estrous) and inactive females. A female is considered estrous on a given day or in a given season if she mated or puckered to a male from ≤2 m away. Puckering is a proceptive behavior that female blue monkeys direct toward males, and it does not occur in any other social context. Most puckering females mated at, or close, to, the time when they puckered. Female blue monkeys use other signals of proceptivity, such as presenting the hindquarters and head-flagging, but they also occur in non-sexual contexts, so were not useful to describe estrus unambiguously.

For data analysis, a mating season is the period from June 15–October 31. Sexual behavior was absent or sporadic in the weeks before and after this period. Records of male presence and female sexual activity from 1995–2001 span this period in every group and year, except for G in 1999, when the mating season ended on October 4. This was the date when G fissioned permanently into Gn and Gs. For October 5–31, I did not combine data from the two daughter groups with those from the original G group because of the enormous change in the social milieu. In comparisons of influx and non-influx mating seasons, I report mean values and associated standard errors of several variables.

Results

Multimale influxes have occurred in all the groups that we have monitored at Kakamega (Fig. 1). Overall, they occurred in 23% (23/98 group-years) of breeding seasons. The frequency of influxes per group varies from 0 to 40% ($n = 10$ groups, with fission products being considered separately from the parent group). It is mainly during the breeding season that multiple males are in the group. When breeding subsides, all but one male leave the group. In only one of 14 group-years of year-round monitoring of our main study groups did we find multiple males in a group for several months after mating activity had subsided.

The number of potentially fertile females in a given year included parous females without young and ones whose most recent offspring were $\geqslant 12$ months old, given that interbirth intervals in the population are $\geqslant 2$ years when the first offspring survives (Cords and Rowell, 1987). Nulliparous females that mated were also included as potentially fertile. The number of potentially fertile females was higher during influx years ($\bar{x} = 13.4 \pm 1.5$, $n = 10$) than during non-influx years ($\bar{x} = 9.6$, $\pm 0.5, n = 24$ years; Mann Whitney U Test, $p = 0.024$). There were even bigger differences in the number of females that actually mated in influx ($\bar{x} = 12.4 \pm 1.7, n = 10$) and non-influx years ($\bar{x} = 6.9 \pm 0.6, n = 24$; Mann Whitney U Test, $p = 0.005$).

From the viewpoint of individual male behavior, these seasonal figures may be less relevant than what is happening on individual days within a given season. In non-influx years, there were more days with no female in estrus ($\bar{x} = 57.4 \pm 5.9\%$ of days, $n = 10$ seasons) than in influx years ($\bar{x} = 20.1 \pm 4.8\%$ of days, $n = 6$ seasons; Mann Whitney U Test, $p < 0.002$, Fig. 2). Across all 16 breeding seasons, there were many days with just one male in the group when there were also many days with no estrous female in the group ($r_s = 0.86$, $p < 0.001$, Fig. 2).

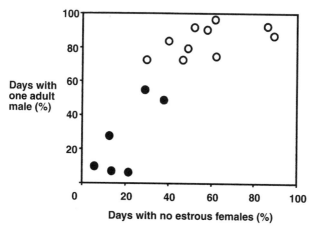

Fig. 2. The number of males in the group as a function of the presence of estrous females. Each data point represents one breeding season in one group. Solid points are influx seasons, while open points are non-influx seasons. Data come from Tw (1995–2001), G (1995–1999), Gn and Gs (2000–2001).

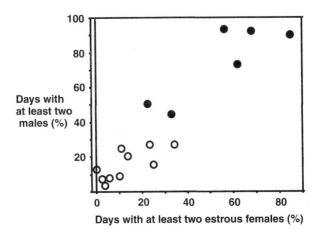

Fig. 3. The number of males in the group as a function of the prevalence of estrus synchrony. Each data point represents one breeding season in one group. Solid points are influx seasons, while open points are non-influx seasons. Data come from Tw (1995–2001), G (1995–1999), Gn and Gs (2000–2001).

Estrus synchrony—two or more females in estrus on a given day—was also more likely to occur during influx ($\bar{x} = 54.5 \pm 9.4\%$ of days, $n = 6$ years) than non-influx years ($\bar{x} = 12.9 \pm 3.5\%$ of days, $n = 10$ years; Mann Whitney U Test, $p < 0.005$, Fig. 3). Across all breeding seasons, the frequent presence of ≥ 2 females in estrus coincided with the frequent presence of >1 male in a group (Fig. 3, $r_s = 0.88$, $p < 0.001$).

If the presence of multiple estrous females makes successful exclusion of other males difficult, one might also expect to see a relationship between the number of males and the number of estrous females on a day-to-day basis. During influx years, there were indeed more males in the group on days when more females were estrous (Fig. 4; Jonckheere Test for Ordered Alternatives, $p < 0.00003$ for Tw and for G groups). During non-influx years, the relationship between the daily number of estrous females and males in the group was less strong (Fig. 4; Jonckheere Test, $p = 0.002$ for Tw, $p = 0.063$ for G, $p = 0.0007$ for Gs, $p = 0.348$ for Gn).

The results presented so far show on various time scales that there is an association between the number of males and number of sexually active females in a group, but such an association does not clarify the direction of the causal arrow. Does female estrus bring males into the group, as the defensibility hypothesis suggests, or does the presence of males in the group bring females into estrus? There is evidence from Kakamega that male presence could stimulate estrus in females: in G, a resident male died outside the breeding season in 1998. He was replaced immediately by a succession of two males, and within three weeks of his death, nearly every female without a suckling infant

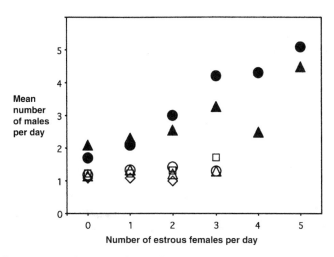

Fig. 4. The mean number of males in the group per day as a function of the number of estrous females present on that day. Data are combined over multiple seasons, with circles representing Tw, triangles representing G, diamonds representing Gn and squares representing Gs. Filled symbols show values for influx breeding seasons (1997, 1999, 2001 in Tw, 1996 and 1998, 1999 in G), while open symbols show values for non-influx seasons (1995, 1996, 1998, 2000 in Tw, 1995, 1997 in G, and 2000, 2001 in Gn and Gs).

began to mate (Pazol and Ziegler, 2000). Because this mating occurred in February and March, it seemed clear that the changed social situation, i.e., presence of new males, triggered the females' sexual behavior out of season.

In principle, observational data on male influx seasons can shed light on how female estrus and male numbers relate causally. If one thing causes the other, it should come first in time. Data are available for four influx seasons in which daily or near-daily observations of the study groups were made for ≥6 months before the onset of the breeding season. In two influx years (Tw 1999, G 1999), female estrus and the presence of additional males coincided very closely in time, and it is not possible to say with confidence which preceded which. In the G 1998 influx year, female estrus during the breeding season preceded the arrival of the first extra male by several weeks. In the Tw 2001 influx year, however, the presence of new males in the group preceded the first female estrus by about two weeks. These results are clearly not conclusive.

We can expand the sample size by considering non-influx years as well, since additional males usually turned up in the group during such years, even if their numbers were smaller and their stays shorter. Six non-influx seasons occurred in which intensive observations of the study groups preceded the advent of the mating season by many months. In five cases, some females became sexually active ≥2–3 weeks before any additional males visited the group, while in the sixth case, female estrus and visits by males coincided too closely in time to determine confidently which came first. Altogether, female estrus clearly preceded the arrival of male visitors in six breeding seasons, male visits preceded female estrus in one, and it is unclear whether estrus or male presence came first in the three remaining seasons.

Discussion

Data from Kakamega blue monkeys support the hypothesis that variation in male numbers is related to defensibility of females. When more females are sexually active, whether over an entire breeding season or on a single day, there are likely to be more males in the group. From the standpoint of the behavior of individual males and females, this pattern makes sense. Males often guard estrous females by following them closely, and chasing or threatening rivals. Because females are widely spaced (Cords, 2002), a single male often cannot guard more than one of them. Even when only one female is estrous, intruding males can gain access to her if a resident temporarily moves away to feed or to investigate the presence of other intruders in another part of the group, which can be spread out over 300 m. The dense foliage of the forest environment means that wide interindividual dispersion goes hand in hand with limited visibility, and over the years we have witnessed many situations in which intruding males behave exactly as if they were taking advantage of an opportunity to sneak copulations undetected by a resident. The facts that intruding males mate with females and that influxes coincide with the breeding season underscore the importance of the defensibility of estrous females in influencing male numbers in a group. This analysis thus confirms conclusions derived from similar analyses of a subset of the data (Cords, 2000).

In a previous analysis (Cords, 2000), the question of whether female estrus brought males into the group or the presence of extra males brought females into estrus was not resolved. With supplementary data, it seems clear that male numbers do, at least sometimes, respond to sexual behavior of females. Such a mechanism must be in place if the female defensibility hypothesis is correct. The existence of such a mechanism does not, however, preclude a stimulating effect of male presence on female estrus. I suspect that both mechanisms operate synergistically in blue monkeys: the presence of males stimulates female sexual activity which in turn ensures the repeated arrival of new males during an influx season.

The importance of female defensibility is supported by comparative data across five populations of *Cercopithecus mitis*. The occurrence of influx breeding seasons in a population is related to the ratio of non-resident males to groups of females (Cords, 2000; Macleod, 2000). Influxes occur in populations wherein this ratio is at least one, but have not been reported in populations in which the ratio is less than one. The ratio of males to female-groups is a measure of intruder pressure, and is independent of population density and sex ratio. Thus cross-population comparisons highlight another aspect of female defensibility that helps to explain variation in male numbers in blue monkeys.

Variable intruder pressure may be the factor that explains why the number of males in a group per day increases much more dramatically with the number of estrous females in the group per day in influx vs. non-influx years (Fig. 4). During non-influx breeding seasons, it appears that there are fewer males around the home range of the focal group, so fewer males may be available to move into groups when opportunity arises, when females are sexually active. During all breeding seasons, we make a point of regularly surveying the home ranges of the study groups and those of neighboring groups for adult males. During influx years, we routinely detected males during the surveys, but during non-influx years, they are harder to find. In thick vegetation, detecting non-resident, often solitary males, whose behavior is probably designed to be inconspicuous, is never an easy task, and would make quantitative estimates of their abundance extremely difficult. From a comparative perspective, however, I am confident that there are differences between influx and non-influx years. Studies of other guenon species, particularly red-tailed monkeys (Jones and Bush, 1988) and patas monkeys (Chism and Rogers, 1997), underscore the importance of variable male densities in facilitating multimale influxes.

If intruder pressure is an important component of female defensibility (along with the number of estrous females), and hence a cause of variation in male numbers in blue monkey groups, the question remains as to what factors bring about variation in intruder pressure. Across populations, demographic factors related to male mortality are likely to be important, as well as ecological factors that influence the size, and consequently the density, of female groups. What causes variable male density across local areas or between years in one site is unknown. Further study that focuses on non-resident male behavior is needed. Jones and Bush's (1988) study of radio-collared male red-tailed monkeys is a first step in this direction. As in Kakamega, they found that the density of non-resident males varied on a local spatial scale, and from year to year within the home range of a single group; however, their sample of individual males is too small to allow an understanding of this variation.

Data from other guenon populations and species, while less extensive, generally support my conclusions. Four groups of samango monkeys (*Cercopithecus mitis labiatus*) in Cape Vidal, South Africa, showed multimale influxes in one year, but non-influx breeding in the next year, when fewer females were receptive (Henzi and Lawes, 1987, 1988). In a later study of the same population, Macleod (2000) found that the number of males present on a single day in one group was correlated with the number of proceptive females on the same day. Among five red-tailed monkey groups in the Kibale Forest, Uganda, the number of adult females per group was negatively correlated with the proportion of time the group had only one adult male, and positively correlated with the number of males seen in the group per observation month over a period of about 2–6 years (Struhsaker, 1988). In two patas monkey groups with adjoining ranges, influxes occurred less often in the group with fewer adult females (Chism and Rowell, 1986; Chism and Rogers, 1997). In one of them in another year, Harding and Olson (1986, their Fig. 2) showed that the number of males in the group on a given day was correlated with the number of sexually receptive females ($r_s = 0.79$, $n = 23$ days, $p = 0.0002$). All these data point to the importance of the number of (receptive) females in determining how many males are in a group.

There are also reports that dispute the importance of female numbers in determining male numbers. In a two-year study of three groups of samango monkeys in the same population previously studied by Henzi and Lawes (1988), Macleod (2000) found that the mean number of males in a group during a single breeding season was not correlated with the number of potentially fertile females as defined by Cords (2000). Similarly, Jones and Bush (1988) found that the frequency of sightings of non-resident males during a breeding season was not related to the number of potentially fertile females in three groups of Kibale red-tailed monkeys. Harding and Olson (1986) noted that influx breeding seasons occurred in patas monkey groups that differed greatly in size.

While these reports may suggest that factors other than the number of receptive females can influence male numbers, it is also possible that measurements of reproductive opportunity were inaccurate. Across all existing studies of male influxes in guenons, the number of males in a group shows the closest relationship to the number of sexually active females per day, whereas measurements of reproductive opportunity that represent a group over an entire season predict male numbers less well. This fact probably reflects the proximate mechanisms that underlie a connection between female availability and male numbers: males seem to join groups in response to the behavior, not simply the number, of females that they detect.

Data from two other guenon populations also suggest that female receptivity determines the number of males in the group rather than vice versa. Henzi and Lawes (1988) seem to have simply assumed this causal direction for the Cape Vidal samango monkeys, perhaps because the reduction in female receptivity from one year to the next was largely related to the presence of young infants born to many of the group's females, thus removing them from the pool of possible mates. Macleod's (2000) follow-up study included more detailed monitoring of the chronology of female sexual behavior and males entering groups, and demonstrated for one group in one (influx) year that female proceptivity preceded males entering the group. Similarly, Harding and Olson (1986) found that males entered the groups a few days after females became sexually receptive, and that increases in the number of receptive females tend to precede increases in the number of intruding males. However, they do not comment on the events that occurred in the weeks before the influx of males, and before the first female became receptive, so it is unclear whether observations began early enough in the season to allow a confident conclusion about whether female receptivity or male intrusions occurred first. However, they described how non-resident male patas monkeys monitor heterosexual groups, apparently looking for proceptive behavior in females; such behavior on the part of males would explain how their numbers respond to the presence of receptive females.

The importance of intruder pressure has been recognized by other researchers who focused on single populations (Struhsaker, 1988; Chism and Rogers, 1997). Only the work on samango monkeys, however, can address the relative importance of intruder pressure and number of receptive females to determine male numbers. Henzi and Lawes (1988) found an almost equal number of non-resident males that remained regularly around their study area in a year with influx breeding and in a year without influxes. While potential intruders thus seemed to be similarly available in both years, they spent more time with females in the year when more females were receptive. Further, Macleod (2000) noted that despite a similar availability of potential intruder males across groups in a single year, influxes occurred only in groups in which females were proceptive to non-residents. Accordingly, in samango monkeys variable intruder pressure does not seem to explain as well as female bahavior why influxes occur in only some groups and some years. It will be a challenge for future researchers of other populations to investigate further the importance of intruder pressure, a variable that is very difficult to measure under most field conditions (Struhsaker, 1988). If realized intruder pressure depends not only on the number of males but also on individual characteristics of particular residents, as suggested by Cords (1984), Struhsaker (1988), and Macleod (2000), documenting the importance of this aspect of female defensibility will be even more difficult.

Summary

Cercopithecus mitis groups show variation in male number during the breeding season. The 23 years of male residence records from multiple groups of blue monkeys in the Kakamega Forest reveal that influxes occurred in 23% of breeding seasons, and coincide with greater number of sexually active females in the group on both per-season and per-day bases. In six of ten seasons with sufficiently detailed records, female sexual behavior preceded the arrival of multiple males in the group, while in one year, male visits came first, and in the remaining three years, estrus and the presence of more than one male occurred essentially simultaneously. It is likely that estrous females bring males to the group, but also that the presence of males feeds back positively on estrus behavior in females. The number of males in the local area, or intruder pressure, may also explain some variation in male numbers. These data, which are the most extensive for any guenon population, are generally supported by other investigations of different populations and species. The number of sexually active females, as well as intruder pressure, are two important components of female defendability that influence the number of males present in the breeding season in at least three guenon species.

ACKNOWLEDGMENTS

I am grateful to the Government of Kenya, Office of the President and Ministry of Education, Science and Technology for permission to conduct research in the Kakamega Forest, to the University of Nairobi Zoology Department and the Institute for Primate Research (National Museums of Kenya) for local sponsorship, and to the local officers of the Forestry Department and Kenya Wildlife Service for cooperation at the field site. For assistance with data gathering, I thank especially Karen Pazol, Simon Mbugua, Praxides Akelo, Steffen Forester, Jasper Kirika, Caroline Okoyo, Julius Omondi, Joel Glick, and Alex Piel, as well as those Kenyan and American students who assisted with observations for shorter periods. Jade Gibson and Katie Ross helped enormously with computer entry of field notes. Mary Glenn, Russ Tuttle and Thelma Rowell provided many useful comments on the manuscript. My study of blue monkeys at Kakamega has been supported by the University of California Research Expeditions Program, Columbia University, the Wenner Gren Foundation, the L.S.B. Leakey Foundation, and the National Science Foundation (Graduate Fellowship, SBR 95–23623, BCS 98–08273).

References

Brooks, T., Pimm, S. L., and Oyugi, J. O. 1999. Time lag between deforestation and bird extinction in tropical forest fragments. *Cons. Biol.* **13**:1140–1150.

Chism, J. B., and Rowell, T. E. 1986. Mating and residence patterns of male patas monkeys. *Ethology* **72**:31–39.

Chism, J., and Rogers, W. 1997. Male competition, mating success and female choice in a seasonally breeding primate (*Erythrocebus patas*). *Ethology* **103**:109–126.

Cords, M. 1984. Mating patterns and social structure in redtail monkeys (*Cercopithecus ascanius*) *Z. Tierpsychol.* **64**:313–239.

Cords, M. 1987a. Forest guenons and patas monkeys: Male-male competition in one-male groups. In: B. B. Smuts, D. L. Cheney, R. M. Seyfarth, R. W. Wrangham, and T. T. Struhsaker (eds.), *Primate Societies*, pp. 98–111. University of Chicago Press, Chicago.

Cords, M. 1987b. Mixed-species association of *Cercopithecus* monkey in the Kakamega Forest, Kenya. *Univ. Calif. Pub. Zool.* **117**:1–109.

Cords, M. 1988. Mating systems of forest guenons: A preliminary review. In: A. Gautier-Hion, F. Bourlière, J.-P. Gautier, and J. Kingdon (eds.), *A Primate Radiation: Evolutionary Biology of the African Guenons*, pp. 323–339. Cambridge University Press, Cambridge.

Cords, M. 2000. The number of males in guenon groups. In: P. Kappeler (ed.), *Primate Males*, pp. 84–96. Cambridge University Press, Cambridge.

Cords, M. 2002. Foraging and safety in adult female blue monkeys in the Kakamega Forest, Kenya. In: L. E. Miller (ed.), *Eat or Be Eaten: Predation-sensitive Foraging in Primates*, pp. 205–221. Cambridge University Press, Cambridge.

Cords, M., and Rowell, T. E. 1987. Birth intervals of *Cercopithecus* monkeys of the Kakamega Forest, Kenya. *Primates* **28**:277–281.

Fashing, P. J., and Cords, M. 2000. Diurnal primate densities and biomass in the Kakamega Forest: An evaluation of census methods and a comparison with other forests. *Am. J. Primatol.* **50**:139–152.

González-Martinez, J. 1998. The ecology of the introduced patas monkey (*Erythrocebus patas*) population of southwestern Puerto Rico. *Am. J. Primatol.* **45**:351–365.

Harding, R. S. O., and Olson, D. K. 1986. Patterns of mating among male patas monkeys (*Erythrocebus patas*) in Kenya. *Am. J. Primatol.* **11**:343–358.

Henzi, S. P. 1988. Many males do not a multimale troop make. *Folia Primatol.* **51**:165–168.

Henzi, S. P., and Lawes, M. J. 1987. Breeding influxes and the behaviour of adult male samango monkeys (*Cercopithecus mitis albogularis*). *Folia Primatol.* **48**:125–136.

Henzi, S. P., and Lawes, M. J. 1988. Strategic responses of male samango monkeys (*Cercopithecus mitis*) to a decline in the number of receptive females. *Int. J. Primatol.* **9**:479–495.

Jones, W. T., and Bush, B. B. 1988. Movement and reproductive behavior of solitary male redtail guenons (*Cercopithecus ascanius*). *Am. J. Primatol.* **14**:203–222.

Kaplin, B. A., Munyaligoga, V., and Moermond, T. C. 1998. The influence of temporal changes in fruit availability on diet composition and seed handling in blue monkeys (*Cercopithecus mitis doggetti*). *Biotropica* **30**:56–71.

Kappeler, P. 2000. Primate males: history and theory. In: P. Kappeler (ed.), *Primate Males*, pp. 3–7. Cambridge University Press, Cambridge.

Macleod, M. C. 2000. *The Reproductive Strategies of Samango Monkeys* (Cercopithecus mitis erythrarchus). Ph.D. Thesis, University of Surrey Roehampton.

Pazol, K., and Ziegler, T. E. 2000. Mating in the absence of ovarian activity following a resident male turnover in a wild blue monkey group, Kakamega, Kenya. *Am. J. Primatol.* **51**: (Suppl. 1) 79.

Pazol, K., Carlson, A. A., and Ziegler, T. E. 2002. Female reproductive endocrinology in wild blue monkeys: A preliminary assessment and discussion of potential adaptive functions. In: M. E. Glenn, and M. Cords (eds.), *The Guenons: Diversity and Adaptation in African Monkeys*, pp. 217–232. Kluwer Academic Publishers, New York.

Struhsaker, T. T. 1988. Male tenure, multimale influxes, and reproductive success in redtail monkeys (*Cercopithecus ascanius*). In: A. Gautier-Hion, F. Bourlière, J.-P. Gautier, and J. Kingdon (eds.), *A Primate Radiation: Evolutionary Biology of the African Guenons*, pp. 340–363. Cambridge University Press, Cambridge.

Costs and Benefits of   15
Alternative Mating Strategies
in Samango Monkey Males

MAIRI C. MACLEOD, CAROLINE ROSS,
and MICHAEL J. LAWES

Introduction

In primates, as in all other mammals, a male's lifetime reproductive success is limited by the number of females he can fertilize at any given time (mating rate) and the length of time for which he can sustain this rate (his reproductive lifespan). Males are therefore expected to act in a way that maximizes their access to fertile females, but how they do this will depend on how females group themselves (Emlen and Oring, 1977; van Schaik and van Hooff, 1983; Dunbar, 1988; Newton, 1988; Clutton-Brock, 1989; Altmann, 1990). Most guenon species have a unimale troop social system. The troop male achieves this status by ousting the incumbent male, and then defending the females against other males. However, although troop males have priority of access to receptive females in their troop, this is not the only way to achieve copulations. Like some other social mammals such as red deer (Clutton-Brock *et al.*, 1982), samango monkey males manage extratroop

MAIRI C. MACLEOD and CAROLINE ROSS • School of Life Sciences, University of Surrey Roehampton, West Hill, London SW15 3SN, UK.   MICHAEL J. LAWES • School of Botany and Zoology, Forest Biodiversity Programme, University of Natal, Pietermaritzburg, South Africa. *The Guenons: Diversity and Adaptation in African Monkeys*, edited by Glenn and Cords. Kluwer Academic/Plenum Publishers, New York, 2002.

copulations (Henzi and Lawes, 1987). Given that extratroop males can thus achieve fitness benefits, the question is why become a troop male? We examine the fitness costs and benefits of the alternative mating strategies of sneaking extratroop copulations and being a resident troop male (i.e., sneakers and residents).

Alternative male mating strategies occur in several primate species including some of the guenons and their relatives. In blue monkeys (Tsingalia and Rowell, 1984; Cords *et al.*, 1986; Henzi and Lawes, 1987; Cords, 2000), red-tailed monkeys (Cords, 1984; Jones and Bush, 1988; Struhsaker, 1988) and patas monkeys (Chism and Rowell, 1986; Harding and Olson, 1986; Ohsawa *et al.*, 1993; Chism and Rodgers, 1997), non-resident or extratroop males enter troops during the mating season and mate with females. Multimale as well as unimale groups have been observed in mona monkeys (Howard, 1977; Glenn *et al.*, 2002), and multimale influxes in Grenada mona monkeys have been reported (based on calls heard; Glenn, 1996). During multimale influxes, intruding males sometimes appear to achieve greater mating success than the resident troop males (Tsingalia and Rowell, 1984; Cords *et al.*, 1986; Henzi and Lawes, 1987). Given that the costs of takeover and defense of a group of females are likely to be high (Henzi and Lawes, 1987), it may pay some males to opt for the extratroop male lifestyle.

Cords *et al.* (1986) and Henzi and Lawes (1987) suggested that the mating behavior of male *Cercopithecus mitis* may represent competing reproductive strategies that, over the long term, could produce equivalent reproductive success. Various lifetime strategies for male *Cercopithecus mitis* may be to (a) become a resident troop male, (b) be a dedicated extratroop male, or (c) adopt both of these strategies sequentially during its reproductive lifespan. Troop males and extratroop male mating strategies coexist in the same populations (Cords, 2000). If the strategies equilibrate in terms of lifetime reproductive success, then the mating strategies may be a mixed evolutionarily stable strategy.

The question we investigate here is: do male *Cercopithecus mitis* have the choice of alternative lifetime reproductive strategies which, on average, yield equivalent lifetime reproductive output, or is the mixture of male strategies simply a consequence of mixed strategy sets in individuals that are at different stages of their reproductive lives?

We first describe the behavior of resident and extratroop male *Cercopithecus mitis* and the strategies that they employ to gain access to females. Second, we use our data to examine the hypothesis that male *Cercopithecus mitis* are following two lifetime strategies of equivalent merit. We predict that if there are two lifetime strategies, then extratroop males should achieve a significant proportion of successful matings compared with troop males. We further predict that any disparity in the instantaneous reproductive success between troop and extratroop males is offset by differences in costs of each strategy, such that the lower-cost, lower-gain strategy can be maintained for a longer period of life.

We conduct a simple cost–benefit analysis of the different strategies by comparing mating success of males following the two strategies, and comparing the level of wounding to determine which strategy produces the highest costs in terms of risk of injury during aggressive episodes. While we acknowledge that mating success does not always give an accurate measure of reproductive success, it is a useful variable to include in our model for heuristic purposes (i.e., to explore the consequences of variation in reproductive rate and lifespan).

Methods

We observed *Cercopithecus mitis labiatus*, known as samango monkeys, at Cape Vidal (28°05′35″S, 32°33′40″E) between January 1995 and February 1997. Cape Vidal lies within the Indian Ocean coastal forest belt and the monkeys' habitat includes three main plant communities: coastal thicket, dune forest, and *Acacia karroo* woodland (Macleod, 2000).

We studied three troops (each containing seven to 11 adult females), and extratroop males within their ranges. The study focused particularly on male behavior and mating rate and success. We collected data during troop and extratroop male follows on four to eight days per month throughout the study. We used instantaneous scan sampling, focal subject sampling, and *ad libitum* observations of rare but important events to collect behavioral data. During troop follows, which were carried out by a single observer, scan samples of all visible group members were conducted every half an hour for ten minutes. During the remaining time, focal subject sampling was performed, and this was biased toward adult males, with half the sampling time being devoted to them, the remaining time being spent on adult females and older juveniles. During troop follows, male sampling included the troop male and any other adult males that associated with the troop more temporarily. Focal sampling time was divided as equally as possible between the males present in the troop at the time. Additional observations were made during separate extratroop male follows and opportunistic contacts.

We recorded the numbers of unique troop males and unique extratroop males observed in the study area each month and calculated the mean total per month for each of them. Troop males were defined as those which spent the majority of time in close proximity with members of a single troop, were vigilant toward intruders into the troop, and gave loud pyow and boom calls (Marler, 1973). Extratroop males were defined as those that had no specific troop affiliation, and had fluid associations with other males that could be long- or short-term. In our experience, extratroop males rarely gave loud calls.

We calculated the number of mating bouts per hour of observer contact with the troop for all males observed to mate with study troop females. A mating bout is defined as a series of sexual mounts separated by ≤15 minutes between mounts during which no mounts occurred between either participant and another individual. Successful bouts occurred when at least one mount ended in visible ejaculation or a postcopulatory hold (a postcopulatory hold occurs where the male holds on to the female in a copulatory position for a few seconds with no thrusting, after mating has taken place). To measure mating success we used mating bouts rather than individual mounts because juveniles interrupted many mounts, so that the mating partners often had to perform several mounts to achieve intromission and an actual mating.

To compare the level of wounding in troop and extratroop males, we recorded the presence of new wounds on each male *ad libitum*. We calculated numbers of recent wounds observed per month for each known troop and extratroop male. Wounds were classified as major, minor, or intermediate. A major wound caused some disablement, at least temporarily, and could be life threatening, while minor wounds were typically small cuts with little significant effect on the activities of the male. We regarded a major wound as more serious than an intermediate and a minor wound combined, so that for analysis, we scored major wounds as four, intermediate ones as two, and minor ones as one. We pooled the scores for each individual male and divided the sum by the number of months in which that male was observed. We compared troop male and extratroop male scores using the Mann-Whitney U test (Zar, 1984).

Results

Observed Male Strategies

We individually recognized 26 adult males in the study area and these assumed the roles of resident males ($n = 6$), extratroop males ($n = 14$), or both ($n = 6$) on separate occasions. Initially, the three study troops, A, B, and C, each had a single resident troop male. Extratroop males traveled alone or in small all-male groups with a modal size of three.

In B, there were two troop male replacements during the study period, and both incoming males were infanticidal. The first male immigrant killed ≥ 3 infants and the second male killed one infant. C troop had the same resident male over the two-year study period. In A troop, the resident male maintained his tenure into the second year of the study, but suffered a gradual replacement as several males entered the troop during the mating season during a multimale influx, and eventually evicted him.

The number of troop males observed in the study area ranged from three to seven over the study period while the number of extratroop males rose dramatically in 1996, when A troop experienced a multimale influx (Fig. 1). The multimale influx was associated with synchronous receptivity among troop A females.

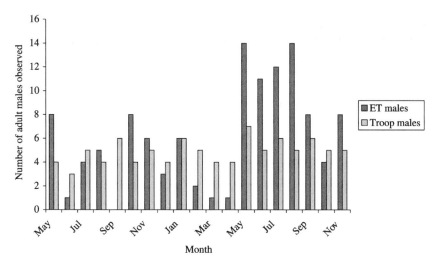

Fig. 1.   Number of unique troop and extratroop males (ET males) observed within the study area (combined home ranges of study troops A, B, and C, overlapping with those of other troops) per month between May 1995 and November 1996.

Costs and Benefits of Alternative Male Strategies

Benefits: mating success

In 1995, only troop males were observed to mate (AMJ was the resident male from a neighboring troop), though only one of them, AMSH, which was newly resident in B, was observed to mate successfully (Table I).

Considerably more mating was observed in 1996, during a multimale influx into A troop, but this still involved only a few males (Table I). The original resident male of A, AMB, had a low mating rate while AMD, which became the dominant male and eventually sole resident of the troop, and AMF, a satellite male, both had relatively high levels of mating success. AMN, the new, infanticidal male of troop B, had the highest level of successful matings. Thus, in both years newly resident males had the highest observed rate of mating success, and only one extratroop male was observed to mate.

Costs: level of wounding

Observations of fights involving male–male contact were confined to the 1996 mating season, and were most apparent near A troop, which had the largest number of males in association. All but three of 32 new wounds (one minor and two intermediate) were sustained during the mating season months (May–November). The relationship between mating competition and wounding suggests that wounding can be regarded as a cost of male competition

Table I. Mating Bouts per Hour of Troop Contact in (a) the 1995 Mating Season, and (b) the 1996 Mating Season. All males observed to mate in the study troops are listed. Male category T = troop male, ET = extratroop male. "Successful" bouts are those where at least one mount ended in visible ejaculation or a postcopulatory hold (see text for description).

Troop	Male ID	Male Category	Bouts/Day	Successful Bouts/Day
1995				
A	AMB	T	0	0
B	AMSH	T	2.14	0.46
B	AMJ	T	0.03	0
C	AMS	T	0.09	0
1996				
A	AMB	T	0.03	0.02
A	AMD	T	1.15	0.09
A	AMF	ET	0.26	0.14
B	AMN	T	1.06	0.54
C	AMS	T	0.55	0.15

for access to females. There was no difference in the amount of wounding between troop and extratroop males (Mann-Whitney test: $Z = 168.5$, $N_1 = 11$, $N_2 = 24$, $p > 0.10$). However, troop males were likely to receive the most serious wounds when a new male ousted them from troop residency. This pattern is expected because a resident male is more likely to escalate a conflict since he has more to lose. Troop males may have a greater risk of serious wounds than extratroop males, but we could not confirm this prediction because troop males often disappeared after they had been ousted and any wounds sustained could not be observed. Of seven males expelled from their troops between January 1995 and July 1997, which includes study troops and neighboring troops, four disappeared, one was badly wounded and disappeared, one survived and traveled with another extratroop male (but had lost three fingers and a thumb), and another was severely wounded and died a few days later. Thus, the risk of severe wounding and death seems high for resident males that defend their troop and it could shorten their reproductive lives.

Cost–Benefit Analysis of Male Mating Behavior

Can the extratroop male strategy of sneaking matings be a viable alternative to securing tenure over a group of females? One way to approach this question is to estimate how many years of sneaked matings would be equivalent to a single period of tenure as a troop male. Here we develop a simple cost-benefit model to answer these questions. We know that the average duration of troop male tenure is 1.67 years, calculated from the rate of troop male replacement in 6 troops over 2.5 years (Macleod, 2000). In addition, by

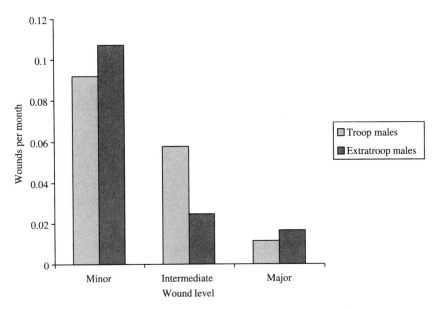

Fig. 2. Mean wound scores for troop males ($n = 12$) and extratroop males ($n = 20$).

estimating the proportion of offspring sired in a troop by the troop male (P_t), the proportion of offspring in a troop sired by extratroop males ($1 - P_t$), and the reproductive lifespan (R) of a troop male, we can begin to estimate the fitness of troop and extratroop males.

We modeled the situation in which for any troop there is: (a) one troop male, (b) one extratroop sneaker, (c) the extratroop sneaker achieves no success in other troops, and (d) therefore, the proportion of offspring in a troop sired by an extratroop male is due to one male's activities, i.e., if 0.1 offspring are due to extratroop copulation, they arise from one male only. These assumptions provide the most optimistic payoffs or fitness benefits for a male adopting a sneaking strategy only.

Two mating strategies are considered: (1) inclusive fitness of a male that has one period of troop tenure and sneaks matings otherwise [S1], and (2) inclusive fitness of a male that sneaks matings in one troop each season and never secures troop tenure [S2]. Essentially, we want to know how many years of S2 are equivalent to S1.

So, when S1 = S2 we get

(total number of infants sired as a troop male (the number of infants
  plus the number sired as an extratroop male) = sired as an extratroop
                                                        male),

and

$$(P_t*1.67) + [(R - 1.67)*(1 - P_t)] = (1 - P_t)*Y_e \qquad \text{Eq. 1}$$

in which Y_e = number of years of extratroop breeding activity (for S2).

To estimate how many years of only the sneaking strategy equate to the fitness returns of a single period of tenure by a troop male we substitute values into the variables in Eq. 1 so that $P_t = 0.9$ (calculated from data in Table I), and therefore $(1-P_t) = 0.1$, and $R = 1.67$ years of reproductive lifespan for a troop male, we get

$$Y_e = [(0.9 *1.67) + (1.67-1.67) * 0.1]/0.1$$
$$Y_e = 15.03 \text{ years}$$

Thus, to achieve the fitness equivalent of a troop male (with the above value of P_t) in one average tenure length, a S2 strategist would have to sneak matings for 15.03 years. Clearly, the benefit to being a troop male is enormous when paternity is assured and accounts for why troop males will escalate conflict and risk serious injury. The model dismisses the sneaking extratroop male strategy since 15 years is probably much longer than their reproductive lifespan. Even if an extratroop male sired 25% of offspring in a troop, he would have to be an S2 strategist for ⩾5 years (Fig. 3a). One could argue that sneaked mating behavior continues to exist because it is the only other strategy available and not because it is a particularly good alternative.

Sneaking is profitable only if there is a good chance the extratroop sneaking male can sire a high proportion (>25%) of the offspring in a troop or if low rates of success are achieved in many troops. The latter explains male influx behavior when breeding opportunities in a troop are suitable, namely that a high proportion of troop females are receptive at the same time (Cords, 2000). In the absence of sufficient data from a number of populations of *Cercopithecus mitis*, the parameter values of P_t and R cannot be reliably fixed. Thus, we examined the effect of variation in these parameter values on the number of years an extratroop male would need to sneak matings (Fig. 3b). The figure shows that as the troop male sires fewer offspring on average, and the more offspring the extratroop male sires, so the number of years of sneaking approaches a value commensurate with troop male tenure. However, for all combinations of P_t and R, extratroop males have to adopt a sneaking strategy for a period considerably longer than the theoretical duration of troop male tenure (Fig. 3b).

Discussion

Behavior of Adult Male Samango Monkeys

Observations of samango monkeys at Cape Vidal over two years indicate that males occur in a variety of grouping patterns, including being associated

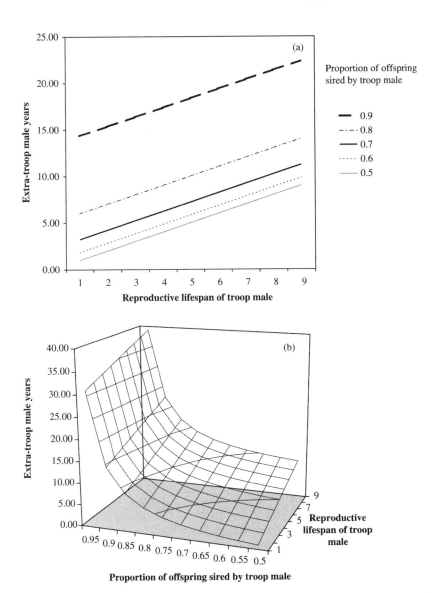

Fig. 3. (a) Number of extratroop male mating years required for an extratroop male to achieve reproductive success equivalent to a single period of tenure as a troop male. This relationship is shown for varying durations of troop male tenure as well as for different conditions of troop male paternity. It can be seen that if an extratroop male sired only 10% of troop infants, it would have to sneak matings for 15.03 years to achieve reproductive success equivalent to a troop male with a 1.67 year period of troop tenure and no further reproductive years. (b) Response surface showing the relationship between extratroop male mating year equivalents and troop male tenure and level of paternity.

with a troop as a resident male, as a temporary resident for days or hours, following a troop as a satellite male for short periods but not interacting with troop members, traveling with other males in an all-male group, or being solitary, either alone or occasionally associated with a vervet monkey (*Cercopithecus aethiops*) troop. The period of troop tenure appears to be a relatively stable state for the male, though troop males may sneak matings in neighboring troops (Macleod, pers. obs.; Henzi and Lawes, 1988; Cords, 2000). Individual extratroop males may be solitary, a member of an all-male group, or a satellite male. This variety of male states also occurs in the population of *Cercopithecus mitis* in Kakamega forest, Kenya (Cords *et al.*, 1986; Rowell, 1997; Cords, 2000). Male patas monkeys also live as residents in groups of females, associate temporarily with groups in the mating season, live in all-male associations or travel singly (Chism and Rodgers, 1997). Mona monkey males have been observed in all-male groups as well as in multimale and unimale bisexual groups (Howard, 1977; Glenn *et al.*, 2002).

Are There Different Lifetime Strategies for Samango Monkey Males?

If extratroop males are pursuing an opportunistic mating strategy in opposition to becoming a resident troop male, we should expect their mating success, measured over several mating seasons, to be significant and similar to that of troop males. During our study of male mating behavior in three troops over two seasons, only one extratroop male was observed to mate successfully, accounting for only 10% of the total observed successful matings.

Since selection pressure operates on lifetime reproductive success, we predict that if extratroop males are following a truly alternative strategy to that of troop males, they should compensate for their low rate of mating success by maintaining their strategy for a longer period (many seasons) or by pursuing more mating opportunities (many matings in different troops in a season). It is perhaps worth noting that under the terms of the model presented here, we assume that an extratroop male mates with females from one group in a season and that, apart from the troop male, no other male than the sneaker achieves any mating success. This assumption provides a best outcome scenario for extratroop males. Although little is known about the number of troops an extratroop male may visit in a season, it is likely to be low (Jones and Bush, 1988), and would probably not compensate for the dilution of mating success in a multi-male influx.

Extratroop males may lead a less costly lifestyle than troop males, taking fewer risks such as those associated with fighting over access to females. However, our data suggest that troop and extratroop males are equally likely to acquire new wounds, though in conflicts between troop and extratroop males it is always the troop male that initiates conflict (Macleod, pers. obs.). Troop males are always dominant except in cases of troop takeover, which are often characterized by simultaneous opposition to the troop male by more than one extratroop male (Macleod, 2000). Although extratroop males that are simply sneaking matings may be unwilling to risk escalating conflicts, when an extratroop male challenges troop male tenure wounding can be severe, especially to the troop male, who appears to defend his position tenaciously since he has more to lose. Termination of tenure could signal the end of a male's reproductive career or even his life. For example, Cords (2002) monitored six troops of *Cercopithecus mitis* at Kakamega for ≤23 years and found no evidence that males have more than one period of troop tenure.

Cost-benefit analysis shows that the extratroop male mating strategy will achieve fitness returns equivalent to the troop male strategy only under strict conditions: (1) if the extratroop male pursues the strategy for a considerable period (>5–9 years), and (2) a relatively large proportion of offspring (≥0.25) are sired by the sneaker in a troop in a season, or the sneaker male achieves limited mating success in several troops. The persistence of the two male mating strategies is frequency-dependent since the more sneakers that breed with a troop's females, the closer the fitness advantage for the troop male converges on the sneaking strategy. This explains the advantage of multi-male influxes to extratroop males.

While there are no data for the duration of male reproductive lifespan, it is unlikely that it is more than five years and hence, the likelihood that the alternative extratroop male mating strategy on average achieves fitness returns equivalent to the troop male strategy is slight. For a relatively short period of tenure (1.67 years), the troop male achieves relatively high reproductive success approaching his lifetime reproductive success. Accordingly, troop tenure is hotly contested among males. But why does extratroop mating behavior persist? Probably it does so because there is no other option and sneaked matings are better than none, but it must also be seen as part of the more important male behavior of attempting to secure troop tenure, rather than as a separate strategy *per se*. In addition, the role of female choice in determining male tenure and ultimately male fitness has been overlooked and extratroop male influxes may function to permit such choice rather than as truly meaningful mating opportunities.

Our findings differ substantially from those reported for other populations of *Cercopithecus mitis* at certain periods. Other observers reported extratroop males having more mating success than the resident male in some troops in some years (Tsingalia and Rowell, 1984; Cords *et al.*, 1986; Henzi and Lawes, 1987; Cords, 2000). For example, Tsingalia and Rowell (1984) report that of 16 males that associated with a troop in Kakamega forest, Kenya, over a six month period, nine were seen to copulate, but the resident male copulated least frequently. In the same troop two years later, the resident and highest ranking male was not seen to mate during a mating season where 12 out of 19 males mated in the troop (Cords *et al.*, 1986). During a previous study of the population described in this chapter, Henzi and Lawes (1987) noted that at least six males mated successfully during a season when at least 25 males invaded a troop. The relative costs and benefits of each strategy thus vary considerably and will depend on the ease with which a single male can monopolize a troop of females.

The number of males associating with a troop of *Cercopithecus mitis* increases with the number of sexually receptive females (Cords, 2000, 2002; Macleod, 2000). An extratroop male's chance of reproductive success in any one year is likely to depend on the number of synchronously receptive females in the troops he encounters. However, the lifetime reproductive success of males following the different strategies will depend on population characteristics that affect the resident troop males' ability to monopolize access to fertile females (Emlen and Oring, 1977). These characteristics are likely to include (1) the operational sex ratio, as this will affect the general level of competition for troop residence, and thus the costs of being a troop male (Rubenstein, 1980; Butynski, 1982), (2) population density and home range size, as they will affect the ease with which a male can move opportunistically from one troop to another in search of mating opportunities, and therefore the relative profitability of being a long-term resident in one troop vs. a roving extratroop male (Srivastava and Dunbar, 1996), (3) female group size, dispersion, and the degree of visibility in their environment, which will affect a male's ability to monitor females and detect intruding males (van Schaik and van Hooff, 1983; Dunbar, 1988; Altmann, 1990; Mitani *et al.*, 1996), and (4) presence or absence of male coalitionary behavior to gain access to females (Terborgh, 1983). The relative costs and benefits of each strategy will be affected by such ecological and demographic factors, which will vary between populations and within populations over time, thus producing a shift in the equilibrium point in the relative frequency of alternative strategies. Clearly, longitudinal data are required from several populations in order to determine the factors responsible for the geographical and temporal variation in the mating system of *Cercopithecus mitis*, as well as genetic studies to establish paternity and obtain valid measures of reproductive success.

Summary

In samango monkeys (*Cercopithecus mitis labiatus*), males compete for residency in unimale troops where they defend females against other males, although extratroop males may sneak copulations with females opportunistically. Here, we attempt to determine whether these are alternative strategies yielding equivalent lifetime reproductive success, or whether the extratroop male component is simply a tactic followed by males who are unable to follow the resident troop male strategy because of their age or physical condition. Observations of mating success and male–male competition support the latter of these hypotheses. Modeling the expected reproductive success of males following the different strategies suggests that dedicated extratroop males in this population would not be able to live for long enough to compensate for their extremely low rate of mating success compared with males achieving troop residency. However, other studies of *Cercopithecus mitis* have revealed a much higher proportion of mating by extratroop males, and it is suggested that ecological and demographic factors may affect the costs and benefits of these two male strategies.

ACKNOWLEDGMENTS

   Institutional support was provided by the University of Surrey Roehampton in the U.K. and by the School of Botany and Zoology, University of Natal, Pietermartizburg, during the fieldwork phase. We would like to thank Paul Carter who helped enormously with data collection and running the research site. The Cape Vidal staff of the KwaZulu-Natal Wildlife service, in particular Ron Joubert, are thanked for their support during the term of the study. Finally, we would like to thank the societies and trusts who assisted this project financially; National Research Foundation (South Africa), the Green Trust, the Boise Fund, the Leakey Trust, the British Federation of Women Graduates, and the Royal Anthropological Institute.

References

Altmann, J. 1990. Primate males go where the females are. *Anim. Behav.* **39**:193–195.
Butynski, T. M. 1982. Harem-male replacement and infanticide in the blue monkey (*Cercopithecus mitis stuhlmanni*) in the Kibale Forest, Uganda. *Am. J. Primatol.* **3**:1–22.
Chism, J. B., and Rogers, W. 1997. Male competition, mating success, and female choice in a seasonally breeding primate (*Erythrocebus patas*). *Ethology* **103**:109–126.
Chism, J. B., and Rowell, T. E. 1986. Mating and residence patterns of male patas monkeys. *Ethology* **72**:31–39.
Clutton-Brock, T. H. 1989. Mammalian mating systems. *Proc. R. Soc. Lond.* **236**:339–372.
Clutton-Brock, T. H., Albon, S. B., and Guiness, F. E. 1982. *Red Deer: Behaviour and Ecology of Two Sexes.* University of Chicago Press, Chicago.
Cords, M. 1984. Mating patterns and social structure in redtail monkeys (*Cercopithecus ascanius*). *Z. Tierpsychol.* **64**:313–329.
Cords, M. 2000. The number of males in guenon groups. In: P. Kappeler (ed.), *Primate Males*, pp. 84–96. Cambridge University Press, Cambridge.
Cords, M. 2002. When are there influxes in blue monkey groups? In: M. E. Glenn, and M. Cords (eds.), *The Guenons: Diversity and Adaptation in African Monkeys*, pp. 189–201. Kluwer Academic Publishers, New York.
Cords, M., Mitchell, B. J., Tsingalia, H. M., and Rowell, T. E. 1986. Promiscuous mating among blue monkeys in the Kakamega Forest, Kenya. *Ethology* **72**:214–226.

Dunbar, R. I. M. 1988. *Primate Social Systems.* Croom Helm Ltd. Kent, England.

Emlen, S. T., and Oring, L. W. 1977. Ecology, sexual selection, and the evolution of mating systems. *Science* **197**:215–223.

Glenn, M. E. 1996. *The Natural History and Ecology of the Mona Monkey* (Cercopithecus mona *Schreber 1774) on the Island of Grenada, West Indies.* Ph.D. Dissertation, Northwestern University, Evanston, Illinois, USA.

Glenn, M. E., Matsuda, R., and Bensen, K. J. 2002. Unique behavior of the mona monkey, *(Cercopithecus mona)*: All-male groups and copulation calls. In: M. E. Glenn, and M. Cords (eds.), *The Guenons: Diversity and Adaptation in African Monkeys,* pp. 133–145. Kluwer Academic Publishers, New York.

Harding, R. S. O., and Olson, D. K. 1986. Patterns of mating among male patas monkeys *(Erythrocebus patas)* in Kenya. *Am. J. Primatol.* **11**:343–358.

Henzi, S. P., and Lawes, M. J. 1987. Breeding season influxes and the behaviour of adult male samango monkeys *(Cercopithecus mitis albogularis).* *Folia Primatol.* **48**:125–136.

Henzi, S. P., and Lawes, M. J. 1988. Strategic responses of male samango monkeys *(Cercopithecus mitis)* to a decline in the number of receptive females. *Int. J. Primatol.* **9**:479–495.

Howard, R. 1977. *Niche Separation among Three Sympatric Species of* Cercopithecus *Monkeys,* Ph.D. Dissertation, The University of Texas at Austin.

Jones, W. T., and Bush, B. B. 1988. Movement and reproductive behavior of solitary male redtail guenons *(Cercopithecus ascanius).* *Am. J. Primatol.* **14**:203–222.

Macleod, M. C. 2000. *The Reproductive Strategies of Samango Monkeys* (Cercopithecus mitis erythrarchus). Ph.D. Dissertation, University of Surrey Roehampton, London, U.K.

Marler, P. 1973. A comparison of vocalizations of red-tailed monkeys and blue monkeys, *Cercopithecus ascanius* and *C. mitis,* in Uganda. *Z. Tierpsychol.* **33**:223–247.

Mitani, J. C., Gros-Louis, J., and Manson, J. H. 1996. Number of males in primate groups: comparative tests of competing hypotheses. *Am. J. Primatol.* **38**:315–332.

Newton, P. N. 1988. The variable social organization of hanuman langurs *(Presbytis entellus),* infanticide, and the monopolization of females. *Int. J. Primatol.* **9**:59–77.

Ohsawa, H., Inoue, M., and Takenaka, O. (1993). Mating strategy and reproductive success of male patas monkeys *(Erythrocebus patas).* *Primates* **34**:533–544.

Rowell, T. E. 1997. *Alternatives for males.* Presentation given for the "Göttinger Freilandtage", at the German Primate Centre, Göttingen.

Rubenstein, D. I. 1980. On the evolution of alternative mating strategies. In: J. E. R. Staddon (ed.), *Limits to Action,* pp. 65–100. Academic Press, London.

Srivastava, A., and Dunbar, R. I. M. 1996. The mating system of Hanuman langurs: a problem in optimal foraging. *Behav. Ecol. Sociobiol.* **39**:219–226.

Struhsaker, T. T. 1988. Male tenure, multi-male influxes, and reproductive success in redtail monkeys *(Cercopithecus ascanius).* In: A. Gautier-Hion, F. Bourlière, J.-P. Gautier, and J. Kingdon (eds.), *A Primate Radiation: Evolutionary Biology of the African Guenons,* pp. 340–363. Cambridge University Press, Cambridge.

Terborgh, J. 1983. *Five New World Primates.* Princeton University Press, Princeton.

Tsingalia, H. M., and Rowell, T. E. 1984. The behaviour of adult male blue monkeys. *Z. Tierpsychol.* **64**:253–268.

van Schaik, C. P., and van Hooff, J. A. R. A. M. 1983. On the ultimate causes of primate social systems. *Behaviour* **85**:91–117.

Zar, J. H. 1984. *Biostatistical Analysis.* 2nd edition. Prentice Hall, Englewood Cliffs.

Female Reproductive Endocrinology in Wild Blue Monkeys: A Preliminary Assessment and Discussion of Potential Adaptive Functions

16

KAREN PAZOL, ANN A. CARLSON,
and TONI E. ZIEGLER

Introduction

Among the best-studied guenons (*Cercopithecus ascanius*, *C. mitis*, and *Erythrocebus patas*) females often engage in sexual behavior at times when conception is unlikely or even impossible. In some cases they remain sexually active for more than one month at a time (Loy, 1981; Cords, 1984; Cords *et al.*, 1986; Macleod, pers. com.) and copulate during pregnancy (Rowell, 1970; Loy, 1981; Cords, 1984; Macleod, 2000; Carlson and Isbell, 2001; Pazol, 2001). On

KAREN PAZOL • Yerkes Regional Primate Research Center, Emory University, Lawrenceville, GA 30043, USA.   ANN A. CARLSON • Large Animal Research Group, Department of Zoology, University of Cambridge, Downing Street, Cambridge CB2 3EJ, UK.   TONI E. ZIEGLER • Department of Psychology and Wisconsin Regional Primate Research Center, University of Wisconsin, 1223 Capitol Court, Madison, WI 53715, USA.
The Guenons: Diversity and Adaptation in African Monkeys, edited by Glenn and Cords. Kluwer Academic/Plenum Publishers, New York, 2002.

a seasonal basis, mating within a population often appears to start before and to continue after the period when conception seems likely (Tsingalia and Rowell, 1984; Chism and Rowell, 1986; Cords *et al.*, 1986), and in some cases the presence of a new male has coincided with a rapid onset of sexual behavior (Fairgrieve, 1995; Carlson, 2000; Macleod, 2000; Pazol and Ziegler, 2000). The apparent disjuncture between mating and conception suggests that sexual behavior in female guenons often serves indirect or non-reproductive functions, including paternity confusion. To test this hypothesis and evaluate the physiological conditions under which females may mate, one needs to have a measure independent of behavior to assess female reproductive condition.

In many species of Old World monkeys, including baboons, mangabeys and several of the macaques, the perineal sex skin shows dramatic fluctuations in size and color across the ovarian cycle (Hrdy and Whitten, 1987). In others, such as Hanuman langurs, visible menstruation provides a clear marker of cyclic stage (Sommer *et al.*, 1992). By contrast, most female guenons have no external morphological indicator of their reproductive state (Hrdy and Whitten, 1987). Therefore, because the collection of serum samples for hormonal analysis is extremely difficult outside captive settings, the development of noninvasive techniques for evaluating reproductive condition is needed to assess the adaptive significance of female mating patterns in wild guenons.

We describe the development of a technique for using fecal steroid assays to monitor female reproductive condition in wild blue monkeys (*Cercopithecus mitis stuhlmanni*). The specific goals of our study were to (1) obtain an estimate of the typical gestation length for *Cercopithecus mitis* in the wild, thereby making it possible to count back from the date of parturition to calibrate fecal profiles against the date of conception; (2) determine if there is a consistent pattern in fecal steroid profiles that can be used reliably to mark ovulation; (3) establish a range of hormonal values within which a female can be considered cycling or non-cycling; and (4) assess whether females typically undergo more than one cycle to conception.

Methods

Study Site and Subjects

We conducted this study from June 1997 through October 1999 in the Kakamega Forest, western Kenya. Cords (1987) has described the study site in detail. Focal subjects were the 14–16 and 17–18 adult females of two habituated social groups (T_w and G, respectively). We obtained behavioral data from every female in the study groups, but we restricted fecal collection to a subset of our subjects. During the breeding season, we collected samples from eight females that were nearing the end of their interbirth interval. For four of these subjects we continued collection through the date of parturition. During the non-breeding season we collected samples from five females that were not pregnant. One female served both as a breeding and a non-breeding season subject.

Behavioral Data Collection and the Estimation of Gestation Length

During each year of the study, we monitored T_w and G intensively from mid-June through the end of August. During this 2.5-month period, which covers the typical peak of the breeding season in the Kakamega Forest, a six- to eight-person research team followed most sexually active females from dawn to dusk. In 1997, 1998, and 1999, two, three, and five researchers, respectively, remained at the study site from the end of August through the end of October. They followed both monkey groups for a minimum of 8 hours per day on most days, though there were occasional days when a group was not observed. We recorded *ad libitum* all copulations and proceptive behavior (persistent follow, headflag, present, pucker lips). We used the presence of semen around a female's vulva to indicate copulation on a given day if its color suggested that it had been freshly deposited.

To estimate gestation length, we followed the procedures outlined in Cross and Martin (1981). We plotted all copulations relative to the date of parturition in a histogram over the entire 80-day period during which conception was considered possible (day -200 to -120 before parturition). We then used the modal value as the best estimate of gestation length: unlike the mean, this figure does not have the potential to be skewed by a few unreliable estimates or unusual pregnancies (Cross and Martin, 1981). For females included in the histogram ($n = 21$) the entire 80-day period occurred when two or more observers were at the study site. We defined the date of parturition as the first day the mother was sighted with her new infant, if she had been sighted on the previous day without the infant. If she had not been sighted on the previous day, we assigned the date of parturition as the midpoint between the first and last sighting of the female with and without her new infant. We included females in the histogram only if the window of potential parturition dates was 5 days or less.

After estimating the typical gestation period, we approximated a potential range of variation. Again following Cross and Martin (1981), we assumed that 95% of all gestation lengths would fall within two standard deviations of the mean. To estimate the standard deviation, we used the formula $SD = (X)(CV)/100$, wherein X is the mean and CV is the coefficient of variation (Thomas, 1986). For the mean we used the modal value from our histogram as the best available estimate. For the coefficient of variation we used the average figure of 4%, which Martin (1992) found in a survey of 15 simian primates. Thus we assumed that *Cercopithecus mitis* does not differ from the typical simian primate in showing more or less variation in gestation length.

Fecal Sample Collection and Analysis

We scheduled fecal collection for every other day during breeding periods and for once a week during non-breeding periods. Upon defecation, we immediately placed the feces in 95% ethanol and stored them at room temperature, but out of direct light. After a maximum of four months, we transported the feces to the Wisconsin Primate Center and stored them at $-20°$C.

To prepare the samples for immunoassay analysis, we evaporated the ethanol from the feces and then lyophilized them to remove any remaining liquid. Next we pulverized the dried material into powder and weighed out 0.1 g aliquots. To extract the steroids from the powder we used 5 ml of a 50:50 distilled water:ethanol solution (Strier and Ziegler, 1997). We then prepared 20% of this extract using a diethyl ether extraction technique (Strier and Zeigler, 1997).

During the assay validation, a sequential hydrolysis/solvolysis experiment (Ziegler *et al.*, 1996) revealed that 90% of the estradiol (E2) and 87% of the pregnanediol (Pd) in *Cercopithecus mitis* feces is found in its free form. Based on these results, we did not consider it necessary to liberate these steroids from their conjugates prior to immunoassay analysis. Also during the assay validation, we used celite chromatography (Ziegler *et al.*, 1989) to separate the steroids into discrete fractions. Values for duplicate samples that had and had not gone through chromatography were significantly correlated ($r^2 = 0.89$, df $= 58$, $p < 0.0001$). Accordingly, we considered celite chromatography unnecessary.

To measure E2 we used the radioimmunoassay (RIA) techniques reported in French *et al.* (1983). We used 30 μl aliquots of the diethyl ether extract. Slopes of the lines created from serial dilution of the standards and from the fecal pool extract did not differ ($t = -0.35$, $p > 0.05$). We assessed accuracy by adding 15 μl of fecal pool extract to the standards and found a mean of $102.2 \pm 1.4\%$. Low pool intra- and interassay coefficients of variation were 2.7 and 19.1%, respectively. High pool intra- and interassay coefficients of variation were 4.2 and 9.1%, respectively. Mean recoveries of tritiated E2 added to the fecal pool were $92.7 \pm 1.8\%$ ($n = 13$).

To measure Pd we used the general enzyme immunoassay (EIA) techniques described in Munro and Stabenfeldt (1984), which we modified following Carlson *et al.* (1996). We used 12 μl aliquots of the diethyl ether extract. To control for non-steroid substances that prevented serial dilutions of the fecal pool from exhibiting parallelism with the standards, we added charcoal-stripped fecal extract (12 μl) to the standard curve points (Strier and Ziegler, 1997). The slopes of the lines created from serial dilution of the charcoal treated standards and from the fecal pool extract did not differ ($t = 1.34$, $p > 0.05$). We assessed accuracy by adding 10 μl of fecal pool extract to the standards and found a mean of $99.3 \pm 3.7\%$. Low pool intra- and interassay coefficients of variation were 8.7 and 16.9%, respectively. High pool intra- and interassay coefficients of variation were 2.6 and 15.6%, respectively. Mean recoveries of tritiated Pd added to the fecal pool were $92.1 \pm 2.7\%$ ($n = 6$).

Interpretation of Fecal Steroid Profiles

To estimate the timing of ovulation, we used an E2 peak $\geqslant 40$ ng/gm feces which occurred when Pd levels were <6 μg/gm feces. We considered conception to have taken place if, over several samples, Pd levels subsequently rose to >6 μg/gm, and E2 levels secondarily rose to >40 ng/gm. We looked for these patterns in the hormonal profiles of individual females with fecal data available from day -200 to -120 before parturition ($n = 4$), and also in a composite profile of females that conceived ($n = 8$). For the composite profile, we divided the data for each subject into 1-week blocks and calculated a weekly average. We then used the averages to calculate a mean and standard error for all females with data in a given week. Because fecal collection began after conception for some females, we aligned the data back from the parturition date, rather than the E2 peak of conception.

To establish a range of hormonal values within which a female can be considered cycling or non-cycling, we compared mean and peak E2 and Pd concentrations among the females most likely to be in the respective reproductive states. In choosing samples for the cycling female group ($n = 4$ females), we used the sample containing the E2 peak of conception and counted back 30 days. For the non-cycling female group ($n = 5$ females), we used samples from sexually inactive females collected at least 1 month before the start of the typical breeding season.

To assess whether females cycled more than one time before conception, we further examined the hormonal profiles of the subjects in the cycling group. For this analysis we took the samples collected from day -60 to -30 before the E2 peak of ovulation. We compared these samples to ones collected during the month of conception, and also to ones collected outside of the breeding season from the subjects in the non-cycling group.

Results

The histogram representing copulating females in relation to their parturition date revealed a large peak in mating with a modal value of 11 females on day -176 before parturition (Fig. 1). Of the females that did not mate on day -176, six copulated either the day before or the day after. The remaining four females copulated either on day -180, -178, -174, or -171. Accordingly, we considered 176 days a good approximation of the typical

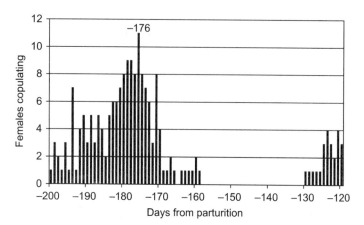

Fig. 1. Females copulating relative to their date of parturition.

gestation length for *Cercopithecus mitis*. Using this value as the best available estimate of the mean, 95% of all pregnancies can be expected to last between 162 and 190 days.

Fecal data were available for four females [Fig. 2(a–d)] over the entire period from day −190 to −162 before parturition. For three of them (Wavy, Angle, and AnG) [Fig. 2(a–c)], we identified an E2 peak ≥40 ng/gm at a time when Pd levels were <6 μg/gm. This peak occurred in conjunction with copulations on day −170, −175, and −178 before parturition (for Wavy, Angle, and AnG, respectively), and presumably marks ovulation. For all three females, the E2 peak was also followed shortly thereafter by a sustained rise in Pd to >6 μg/gm and a secondary increase in E2 to >40 ng/gm. The sustained rise in both hormones over several successive samples following the initial E2 peak presumably marks conception.

For the fourth female [Kitty; Fig. 2(d)], an E2 peak during a period when Pd levels were <6 μg/gm occurred on day −185 and −176 before parturition. Although the E2 peak on day −185 was ≥40 ng/gm, we did not consider it a likely marker of ovulation: we observed no copulations, and Pd levels had fallen to <6 μg/gm just 9 days prior (versus 28–39 days for the other females). The E2 peak on day −176 was only 26.9 ng/gm. However, this peak occurred in conjunction with copulations and was followed shortly thereafter by a rise in Pd to >6 μg/gm and a secondary increase in E2 to >40 ng/gm. Thus we considered the E2 peak on day −176 the most likely marker of conception, though we are uncertain.

The composite profile (Fig. 3) revealed an initial E2 peak at week −26 before parturition (day −182 to −175). That it occurred when Pd levels were low suggests ovulation. Following this peak there was also a sustained rise in Pd to >6 μg/gm and a secondary increase in E2 to >40 ng/gm. The sustained

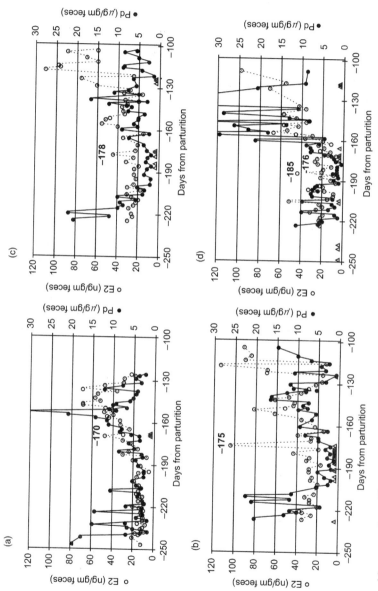

Fig. 2. Fecal hormonal profiles for four females in relation to their date of parturition. Triangles denote days on which copulations occurred. (a) Fecal profile for Wavy from day −247 to −126. (b) Fecal profile for Angle from day −227 to −105. (c) Fecal profile for AnG from day −225 to −104. (d) Fecal profile for Kitty from day −224 to −109.

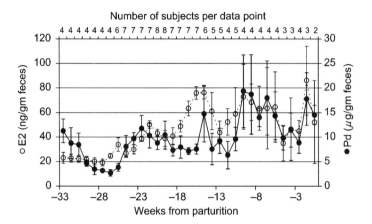

Fig. 3. Composite fecal hormonal profile.

rise in both hormones suggests conception. Hence, the composite data suggest a gestation length similar to that which we found among individual subjects and which also concurs with our estimate based on copulations observed in relation to parturition dates.

The comparison of hormone levels among subjects in the cycling and the non-cycling groups revealed that the cycling females had elevated E2 concentrations (Mann–Whitney U Test, two-tailed, $p = 0.016$; $\bar{x}_{cycling} = 25.2 \pm 2.9$ ng/gm, $n = 4$; $\bar{x}_{non-cycling} = 14.6 \pm 1.3$ ng/gm, $n = 5$). Moreover, reflecting the mid-cycle ovulatory peak, subjects in the cycling group had much higher peak E2 values than subjects in the non-cycling group (Mann–Whitney U Test, two-tailed, $p = 0.016$; $\bar{x}_{cycling} = 62.0 \pm 14.0$ ng/gm, $n = 4$; $\bar{x}_{non-cycling} = 23.1 \pm 2.2$ ng/gm, $n = 5$). An examination of the mean and peak E2 values among individual subjects (Table I) revealed no overlap in values between females in the cycling and the non-cycling groups. This pattern of non-overlapping values allowed us to establish a range within which a female can be classified unambiguously as cycling (peak E2 $\geqslant 44.5$ ng/gm; mean E2 $\geqslant 20.7$ ng/gm) or non-cycling (peak E2 $\leqslant 26.0$ ng/gm; mean E2 $\leqslant 16.7$ ng/gm).

In contrast to E2, the comparison of Pd concentrations revealed that subjects in the cycling group had lower mean levels than subjects in the non-cycling group (Mann–Whitney U Test, two-tailed, $p = 0.016$; $\bar{x}_{cycling} = 3.5 \pm 0.3$ µg/gm, $n = 4$; $\bar{x}_{non-cycling} = 8.9 \pm 2.0$ µg/gm, $n = 5$). Females in the cycling group also had somewhat lower peak Pd values, but this difference was not significant (Mann–Whitney U Test, two-tailed, $p = 0.111$; $\bar{x}_{cycling} = 8.3 \pm 1.1$ µg/gm, $n = 4$; $\bar{x}_{non-cycling} = 14.0 \pm 1.9$ µg/gm, $n = 5$). An examination of the mean Pd levels among individual subjects (Table I) revealed no overlap in values between females in the cycling and the non-cycling groups. This pattern of non-overlapping values again allowed us to establish a range within which a

Table I. Mean and Peak Hormonal Concentrations for Subjects Used to Establish a Definition of a Cycling and a Non-Cycling Female

Female	Condition	E2 (ng/gm feces)		Pd (μg/gm feces)	
		Mean	Peak	Mean	Peak
AnG	Cycling	26.4	**44.5**	3.5	9.9
Angle	Cycling	32.8	104.0	2.8	5.2
Kitty	Cycling	**20.7**	51.4	3.8	9.7
Wavy	Cycling	**20.7**	48.0	**4.0**	8.5
Gaunt	Non-cycling	15.6	**26.0**	8.2	12.5
Kamba	Non-cycling	**16.7**	24.6	8.2	13.7
SF	Non-cycling	16.2	24.7	16.5	18.5
Spanner	Non-cycling	9.8	14.6	**4.7**	8.0
Wavy	Non-cycling	14.9	25.7	6.8	17.3

Values that define the upper and lower limits for a given reproductive condition are in bold. For E2, values in bold represent the low end of the established range for cycling females and the high end of the established range for non-cycling females. For Pd, values in bold represent the high end of the established range for cycling females and the low end of the established range for non-cycling females.

female can be classified unambiguously as cycling (mean Pd ≤ 4.0 μg/gm) or non-cycling (mean Pd ≥ 4.7 μg/gm). The slight overlap in peak Pd values prevented us from establishing such a range with this measure.

Hormonal levels for subjects in the cycling group during the month before conception (day −60 to −30 before the E2 peak) indicated that none of the females was clearly cycling (Table II). One female (Wavy) had E2 and Pd values well within the established range for a non-cycling female. The remaining females (AnG, Angle, and Kitty) also had Pd values within the established range for a non-cycling female, though their E2 levels were elevated.

Table II. Mean and Peak Hormonal Concentrations for Cycling Females during the Period from Day −60 to −30 before the E2 Peak of Conception

Female	E2 (ng/gm feces)				Pd (μg/gm feces)	
	Mean	Classification	Peak	Classification	Mean	Classification
AnG	24.8	C (E2 > 20.7)	30.4	I (26.0 < E2 < 44.5)	10.4	N (Pd > 4.7)
Angle	29.3	C (E2 > 20.7)	37.0	I (26.0 < E2 < 44.5)	13.9	N (Pd > 4.7)
Kitty	25.3	C (E2 > 20.7)	51.4	C (E2 > 44.5)	5.2	N (Pd > 4.7)
Wavy	13.3	N (E2 < 16.7)	20.0	N (E2 < 26.0)	4.7	N (Pd > 4.7)

Abbreviations: C = value falls within the established range for a cycling female; N = value falls within the established range for a non-cycling female; I = value falls between the established ranges for cycling and non-cycling females.

Discussion

Our findings suggest that fecal steroid assays can be used reliably to assess reproductive function in *Cercopithecus mitis*, and they offer a number of insights into guenon reproduction. Based on mating and birth records from 21 females, we determined that 95% of all pregnancies in *Cercopithecus mitis* can be expected to last between 162 and 190 days. This estimate is somewhat longer than the 140-day period that has been calculated previously for *Cercopithecus mitis* (Rowell, 1970; Bramblett *et al.*, 1975). However, the old estimate is based on only five females, all of which were maintained in captivity where the relaxation of nutritional constraints may result in shorter gestation periods (Riopelle and Hale, 1975; Silk, 1986; Borries *et al.*, 2001). Moreover, among the other members of the Cercopithecini, gestation length has been calculated between 165 and 213 days (Ardito, 1976; Butynski, 1988). Hence, with a 140-day gestation period, *Cercopithecus mitis* would be a clear outlier, with no discernible reason for the difference.

Within the period from day -162 to -190 before parturition, we found a hormonal pattern consistent with ovulation and conception in the profiles of three out of four individual females and in a composite profile composed of data from eight females. That we could not identify ovulation in every case illustrates the limitations of fecal sampling. These limitations may be especially great among the forest guenons because of their diverse and variable diets that contain a high percentage of high-fiber foods, especially during times of low fruit availability (Lambert, 2002a). Dietary fiber accelerates gut passage rate (Goldin *et al.*, 1982; Milton and Demment, 1988; Maisels, 1993) and increases overall fecal mass (Goldin *et al.*, 1982), thereby altering the lag-time to excretion and the final concentration of hormones eliminated in the feces. Although the extent of these alterations can be assessed experimentally, estimates obtained from captivity must be applied with caution to wild guenons because of the considerable variation within and between subjects and populations in the foods consumed (Cords, 1986; Gautier-Hion, 1988; Chapman *et al.*, 2002; Lambert, 2002b).

Although caution must be taken when interpreting values from individual fecal samples, we identified clear differences in the overall hormonal profiles of cycling and non-cycling females. The complete lack of overlap in mean E2 and Pd as well as peak E2 values during the month of conception versus non-breeding periods allowed us to establish a range within which a female can be classified unambiguously as cycling or non-cycling. Thus, fecal steroid assays can provide a powerful tool in *Cercopithecus mitis* for evaluating ovarian function over a given time period. Indeed, an examination of the overall profiles suggests that some unusual features may characterize the reproductive biology of *Cercopithecus mitis*. Many prior endocrine studies on seasonally breeding primates have shown that females have low and unvarying estrogen and progesterone levels during acyclic periods (*Brachyteles arachnoides*: Strier and Ziegler, 1994, 1997; *Macaca fuscata*: Nozaki and Oshima, 1987; *M. mulatta*: Walker *et al.*, 1984; *Papio cynocephalus*: Wasser *et al.*, 1988; Stavisky *et al.*, 1995; *Presbytis entellus*: Ziegler *et al.*, 2000; *Propithecus verreauxi*: Brockman and Whitten, 1996). Hence, although the low E2 levels we found outside the mating season are consistent with previous knowledge of female reproductive endocrinology, the high Pd levels we found are anomalous. Nonetheless, high levels of progesterone during the luteal phase of the ovarian cycle block the positive feedback effects of E2 that are responsible for the mid-cycle LH surge, and they also enhance the negative feedback effects that E2 has over most of the ovarian cycle (Hotchkiss and Knobil, 1994). Therefore, high progesterone levels would be expected to provide a powerful block on ovulation outside seasonal breeding periods as well as during the luteal phase of the ovarian cycle.

A system for maintaining anovulatory periods with high progesterone levels is quite distinct from that seen in ewes (*Ovis* spp.) and rhesus macaques (*Macaca mulatta*). In these well-studied species, seasonal anovulation is maintained by a dramatic increase in hypothalamic sensitivity to the negative feedback effects of E2. Consequently, even small amounts of E2 suppress LH secretion (Legan *et al.*, 1977; Karsch *et al.*, 1984; Wilson *et al.*, 1987). In ewes, it has been well documented that this heightened sensitivity is regulated by photoperiod (Karsch *et al.*, 1984; Goodman, 1994). In rhesus monkeys, the importance of day-length is strongly suggested by experimental manipulations in which exposure to light for only a short period each day, or the administration of melatonin to mimic this pattern, has led to the activation of reproductive function in adult males and prepubertal females (Wilson and Gordon, 1989; Chik *et al.*, 1992). Because guenons are primarily equatorial (Lernould, 1988), it is possible that a constant day-length forced the ancestral guenon to evolve an alternative mechanism to regulate seasonal reproduction. However, it is tempting to speculate that a system driven by high progesterone levels has certain adaptive advantages. Although high luteal phase progesterone levels inhibit the positive and enhance the negative feedback effects of E2, prior exposure to progesterone may have an important priming effect that is necessary for the full expression of the mid-cycle LH surge (Hotchkiss and Knobil, 1994; Caraty and Skinner, 1999). Hence, the maintenance of high progesterone levels followed by a rapid decline at the start of annual breeding periods may result in an LH surge sufficient to induce ovulation on the first cycle of the season. Given that the timing of mating and births among many of the guenons appears to be related to rainfall and its concomitant effects on food availability (Butynski, 1988), a system driven by high progesterone levels could be advantageous in allowing females to respond quickly to annual differences in precipitation patterns.

Our findings suggest that conception in *Cercopithecus mitis* may, in fact, typically occur on the first ovulation of the breeding season. All four females with complete fecal records from day -60 to -30 before conception continued to have Pd levels well within the range established for a non-cycling female. These high Pd levels most likely blocked ovulation. Nonetheless, three of the females (Angle, AnG, Kitty; Table II) had heightened E2 levels at the time. Because E2 enhances female sexual motivation (Wallen, 2000), one might speculate that a rise in E2 before the decline in Pd levels may serve as a mechanism to promote sexual behavior at a time when conception is unlikely. Indeed, two of three females with elevated E2 levels [Angle and Kitty; Fig. 2(a) and (d)] copulated at this time.

Hrdy (1979) originally developed the hypothesis that mating at times when conception is unlikely can function to confuse paternity. However, in her original formulation of the hypothesis, Hrdy focused primarily on postconception mating and the extension of sexually active periods. In a recent resurgence of the discussion of her hypothesis, it has also been recognized that females could potentially manipulate paternity assessment by mating before their conception cycle (van Schaik *et al.*, 1999, 2000; van Noordwijk and van Schaik, 2000). Based on our findings, one might speculate that females have actually evolved hormonal mechanisms to promote the early onset of sexual behavior. However, to fully support this hypothesis, we must confirm that the patterns we observed are in fact typical by collecting endocrine data on a much larger sample of females. Moreover, it will be necessary to determine the mechanism by which high progesterone levels are maintained during seasonal anovulatory periods and whether similar endocrine systems are present in other species in which females are vulnerable to infanticide.

Summary

In this paper we demonstrate that fecal steroid assays can be used to monitor female reproductive function in wild blue monkeys. Based on mating and birth records from 21 females, we determined that 95% of all pregnancies last between 162 and 190 days. Within the period when conception is likely we identified a clear estrogen peak marking ovulation for three out of four subjects with complete fecal records. We also examined hormonal levels among females of known reproductive state and established a range of values within which a female can be classified unambiguously as cycling or non-cycling. Our data tentatively suggest that blue monkeys typically ovulate just once before conception. Nonetheless, estrogen levels seem to rise before the conception cycle and may promote the early onset of sexual behavior for purposes, such as paternity confusion, that are not directly related to reproduction.

ACKNOWLEDGMENTS

We are grateful to the Government of Kenya for permission to conduct research in the Kakamega Forest and to the Institute for Primate Research (National Museums of Kenya) for local sponsorship. We thank Praxides Akelo and Benjamin Okalo for their help with the fecal sample collection, and Guenther Scheffler and Dan Wittwer for their technical assistance with the fecal analysis. We are grateful to Marina Cords and her many students for helping us to monitor the study groups for sexual behavior. Marina Cords, Mary Glenn, and Russell Tuttle provided helpful suggestions on earlier versions of this manuscript. This research was funded by the Anthropology Department of the University of Pennsylvania, the Wenner Gren Foundation, the L. S. B. Leakey Foundation, two National Science Foundation grants to Marina Cords (SBR 95-23623, BCS 98-08273), and the Wisconsin Regional Primate Research Center (NIH grant RR00167). This is publication #41-010 of the WRPRC.

References

Ardito, G. 1976. Check-list of the data on the gestation length of primates. *J. Hum. Evol.* **5**:213–222.

Borries, C., Koenig, A., and Winkler, P. 2001. Variation of life history traits and mating patterns in female langur monkeys (*Semnopithecus entellus*). *Behav. Ecol. Sociobiol.* **50**:391–402.

Bramblett, C. A., Pejaver, L. D., and Drickman, D. J. 1975. Reproduction in captive vervet and Sykes' monkeys. *J. Mammal.* **56**:940–946.

Brockman, D. K., and Whitten, P. L. 1996. Reproduction in free-ranging *Propithecus verreauxi*: Estrus and the relationship between multiple partner matings and fertilization. *Am. J. Phys. Anthropol.* **100**:57–69.

Butynski, T. M. 1988. Guenon birth seasons and correlates with rainfall and food. In: A. Gautier-Hion, F. Bourlière, J.-P. Gautier, and J. Kingdon (eds.) *A Primate Radiation: Evolutionary Biology of the African Guenons*, pp. 284–322. Cambridge University Press, Cambridge.

Caraty, A., and Skinner, D. C. 1999. Progesterone priming is essential for the full expression of the positive feedback effect of estradiol in inducing the preovulatory gonadotropin-releasing hormone surge in the ewe. *Endocrinology* **140**:165–170.

Carlson, A. A. 2000. *Social Relationships and Mating Patterns of Patas Monkeys in Laikipia, Kenya.* Ph.D. Dissertation. University of Wisconsin, Madison.

Carlson, A. A., Ginther, A. J., Scheffler, G. R., and Snowdon, C. T. 1996. The effects of infant births on the sociosexual behavior and hormonal patterns of a cooperatively breeding primate (*Cebuella pygmaea*). *Am. J. Primatol.* **40**:23–39.

Carlson, A. A., and Isbell, L. Y. (2001). Causes and consequences of single- and multi-male mating in free-ranging patas monkeys (*Erythrocebus patas*). *Anim. Behav.* **62**; 1047–1058.

Chapman, C. A., Chapman, L. J., Cords, M., Gathua, J. M., Gautier-Hion, A., Lambert, J. E., Rode, K., Tutin, C. E. G., and White, L. J. T. 2002. Variation in the diets of *Cercopithecus* species: Differences within forests, among forests, and across species. In: M. E., Glenn, and M. Cords (eds.), *The Guenons: Diversity and Adaptation in African Monkeys*, pp. 325–350. Kluwer Academic Publishers, New York.

Chik, C. L., Almeida, O. F., Libre, E. A., Booth, J. D., Renquist, D., and Merriam, G. R. 1992. Photoperiod-driven changes in reproductive function in male rhesus monkeys. *J. Clin. Endocrinol. Metab.* **75**:1068–1074.

Chism, J. B., and Rowell, T. E. 1986. Mating and residence patterns of male patas monkeys. *Ethology* **72**:31–39.

Cords, M. 1984. Mating patterns and social structure in redtail monkeys (*Cercopithecus ascanius*). *Z. Tierpsychol.* **64**:313–329.

Cords, M. 1987. Mixed-species association of *Cercopithecus* monkeys in the Kakamega Forest, Kenya. *University of California: Publications in Zoology* **117**:1–109.

Cords, M. 1986. Interspecific and intraspecific variation in diet of two forest guenons, *Cercopithecus ascanius* and *C. mitis*. *J. Anim. Ecol.* **55**:811–827.

Cords, M., Mitchell, B. J., Tsingalia, H. M., and Rowell, T. E. 1986. Promiscuous mating among blue monkeys in the Kakamega Forest, Kenya. *Ethology* **72**:214–226.

Cross, J. F., and Martin, R. D. 1981. Calculation of gestation period and other reproductive parameters for primates. *Dodo, J. Jersey Wildl. Preserv. Trust* **18**:30–43.

Fairgrieve, C. 1995. Infanticide and infant eating in the blue monkey (*Cercopithecus mitis stuhlmanni*) in the Budongo Forest Reserve, Uganda. *Folia Primatol.* **64**:69–72.

French, J. A., Abbott, D. H., Scheffler, G., Robinson, J. A., and Goy, R. W. 1983. Cyclic excretion of urinary oestrogens in female tamarins (*Saguinus oedipus*). *J. Reprod. Fertil.* **68**: 177–184.

Gautier-Hion, A. 1988. The diet and dietary habits of forest guenons. In: A. Gautier-Hion, F. Bourlière, J.-P. Gautier, and J. Kingdon (eds.), *A Primate Radiation: Evolutionary Biology of the African Guenons*, pp. 257–283. Cambridge University Press, New York.

Goldin, B. R., Adlercreutz, H., Gorbach, S. L., Warram, J. H., Dwyer, J. T., Swenson, L., and Woods, M. N. 1982. Estrogen excretion patterns and plasma levels in vegetarian and omnivorous women. *N. Engl. J. Med.* **307**:1542–1547.

Goodman, R. L. 1994. Neuroendocrine control of the ovine estrous cycle. In: E. Knobil, and J. D. Neill (eds.), *The Physiology of Reproduction*, pp. 659–709. Raven Press, New York.

Hotchkiss, J., and Knobil, E. 1994. The menstrual cycle and its neuroendocrine control. In: E. Knobil, and J. D. Neill (eds.), *The Physiology of Reproduction*, pp. 711–749. Raven Press, New York.

Hrdy, S. B. 1979. Infanticide among animals: A review, classification, and examination of the implications for the reproductive strategies of females. *Ethol. Sociobiol.* **1**:13–40.

Hrdy, S. B., and Whitten, P. L. 1987. Patterning of sexual activity. In: B. B. Smuts, D. L. Cheney, R. M. Seyfarth, R. W. Wrangham, and T. T. Struhsaker (eds.), *Primate Societies*, pp. 370–384. University of Chicago Press, Chicago.

Karsch, F. J., Bittman, E. L., Foster, D. L., Goodman, R. L., Legan, S. J., and Robinson, J. E. 1984. Neuroendocrine basis of seasonal reproduction. *Recent Prog. Horm. Res.* **40**:185–225.

Lambert, J. E. 2002a. Digestive retention times in forest guenons with reference to chimpanzees. *Int. J. Primat.* **23** (6).

Lambert, J. E. 2002b. Resource switching and species coexistence in guenons: A community analysis of dietary flexibility. In: M. E. Glenn, and M. Cords (eds.), *The Guenons: Diversity and Adaptation in African Monkeys*, pp. 309–323. Kluwer Academic Publishers, New York.

Legan, S. J., Karsch, F. J., and Foster, D. L. 1977. The endocrine control of seasonal reproductive function in the ewe: A marked change in response to the negative feedback action of estradiol on luteinizing hormone secretion. *Endocrinology* **101**:818–824.

Lernould, J. 1988. Classification and geographical distribution of guenons: A review. In: A. Gautier-Hion, F. Bourlière, J.-P. Gautier, and J. Kingdon (eds.), *A Primate Radiation: Evolutionary Biology of the African Guenons*, pp. 54–78. Cambridge University Press, New York.

Loy, J. 1981. The reproductive and heterosexual behaviours of adult patas monkeys in captivity. *Anim. Behav.* **29**:714–726.

Macleod, M. C. 2000. *The Reproductive Strategies of Samango Monkeys* (Cercopithecus mitis erythrarchus). Ph.D. Dissertation, University of Surrey, Roehampton.

Maisels, F. 1993. Gut passage rate in guenons and mangabeys: Another indicator of a flexible feeding niche? *Folia Primatol.* **61**:35–37.

Martin, R. D. 1992. Female cycles in relation to paternity in primate societies. In: R. D. Martin, A. F. Dixson, and E. J. Wickings (eds.), *Paternity in Primates: Genetic Tests and Theories*, pp. 238–274. Karger, Basel.

Milton, K., and Demment, M. W. 1988. Digestion and passage kinetics of chimpanzees fed high and low fiber diets and comparison with human data. *J. Nutr.* **118**:1082–1088.

Munro, C., and Stabenfeldt, G. 1984. Development of a microtitre plate enzyme immunoassay for the determination of progesterone. *J. Endocrinol.* **101**:41–49.

Nozaki, M., and Oshima, K. 1987. Seasonal changes of the gonadotropic function in the female Japanese monkey. *Prog. Biometeorol* **5**:41–49.

Pazol, K. 2001. *Social, Ecological, and Endocrine Influences on Female Relationships in Blue Monkeys* (Cercopithecus mitis stuhlmanni). Ph.D. Dissertation, University of Pennsylvania, Philadelphia.

Pazol, K., and Ziegler, T. E. 2000. Mating in the absence of ovarian activity following a resident male turnover in a wild blue monkey group, Kakamega, Kenya. *Am. J. Primatol.* **51** (Suppl. 1): 79.

Riopelle, A. J., and Hale, P. A. 1975. Nutritional and environmental factors affecting gestation length in rhesus monkeys. *Am. J. Clin. Nutr.* **28**:1170–1176.

Rowell, T. E. 1970. Reproductive cycles of two *Cercopithecus* monkeys. *J. Reprod. Fertil.* **22**:321–338.

Silk, J. B. 1986. Eating for two: Behavioral and environmental correlates of gestation length among free-ranging baboons (*Papio cynocephalus*). *Int. J. Primatol.* **7**:583–602.

Sommer, V., Srivastava, A., and Borries, C. 1992. Cycles, sexuality, and conception in free-ranging langurs (*Presbytis entellus*). *Am. J. Primatol.* **28**:1–27.

Stavisky, R., Russell, E., Stallings, J., Smith, E. O., Worthman, C., and Whitten, P. L. 1995. Fecal steroid analysis of ovarian cycles in free-ranging baboons. *Am. J. Primatol.* **36**:285–297.

Strier, K. B., and Ziegler, T. E. 1994. Insights into ovarian function in wild muriqui monkeys (*Brachyteles arachnoides*). *Am. J. Primatol.* **32**:31–40.

Strier, K. B., and Ziegler, T. E. 1997. Behavioral and endocrine characteristics of the reproductive cycle in wild muriqui monkeys, *Brachyteles arachnoides*. *Am. J. Primatol.* **42**:299–310.

Thomas, D. H. 1986. *Refiguring Anthropology: First Principles of Probability and Statistics*. Waveland Press, Inc., Prospect Heights.

Tsingalia, H. M., and Rowell, T. E. 1984. The behaviour of adult male blue monkeys. *Z. Tierpsychol.* **64**:253–268.

van Noordwijk, M. A., and van Schaik, C. P. 2000. Reproductive patterns in eutherian mammals: Adaptations against infanticide? In: C. P. van Schaik, and C. H. Janson (eds.), *Infanticide by Males and Its Implications*, pp. 322–360. Cambridge University Press, Cambridge.

van Schaik, C. P., Hodges, J. K., and Nunn, C. L. 2000. Paternity confusion and the ovarian cycles of female primates. In: C. P. van Schaik, and C. H. Janson (eds.), *Infanticide by Males and Its Implications*, pp. 361–387. Cambridge University Press, Cambridge.

van Schaik, C. P., van Noordwijk, M. A., and Nunn, C. L. 1999. Sex and social evolution in primates. In: P. C. Lee (ed.), *Comparative Primate Socioecology*, pp. 204–231. Cambridge University Press, Cambridge.

Walker, M. L., Wilson, M. E., and Gordon, T. P. 1984. Endocrine control of the seasonal occurrence of ovulation in rhesus monkeys housed outdoors. *Endocrinology* **114**:1074–1081.

Wallen, K. 2000. Risky business: Social context and hormonal modulation of primate sexual desire. In: K. Wallen, and J. E. Schneider (eds.), *Reproduction in Context: Social and Environmental Influences on Reproductive Physiology and Behavior*, pp. 289–323. The MIT Press, Cambridge.

Wasser, S. K., Risler, L., and Steiner, R. A. 1988. Excreted steroids in primate feces over the menstrual cycle and pregnancy. *Biol. Reprod.* **39**:862–872.

Wilson, M. E., and Gordon, T. P. 1989. Short-day melatonin pattern advances puberty in seasonally breeding rhesus monkeys (*Macaca mulatta*). *J. Reprod. Fertil.* **86**:435–444.

Wilson, M. E., Pope, N. S., and Gordon, T. P. 1987. Seasonal modulation of luteinizing-hormone secretion in female rhesus monkeys. *Biol. Reprod.* **36**:975–984.

Ziegler, T., Hodges, K., Winkler, P., and Heistermann, M. 2000. Hormonal correlates of reproductive seasonality in wild female Hanuman langurs (*Presbytis entellus*). *Am. J. Primatol.* **51**:119–134.

Ziegler, T. E., Scheffler, G., Wittwer, D. J., Schultz-Darken, N., Snowdon, C. T., and Abbott, D. H. 1996. Metabolism of reproductive steroids during the ovarian cycle in two species of callitrichids, *Saguinus oedipus* and *Callithrix jacchus*, and estimation of the ovulatory period from fecal steroids. *Biol. Reprod.* **54**:91–99.

Ziegler, T. E., Sholl, S. A., Scheffler, G., Haggerty, M. A., and Lasley, B. L. 1989. Excretion of estrone, estradiol, and progesterone in the urine and feces of the female cotton-top tamarin (*Saguinus oedipus oedipus*). *Am. J. Primatol.* **17**:185–195.

Grooming and Social Cohesion in Patas Monkeys and Other Guenons

17

JANICE CHISM and WILLIAM ROGERS

Introduction

Patas monkeys (*Erythrocebus patas*) and vervet monkeys (*Cercopithecus aethiops*) are phylogenetically closely allied with the forest guenons. Their ecologies, however, have been regarded as setting them apart from their forest-living relatives. In addition to ecological differences, the social organization of vervet monkeys differs from those of most forest guenons in that they have relatively rigid matrilineally based dominance hierarchies (Gouzoules and Gouzoules, 1986). Over the last two decades, several field studies of patas monkeys, and more detailed information on a few forest guenons, have shown that in many respects, patas monkeys closely resemble their guenon relatives in patterns of female–female social relations (Rowell *et al.*, 1991; Chism, 1999a, 2000; Cords, 2000; Carlson and Isbell, in prep.). For example, a feature of social organization common to patas monkeys and most of the guenons is that females have less rigidly hierarchical dominance relationships than those of most other cercopithecines, notably baboons, macaques, and vervet monkeys (Rowell and Olson, 1983; Isbell and Pruetz, 1998; Chism, 2000; Cords, 2000).

JANICE CHISM and WILLIAM ROGERS • Department of Biology, Winthrop University, Rock Hill, SC 29733, USA.
The Guenons: Diversity and Adaptation in African Monkeys, edited by Glenn and Cords. Kluwer Academic/Plenum Publishers, New York, 2002.

Like most guenons, patas monkeys are female philopatric and live in what Wrangham (1980, 1987) described as female-bonded groups. More recently, this form of social organization has been classified as Resident-Egalitarian (Sterck *et al.*, 1997). Females typically remain in their natal groups throughout their lives while males disperse at or near puberty (Chism *et al.*, 1984; Chism, 1999a). Throughout most of the year, patas monkey groups typically contain a single adult male with multimale influxes during some annual short conception periods (Chism and Rowell, 1986; Harding and Olson, 1986; Ohsawa *et al.*, 1993; Chism and Rogers, 1997; Carlson and Isbell, 2001). Adult males typically have short tenures in groups; at our study site these ranged from a few days to nine months and averaged just three months (Chism and Rowell, 1986). Thus, patas monkeys have retained a social organization typical of the forest guenons although they occupy savanna habitats more similar to those of the baboons and vervet monkeys with their multimale groups. In the absence of additional males, females must assume more of the responsibility for resource defense and predator protection. In consequence, patas females need to develop and maintain cohesive relationships that allow them to act together to defend themselves, their offspring, and the group's resources.

Patas females and juveniles studied at ADC Mutara Ranch in Kenya's Laikipia District typically initiated and were active participants in the frequent intergroup encounters while males often acted as bystanders, taking active roles only during conception periods (Chism, 1999b). Females cooperate in caring for young infants (Chism, 1986, 2000) and in defending immatures of the group against aggression by adult males (Chism and Rowell, 1986; Chism and Rogers, 1997; Carlson, 2000) and predator attacks (Chism *et al.*, 1984; Carlson, 2000). At ADC Mutara, females as well as males acted as lookouts when the group drank at water tanks or crossed areas of open grassland (Chism and Rowell, 1988).

Dunbar (1988) and others have suggested that grooming is a major contributor to social bond-formation among primates and that such social bonds facilitate many other kinds of interactions. Given the central roles of females in all aspects of patas monkey society combined with the lack of matrilineally based dominance hierarchies, patas females would seem to have great need of strong social bonds. In this respect, patas monkeys appear to resemble other guenons closely. Although we do not have detailed information on the social relations of many guenons living under natural conditions, for most of those species that have been studied, females appear to have similarly central roles in group coordination and defense (Rowell *et al.*, 1991; Cords, 2002). This suggests that guenon females need a mechanism, some form of social glue, to help develop and maintain close relationships. Grooming seems the obvious candidate for such a mechanism because (1) it provides a hygienic benefit to the receiver, (2) it may provide psychological benefits to both giver and receiver, and (3) its energetic costs are low. We tested this idea with a series of predictions about the nature of grooming relations among patas females.

First, we examined the effect of group size on female grooming patterns. Dunbar's (1991) analysis of grooming data from a large number of primate species indicated that grooming (measured as percent of observation time) increased as group size increased. He suggested that primates in larger groups have a greater social load that they support via increased grooming. As groups get larger, tensions among members may increase leading individuals to have more need of alliances to mitigate aggression or feeding interference. We tested Dunbar's prediction that there would be more grooming in larger groups by comparing grooming frequency and rates for adult female patas monkeys in a small and a large group.

To examine the effects of the need for cooperation in group and resource defense on grooming relations among patas females, we looked for changes in the amount of time spent grooming on days with intergroup encounters and during a period when a rapid loss of adult females occurred.

During intergroup encounters, patas females alternated grooming their own group members with bouts of chasing and vocalizing at the members of the other. We therefore expected time spent grooming to increase on days with encounters if females needed to reinforce social bonds so that they could act together to defend the group's resources. We predicted that females in the smaller group would be more likely to increase grooming on intergroup encounter days than would females in the large group because, with fewer females to defend their resources, we expected the smaller group's females to have more need for close cooperation and reinforcement of social bonds in these encounters. We also predicted more frequent but shorter grooming bouts on days with encounters because we thought females would frequently interrupt their grooming to monitor the proximity and activities of members of the other group.

We predicted that if grooming acts as a form of social glue, then grooming frequency might also increase when many females died over a brief period because surviving females would need to reestablish the social fabric of the group and to restructure the group's decision-making process. This would be particularly likely if, as was the case in our study group, group leaders were among those lost.

During and after the period when many females were lost, it seemed possible that surviving females would need to increase their levels of vigilance to make up for there being fewer eyes to watch out for predators and the encroachment of other patas monkey groups. For example, Cords (1995, 1997) found that grooming is associated with lowered levels of vigilance in wild blue monkeys. Females might reconcile their need to groom more to reestablish social bonds with their need to increase their vigilance by more frequently interrupting their grooming to scan the environment. While brief visual scanning need not terminate a grooming bout, patas females often stood to scan (in trees) or bipedally scanned (on the ground) resulting in longer interruptions that terminated bouts.

Methods

We observed two study groups during a 27-month field study of patas monkeys at ADC Mutara Ranch in Kenya's Laikipia District from May 1979 through July 1981. For this analysis, we used data derived from 209 hours of focal samples on 17 adult females, 126 hours on ten females in Mutara I (MI), the smaller of the two study groups and 83 hours on seven females in the larger group, Mutara II (MII). We collected focal samples in the second year of the study when subjects were well habituated and we reliably recognized individuals. MI averaged *ca.* 20 individuals and MII averaged *ca.* 50 individuals over this period. Our analysis includes observations from one complete annual ecological and reproductive cycle. We computerized the data and initially analyzed it via SAS version 7 software. We used one and two-way ANOVAs and the Student *t*-Test (Heath, 1995) to test for differences. Percentages were first arc-sin transformed to approximate a normal distribution (Sokal and Rohlf, 1981). All significant differences are at $p < 0.05$.

Our analysis focused on the following measures: time adult females spent grooming all other group members as a percent of all time observed, the percent of time females spent grooming other females only, the rate of grooming per available partner (either all other group members or other adult females only), the rate of grooming bouts per hour of observation, and the duration of grooming bouts. Interruptions in grooming that lasted at least 30 seconds were treated as terminating bouts.

To examine Dunbar's (1991) prediction about the effect of group size on grooming, we compared the percent of observed time that females in the small and large group spent grooming. Because Altmann (1999) suggested that if Dunbar's data were adjusted for different numbers of potential partners they would show that grooming was actually more frequent in smaller groups than in larger ones, we also looked at time spent grooming as a rate adjusted for numbers of possible partners.

To test our hypothesis about the effects of intergroup encounters on social bonds we examined a subset of 61.2 hours of focal sample data collected from 1 December 1980 through 20 January 1981, a period when encounters were frequent and before the loss of several females from the small study group (Table I). We selected this period for analysis because it represented a time when numbers of adult females were stable and it coincided with the time of the year in the early, long dry season when all of the two groups' females were in late pregnancy or early lactation. Thus, it was a time when access to resources should have been critical to the females' ability to reproduce successfully. In addition, this period included a sufficient number of focal samples so that we could compare grooming for most of the group's adult females on days when encounters occurred with samples for days without encounters. We chose to use days with or without encounters as our unit of analysis because at Mutara

Table I. Distribution of Observations of Grooming on Days with and without Intergroup Encounters

	Mutara I	Mutara II
Days with Intergroup Encounters	7 adult females 6 days 8.8 hours	6 adult females 7 days 13.7 hours
Days without Intergroup Encounters	9 adult females 13 days 24 hours	4 adult females 8 days 14.7 hours

intergroup encounters frequently lasted for several hours and often groups were in visual or auditory contact with each other for long periods before and after they got close enough to engage each other in chasing or threatening. Given the difficulty of defining exactly when an encounter began or ended, we did not try to compare samples taken on the same day before and after an encounter.

To test our hypothesis about the effects of loss of many adult females on grooming, we used 35.3 hours of focal samples collected over an approximately six-week period in January and February 1981 when four out of the ten adult females in MI died, apparently as a result of predation (Chism *et al.*, 1984). These data were divided into a preloss period (6 January 1981–20 January 1981), the period during which the losses occurred (21 January 1981–31 January 1981) and a postloss period (1 February 1981–13 February 1981). Table II shows the distribution of observations over these periods.

Results

Effect of Group Size

Contrary to our prediction based on Dunbar's findings (1991), grooming did not increase with group size. In fact, females in the smaller group, MI, spent more time grooming overall and grooming other adult females than did adult females in the larger group, MII. Females in MI groomed all others more frequently ($\bar{x} = 9.5\%$ of time observed) than females in MII ($\bar{x} = 3.1\%$ of time

Table II. Distribution of Observations over the Preloss, Loss, and Postloss Periods in Mutara I

	PreLoss	Loss	PostLoss
Dates	6–20 Jan.	21–31 Jan.	1–13 Feb.
Sample Sizes	6 Adult Females 5 sample days 11.4 hours	9 Adult Females 9 sample days 11.3 hours	6 Adult Females 9 sample days 12.7 hours

observed) and these differences were significant ($t = 2.570$, $t_{(15)} = 2.131$). Adult females in MI also groomed other females more ($\bar{x} = 4.94\%$ of time observed) than did females in the larger group, MII ($\bar{x} = 2.23\%$), but this difference was not significant ($t = 1.568$, $t_{(15)} = 2.131$). When we adjusted for numbers of potential partners available to females in the two groups both time spent grooming all others ($\bar{x} = 0.51\%$ for MI, $\bar{x} = 0.06\%$ for MII, $t = 4.769$, $t_{(11)} = 2.201$) and time spent grooming females only ($\bar{x} = 0.73\%$ for MI, $\bar{x} = 0.14\%$ for MII, $t = 2.250$, $t_{(15)} = 2.131$) were significantly greater for the small group also.

Effect of Intergroup Encounters

Our data did not support the prediction that adult female patas would increase time spent grooming on days with intergroup encounters. Neither group's females significantly increased the rate of grooming of other adult females (Two-way ANOVA, $F = 2.681$ for condition, $F = 2.942$ for group, $F_{(1,12)} = 6.55$) or of all group members ($F = 3.560$ for condition, $F = 0.501$ for group, $F_{(1,12)} = 6.55$) on encounter days over days when no encounters occurred.

Grooming bout length in the smaller group (MI) shifted in the predicted direction: Bouts were shorter on days with encounters than on days without encounters, but this difference failed to reach significance. Days with and without encounters did not differ significantly in grooming bout rates (Table III).

Effect of Loss of Group Females

While we expected adult females in the smaller group (MI) to increase the amount of time they spent grooming other adult females in the period following the loss of a substantial fraction of the group's adult females, we found a somewhat more complicated response. Mean rates for the amount of time

Table III. Bout Rate and Bout Duration for Days with and without Intergroup Encounters for MI

	Bout Rate	Bout Duration
Days with Encounters	$\bar{x} = 6.60$ bouts/hour	$\bar{x} = 0.59$ minute
	s. d. $= 7.93$	s. d. $= 0.26$
	$n = 7$	$n = 4$
Days without Encounters	$\bar{x} = 8.57$ bouts/hour	$\bar{x} = 0.87$ minute
	s. d. $= 9.06$	s. d. $= 0.47$
	$n = 9$	$n = 8$
t-Test Results	$t = 0.454$	$t = 1.057$
	$t_{(14)} = 2.145$	$t_{(10)} = 2.228$

Table IV. Mean Grooming Bout Rate for All Grooming during Preloss, Loss and Postloss Periods[a]

	Mean	Standard dev.	N
Pre-loss	0.16 bouts/hour	S = 0.135	5
Loss	0.42 bouts/hour	S = 0.375	9
Post-loss	0.36 bouts/hour	S = 0.472	6

ANOVA: $F = 0.811$, $F_{(2,17)} = 3.59$, n.s.
[a] Means adjusted for numbers of potential grooming partners.

Table V. Mean Grooming Bout Duration for All Grooming during Preloss, Loss and Postloss Periods

	Mean	Standard dev.	N
Pre-loss	0.47 minute	S = 0.138	4
Loss	0.61 minute	S = 0.539	9
Post-loss	0.42 minute	S = 0.238	4

ANOVA: $F = 0.389$, $F_{(2,17)} = 3.93$, n.s.

females groomed other females actually decreased from 0.16 minute/partner in the preloss period to 0.02 minute/partner in the loss period, then increased to 0.48 minute/partner in the postloss period. These differences were not significant (ANOVA, $F = 2.088$, $F_{(2,18)} = 3.55$), nor were differences in time spent grooming all group members ($F = 0.250$, $F_{(2,18)} = 3.55$).

Grooming bout rate increased between the preloss and loss periods then decreased during the postloss period, but these differences were also not significant (Table IV). Contrary to our predictions, grooming bout length increased from the preloss to the loss period, then during the postloss period, bout length returned to approximately the pre loss period level although, once more, these differences were not significant (Table V).

Discussion

Our analysis of grooming among females in two wild patas monkey groups indicates that contrary to Dunbar's (1991) findings for primates in general, females in the small group groomed more frequently than females in the larger group did, both with all group members and with other adult females only. van Schaik (1989) characterized patas monkeys as having egalitarian, individualistic relationships among females based on their ecology, which includes significant between-group competition. Patas females in small groups might well have more difficulty defending their group's resources during intergroup competitions than females in larger groups do. If grooming acts as the primary means to maintain social bonds, it might be expected that females in small groups would need to work harder at bond maintenance than females in larger groups. Since patas monkey groups in our study area had frequent intergroup encounters (Chism, 1999b), the greater frequency of grooming by females in the smaller group may be directly related to a greater need to maintain cohesion during resource defense.

Our finding that females in the small group did not increase their grooming on days where intergroup encounters occurred, however, appears to contradict this interpretation. We think the most likely explanation is that patas females have relatively inflexible time budgets, particularly during the dry season period, which overlaps with late pregnancy and early lactation, the interval we used for our analysis of the effects of intergroup encounters. Thus, females may have limited time for social interactions each day and on days with intergroup encounters they spend some of it fighting with neighbors instead of grooming. If that is the explanation one might actually expect a more dramatic decrease in grooming on days with encounters, but, in fact, there is no significant difference. In addition, the large and small study groups did not differ in grooming frequency on days with and without encounters. We believe these findings actually provide support for the role of grooming in maintaining social bonds, and that in the case of intergroup encounters at least it appears to do so without regard to group size.

In contrast, *Cercopithecus mitis*, including blue monkeys in the Kakamega Forest in Kenya (Rowell *et al.*, 1991; Cords, 2002) and samango monkeys near the southern limit of their range in South Africa, experienced frequent intergroup encounters and females groomed more often than expected during and immediately after encounters (Payne *et al.*, in prep.). Cheney (1992), too, reported intense grooming during vervet monkeys' intergroup encounters but found that the patterning of grooming (who groomed whom and at what rate) did not change.

As predicted, grooming bout rates increased, though not significantly, on days with intergroup encounters. These encounters often created apparently chaotic situations. On several occasions, predators appeared to take advantage of the monkeys' distracted states to approach them without being detected. The distraction may have negated any possible advantage to females in the larger group of having more eyes available to look for predators, which under-scores the idea that predator protection does not depend solely on the numbers of watchers but instead on their coordinated efforts.

The rapid loss of four of the small patas monkey group's adult females, the second set of circumstances in which we expected active reaffirmation of bonds, produced no significant overall increase in grooming among the surviving females generally. The two youngest adult females, however, increased time spent grooming in the period that followed the loss. (One of the young females, BG, increased her grooming from 2.42 to 3.67% of time; the second, RO, increased her grooming from 0 to 7.52% of time). The older, surviving females did not show this pattern. Although we did not know genealogical relationships of all of the adult females, we are reasonably certain that the mothers of both of the young females were among the females that died during this period.

We detected small changes in the predicted directions in time spent grooming and grooming bout rates in response to loss of females, and in grooming bout duration on days with intergroup encounters, but despite reasonably large field sample sizes none of the changes reached significance. In every case, group patterns were swamped by enormous individual variation. Our findings suggest that it may be unreasonable to expect identical responses to events such as encounters and loss of group females for all females in a group. For species with strongly defined linear dominance relations, there has been discussion of whether low- and high-ranked females ought to participate equally in defense of their group's resources (van Schaik, 1989). Patas females do not have strongly linear dominance relations (Rowell and Olson, 1983; Carlson and Isbell, in prep.). Even in such an egalitarian species, however, differences in age, parity, and kinship may influence female responses to intra- and intergroup social challenges. For example, in our study, young females experienced the most drastic change in their social relationships and so could be expected to have had the most need to use grooming to restructure these relationships, as they appeared to do.

We expected that having fewer females in the group would result in shorter grooming bouts if surviving females had to increase vigilance to make up for fewer individuals to watch for danger. Paradoxically, bout length showed a tendency to increase during a period when we thought predation was especially high since we suspected that all four of the females that disappeared were taken by leopards. Grooming bout duration returned quickly to preloss levels, indicating that grooming patterns overall were not adjusted to compensate for several fewer females. The increase in grooming after the loss by the two young females suggests that an event such as this may be very disruptive to the relationships of some females but not to others. Perhaps this should not really be a surprising conclusion. After all, it is individual monkeys who interact in these groups, not classes of animals.

Our analysis of grooming provides only limited support for the idea that grooming is a form of social glue in patas monkeys. Although patas show many similarities in social organization with other guenons, it is possible that grooming's contribution to social cohesion may be less important for them, at least at some times during the annual reproductive cycle. The finding that intergroup encounters did not cause an increase in grooming as they do in blue, samango, and vervet monkeys suggests that patas females may have less flexible time budgets than these other species, but not necessarily that grooming is less important to them. In addition, since patas monkeys spend most of their day on the ground walking through grass they may pick up more ectoparasites and have more constant primary (hygienic) grooming require-ments than other guenons.

Finally, it may be that the lack of a strong linear dominance hierarchy and the seeming unresponsiveness of grooming frequency to such events as intergroup encounters and loss of female group members may simply reflect a high degree of relatedness among patas females resulting from female philopatry and relatively small group size in Laikipia. Thus, since females are supporting close female relatives during defense of group members and resources they may not need to increase social cohesion.

Summary

Patas monkeys (*Erythrocebus patas*), although ecologically distinct, share many features of social organization with the forest guenons including similar patterns of female–female relations and less rigid dominance hierarchies than is typical of other cercopithecines, notably baboons and macaques. Like many guenons, patas monkeys are female philopatric and females need to cooperate closely to defend their offspring and their group's range and its resources. As patas monkeys and at least some guenons also cooperate in caring for infants, these species would seem to need to maintain strong social bonds. We examined the idea that grooming acts to maintain and reinforce social bonds in species like patas monkeys with rather loosely defined dominance relations. We analyzed grooming patterns of adult female patas monkeys in two wild groups, one small and one large, in Kenya's Laikipia District. We looked at the effect of group size on grooming frequency. We found that, contrary to earlier predictions, females in the smaller group groomed all other group members and other adult females significantly more frequently than did females in the larger group. We also assessed the role of grooming in reaffirming social bonds under two sets of circumstances we believed would create a need for strong cooperation among group females: during intergroup encounters and also during a period when several females in the small study group disappeared. We found no significant effect of either of these potential causes of social stress on patas female grooming relations generally. Our analysis provided partial support for the idea that grooming acts in a general way to maintain or reaffirm social bonds in female philopatric species.

ACKNOWLEDGMENTS

We thank the Government of Kenya, the Office of the President, and the Agricultural Development Corporation for permission to conduct research at ADC Mutara Ranch. The Bower and Slade families provided us with assistance and friendship during our stays in Kenya. Dana Olson participated in data collection during the original 1979–1981 study. Anne Carlson, Lynne Isbell, Hallam Payne, and Peter Henzi generously shared unpublished manuscripts with us. Keira Chism and Will and Alexandra Rogers kindly put up with the many intrusions of this research into their lives. We thank Marina Cords for reading an earlier version of this paper for us at the XVIIIth Congress of the International Primatological Society in Adelaide, Australia and, along with Mary Glenn, for many helpful suggestions that aided our thinking in preparation of this manuscript.

References

Altmann, J. 1999. The role of social grooming in a primate society. Keynote address, 22nd Annual Meeting of the American Primate Society, New Orleans, Aug. 1999.

Carlson, A. A. 2000. *Social Relationships and Mating Patterns of Patas Monkeys* (Erythrocebus patas) *in Laikipia, Kenya*. Ph.D. Dissertation, University of Wisconsin, Madison.

Carlson, A. A., and Isbell, L. A. 2001. Causes and consequences of single-male and multimale mating in free-ranging patas monkeys, *Erythrocebus patas*. *Anim. Behav.* **62**:1047–1058.

Carlson, A., and Isbell, L. (in prep.). Social relationships of patas monkeys (*Erythrocebus patas*) in Laikipia, Kenya, I. Female–female interactions.

Cheney, D. 1992. Intragroup cohesion and intergroup hostility: the relation between grooming distributions and intergroup competition among female primates. *Behav. Ecol.* **3**:334–345.

Chism, J. 1986. Development and mother-infant relations among captive patas monkeys. *Int. J. Primatol.* **7**:49–81.

Chism J. 1999a. Decoding patas social organization. In: P. Dolhinow, and A. Fuentes (eds.), *The Nonhuman Primates*, pp. 86–92. Mayfield, Mountain View, CA.

Chism J. 1999b. Intergroup encounters in wild patas monkeys (*Erythrocebus patas*) in Kenya. *Am. J. Primatol.* **49**:43.

Chism, J. 2000. Allocare patterns among cercopithecines. *Folia Primatol.* **71**:55–66.

Chism, J., and Rogers, W. 1997. Male competition, mating success and female choice in a seasonally-breeding primate (*Erythrocebus patas*). *Ethology* **103**:109–126.

Chism, J., and Rowell, T. E. 1986. Mating and residence patterns of male patas monkeys. *Ethology* **72**:31–39.

Chism, J., and Rowell, T. E. 1988. The natural history of patas monkeys. In: A. Gautier-Hion, F. Bourlière, J.-P. Gautier, and J. Kingdon (eds.), *A Primate Radiation: Evolutionary Biology of the African Guenons*, pp. 412–438. Cambridge University Press, Cambridge.

Chism, J., Rowell, T. E., and Olson, D. K. 1984. Life history patterns of female patas monkeys. In: M. Small (ed.), *Female Primates: Studies by Women Primatologists*, pp. 175–190. Liss, New York.

Cords, M. 1995. Predator vigilance costs of allogrooming in wild blue monkeys. *Behaviour* **132**:559–569.

Cords, M. 1997. Friendships, alliances, reciprocity and repair. In: A. Whiten, and R. W. Byrne (eds.), *Machiavellian Intelligence II: Extensions and Evaluations*, pp. 24–49. Cambridge University Press, Cambridge.

Cords, M. 2000. Agonistic and affiliative relationships of adult females in a blue monkey group. In: P. F. Whitehead, and C. J. Jolly (eds.), *Old World Monkeys*, pp. 453–479. Cambridge University Press, Cambridge.

Cords, M. 2002. Friendship among adult female blue monkeys (*Cercopithecus mitis*). *Behaviour* **139**:291–314.

Dunbar, R.I.M. 1988. *Primate Social Systems*. Comstock/Cornell University Press, Ithaca, NY.

Dunbar, R. I. M. 1991. Functional significance of social grooming in primates. *Folia Primatol.* **57**:121–131.

Gouzoules, S., and Gouzoules, H. 1987. Kinship. In: B. B. Smuts, D. L. Cheney, R. M. Seyfarth, R. W. Wrangham, and T. T. Struhsaker (eds.), *Primate Societies*, pp. 299–305. University of Chicago Press, Chicago.

Harding, R. S. O., and Olson, D. K. 1986. Patterns of mating among male patas monkeys (*Erythrocebus patas*) in Kenya. *Am. J. Primatol.* **11**:343–358.

Heath, D. 1995. *An Introduction to Experimental Design and Statistics for Biology.* UCL Press, London.

Isbell, L., and Pruetz, J. 1998. Differences between vervets (*Cercopithecus aethiops*) and patas monkeys (*Erythrocebus patas*) in agonistic interactions between adult females. *Int. J. Primatol.* **19**:837–855.

Payne, H., Lawes, M. J., and Henzi. S. P. (in prep.). Competition and the exchange of grooming among female samango monkeys (*Cercopithecus mitis*).

Ohsawa, H., Inoue, M., and Takenaka, O. 1993. Male mating strategy and reproductive success of male patas monkeys (*Erythrocebus patas*). *Primates* **34**:533–544.

Rowell, T. E., and Olson, D.K. 1983. Alternative mechanisms of social organization in monkeys. *Behaviour* **86**:31–54.

Rowell, T. E., Wilson, C., and Cords, M. 1991. Reciprocity and partner preference in grooming of female blue monkeys. *Int. J. Primatol.* **12**:319–336.

Sokal, R., and Rohlf, F. 1981. *Biometry, 2nd ed.* W. H. Freeman, San Francisco.

Sterck, E., Watts, D., and van Schaik, C. P. 1997. The evolution of female social relationships in nonhuman primates. *Behav. Ecol. Sociobiol.* **41**:291–309.

van Schaik, C. P. 1989. The ecology of social relationships amongst female primates. In: V. Standen, and R. Foley (eds.), *Comparative Socioecology, V*, pp. 195–218. Blackwell Scientific, Oxford.

Wrangham, R. W. 1980. An ecological model of female-bonded primate groups. *Behaviour* **75**:262–300.

Wrangham, R. W. 1987. Evolution of social structure. In: B. B. Smuts, D. L. Cheney, R. M. Seyfarth, R. W. Wrangham, and T. T. Struhsaker (eds.), *Primate Societies*, pp. 282–296. University of Chicago Press, Chicago.

Development of Mother–Infant Relationships and Infant Behavior in Wild Blue Monkeys (*Cercopithecus mitis stuhlmanni*)

18

STEFFEN FÖRSTER and MARINA CORDS

Introduction

A characteristic feature of primate life histories is the prolonged period of postnatal development in which newborns depend on adults, especially their mothers, for nutrition, transport, and protection, and develop social skills that enable them to become fully integrated members of society. Many researchers have focused on the development of infant nonhuman primates, attempting to identify factors that influence the ontogeny of

STEFFEN FÖRSTER • Department of Zoology, Technische Universität Braunschweig, Braunschweig, Germany.   MARINA CORDS • Department of Ecology, Evolution and Environmental Biology, Columbia University, New York, NY 10027, USA.
The Guenons: Diversity and Adaptation in African Monkeys, edited by Glenn and Cords. Kluwer Academic/Plenum Publishers, New York, 2002.

social behavior, with strong emphasis on the mother–infant relationship (Nicolson, 1987). Studies on infant development in Old World monkeys have concentrated mainly on macaques, especially *Macaca mulatta* (Seay, 1966; Hinde and Spencer-Booth, 1967, 1971; Harlow and Harlow, 1969; Hinde, 1969; Sackett, 1972; Anderson and Mason, 1974; Berman, 1980a; Stevenson-Hinde and Simpson, 1981; Simpson, 1985). Other catarrhine taxa for which data on infant or juvenile behavioral and social development are available include *Papio* spp. (Ransom and Rowell, 1972; Nash, 1978; Altmann, 1980), *Cercopithecus aethiops*, (Struhsaker, 1971; Lee, 1984a; Fairbanks and McGuire, 1985; Fairbanks, 1989), *Erythrocebus patas*, (Chism, 1986; Loy and Loy, 1987), and a few other cercopithecines (*C. neglectus:* Chalmers, 1972; Kirkevold and Crockett, 1987; *C. mitis albogularis:* Bramblett and Coelho, 1987).

Most of the studies were focused on terrestrial or semiterrestrial species and were carried out in captivity or with provisioned colonies. Whether their results can be generalized across the various social and ecological conditions in which cercopithecine species live in the wild is moot. Indeed, social, demographic, and ecological factors have strong and often long-lasting effects on infant and juvenile monkeys (Berman, 1980b; Johnson and Southwick, 1984; Lee, 1984b; Fairbanks and McGuire, 1987; Fairbanks, 1988; Maestripieri, 1994a; Berman *et al.*, 1997). Accordingly, we need to extend the research on infant development to other species with different social organizations or living in different environments.

We aimed to begin filling the gap in our knowledge by providing information on infant development in a highly arboreal cercopithecine species in the wild. Blue monkeys (*Cercopithecus mitis*) differ from macaques, vervet monkeys, and baboons not only in their arboreality but also in their social organization. They live in one-male multifemale groups most of the time and have a weakly differentiated dominance hierarchy among females whose ranks seem unrelated to affiliative behavior or reproductive success (Cords, 1987a, 2000a).

We characterize behavioral development of infant blue monkeys during their first six months to assess how an infant's relationship with its mother influences its behavior, paying special attention to factors that others have found to influence various aspects of infant development, including infant sex (Mitehell, 1979; Meaney *et al.*, 1985; Brown and Dixson, 2000), infant-rearing experience of the mother (Hooley and Simpson, 1981; Maestripieri, 1998), and maternal dominance status (Ransom and Rowell, 1972; Cheney, 1978). We also compare blue monkey infant development, especially the infant's relationship with its mother, with other cercopithecine species that are more terrestrial and have different social organizations. We evaluate the hypothesis that arboreality slows infant development because of the physical dangers of life in a three-dimensional habitat (Chalmers, 1972; Sussman, 1978).

Methods

Study Site

The study site is the Kakamega Forest (Isecheno section), western Kenya (0° 14′ N, 34° 52′ E) at an altitude of *ca*. 1580 m. The indigenous vegetation of the forest is a semideciduous type of Guineo–Congolian lowland rain forest. The site included areas with some human influence, such as enrichment planting of nonindigenous trees, indigenous plantation forest, and a village area that was used on a seasonal basis by one of the study groups, though not during our study. Cords (1987b) provided a detailed description of the site.

Subjects

The blue monkey population (*Cercopithecus mitis stuhlmanni*) at Kakamega is a high-density one for the species, with *ca*. 220 individuals per km^2 (Fashing and Cords, 2000). Blue monkeys live in groups with females as permanent members. Young males leave their natal groups before reaching full bodily size at an age of about seven years. For most of the year, only one adult male is present in the group, though additional males may join the group during the mating season (Cords, 2000b). Our study took place when only one adult male was present in each study group. Females defend territories against neighboring groups, whereas males are rarely involved in intergroup aggression (Cords, 2002).

We focused on 12 infant blue monkeys from three well-habituated groups with 17–37 noninfant members, which occupied adjacent home ranges of *ca*. eight to 17 ha. We could distinguish individuals via facial and other features like the shape of tail hairs. Table I is a summary of the ages, sexes, approximate sizes at birth, and sampling durations of the subjects, and characteristics of their mothers.

Data Collection

Förster conducted focal samples of the infants between November 1999 and June 2000. He sampled each subject for 336 ± 58 minutes ($n = 60$ 2-week sample blocks) per two-week period for the first three months of life, and for an average of 148 ± 15 minutes ($n = 51$ 2-week sample blocks) for the second three months. The total duration of focal samples for each infant over the whole study period averaged 38.5 ± 10.3 hours ($n = 12$).

Predetermined sampling schedules were impractical given the dispersion of group members over several hundred meters in dense vegetation, but we distributed samples of each subject as evenly as possible over each two-week

Table I. Birthdate, Sex and Relative Size of the Subjects, the Age up to Which They Were Sampled, and Parity and Relative Rank of Their Mothers

Mother's name	Sex	Birthdate (MM-DD-YY)	Relative size[a]	Maximum age (weeks)	Parity	Relative rank[b]
		Infant characteristics			Maternal characteristics	
Fletcher	male	12-13-99	medium big	24	multiparous	medium
Tap	male	01-04-00	medium big	24	multiparous	low
Bibi	male	01-10-00	medium big	24	multiparous	low
Pody	male	02-01-00	big	20	multiparous	medium
Spanner	male	02-07-00	medium big	22	multiparous	medium
Plume	male	02-09-00	medium small	12	multiparous	low
Lolita	female	11-16-99	small	24	primiparous	medium
Dingle	female	11-26-99	medium small	24	primiparous	low
Bow	female	12-13-99	big	24	primiparous	medium
Mustache	female	01-30-00	medium small	24	multiparous	low
Angle	female	02-04-00	medium big	20	multiparous	high
Xmas	female	02-27-00	medium big	20	multiparous	high

[a] Relative size of an infant was assigned to one of four classes during the first week of life.
[b] Females in each group were assigned rank classes (high, medium, low) according to their position in the upper, middle, or lower third of the hierarchy. If the total number of females was not divisible by 3, the 'medium' rank class was made the largest (exception: Xmas, who was assigned high rank because of her distinct dominant behavior towards other medium ranked females and an unstable relationship with the 3rd highest ranking female).

time block. To adjust for possible diurnal differences in activity, we divided the day into three periods (0730–1100 h, 1100–1430 h, 1430–1800 h) and sampled all infants nearly equally in these three periods during the first month of the study. Older infants grew increasingly difficult to locate as focal subjects, and we could not maintain an even distribution of samples across the three periods. Therefore, from each infant's third month we combined the morning and afternoon hours as one period, which is distinguished from the midday hours when general activity levels were usually lower. Samples for all subjects are similarly distributed between these more and less active periods.

Subjects quite often disappeared from view because of the dense vegetation. If we lost a subject for <15 minutes, we extended the sample until it had been in view for about 75 minutes. If the infant remained out of view for >15 minutes, we terminated the sample at the moment of disappearance and continued at the next opportunity within the same diurnal period, usually on the same day or on the following day.

Focal samples include both point and one/zero recording methods at 30 second intervals. We noted all occurrences of approaches and leaves, as well as some other rare behavior. Table II is a list of the behavioral categories recorded and recording methods. Several qualifications should be

Table II. Behavioral Categories and Method of Recording

Behavior	Description	Recording method
On nipple	Infant has nipple in mouth (whether clinging or not)	Point and 1/0 samples
On ventrum	Infant on mother's ventrum, without nipple in mouth	Point and 1/0 samples
Clinging	Infant on mother's ventrum but it is unclear whether it is on the nipple	Point and 1/0 samples
Exploring ventrum	Infant crawling on mother's ventrum, often to her sides, with eyes open (no nipple seeking movements)	Point and 1/0 samples
Independent locomotion	Infant locomoting off mother (or caretaker)	Point samples
Explorative feeding	Handling and mouthing objects (includes gnawing) in a playful manner	Point samples
Feeding	Searching for, manipulating, and swallowing food objects	Point samples
Resting	Sitting off mother, not engaged in social interaction, nor feeding (includes sleeping)	Point samples
Playing	Social play with peers	1/0 samples
Groomed	Infant groomed by mother	Point and 1/0 samples
Approach	Moving from >1 m to <1 m	Event recording
Leave	Moving from <1 m to >1 m	Event recording
Following	Infant actively follows mother while remaining in proximity (1 m) to her	1/0 samples
Restricting	Mother prevents infant from moving away from her	1/0 samples
Retrieving	Mother approaches infant and picks it up	Event recording
Rejecting	Mother denies nipple access	Event recording

kept in mind. First, because of limited visibility, it was often unclear whether an infant was suckling or just mouthing the nipple and whether it was sleeping. Therefore, we used three categories that reflect different grades of mother–infant contact behavior and nursing (Table II). Second, we called any kind of locomotion by the infant independent locomotion, including movements during play, which we scored as independent locomotion with the point sampling method but also as play with the 1/0 sampling method. Third, we divided foraging behavior into two classes: explorative feeding (including mouthing) and feeding (ingestion and swallowing). As it was difficult to tell whether infants actually swallowed food, especially until the fourth month of life, we recorded all handling and mouthing of food objects as explorative feeding until we saw the infant actually swallowing food for the first time. From then on, feeding was usually distinguishable from explorative feeding. Fourth, both solitary object play and social play occurred, but we recorded only the latter. We could not reliably distinguish object play, which included handling and mouthing twigs, leaves or other small objects, from explorative feeding and therefore we did not record it separately. Social play was a distinctive behavioral category, involving chases or rough-and-tumble body contact. Fifth, rejection rates in our analyses represent the frequency of rejections per total sample time, not the rejection rate per number of attempted nipple contacts.

Data Analysis

We adjusted all parameters in each sample for sample duration by dividing the number of points/intervals in which the parameter was recorded by the number of points/intervals sampled. Then we averaged the time budget values for each infant over two- or four-week periods depending on the analysis.

We used repeated measure ANOVA to analyze the development of behavioral parameters over the first five months, with four-week periods as the intra-subject factor (repeated measure), while inter-subject factors include sex, parity of mother (primiparous vs. multiparous), group membership, rank of mother (high, medium, or low), and time of birth relative to birth season (early vs. late). In each group, early and late cohorts were separated by several weeks or months (Table I). For each inter-subject factor, we ran a separate repeated measures ANOVA because of the small number of subjects. We pooled data over four-week periods to achieve a normal distribution, which could not be obtained with data from two-week blocks, even with transformations. As the repeated measure procedure is sensitive to missing values, we included no data from the sixth month of life because we had sampled only eight infants through all six months. If significant main effects occurred, we usually mention them only when interactions with age are not significant. We scrutinized all interaction effects carefully and ran follow-up one-way ANOVAs separately for each 4-week block to determine when significant differences occurred. In all analyses of temporal changes of a given measure, we searched for interindividual variation to be sure that mean values are representative. We report only statistically significant differences between individuals.

We compared the results of the repeated-measures ANOVA with linear multiple regression analyses to test for combined influences of different independent variables on a given dependent variable (usually a behavioral parameter). We entered independent variables in a stepwise procedure, with the probability of F to enter ≤ 0.05, and the probability of F to remove ≥ 0.1 (standard setting: Sokal and Rohlf, 1981) and with pairwise exclusion of cases. We examined data points that were more than three times the standard deviation from the mean and excluded them from the model if there were reasonable explanations for their aberrant values, e.g., exceptionally short sampling time, or infant with injury. We used the Durbin–Watson statistic to assess the degree of autocorrelation between adjacent residuals, accepting values from 1.5 to 2.5. We tested the assumption of equal variances across variables by inspecting residual plots: standardized residuals against standardized predicted values. For all extracted factors we report the regression coefficient B together with its p-value and associated adjusted R^2 (or ΔR^2_{adj}) value.

In correlational analyses we used the Spearman rank correlation for two-week-block data, which were not normally distributed, or Pearson correlation for four-week-block data, which were normally distributed.

Generally, we report means and standard deviations to summarize scores; however, we use medians (and ranges) when summarizing ≤ 12 scores with outlying values.

To reduce the number of intercorrelated variables, especially in measures of mother–infant contact behavior, we ran a principal components analysis. We extracted one factor containing the following measures: clinging, time on nipple, independent locomotion, feeding/explorative feeding, resting, time off mother, time in contact to mother, time in proximity to mother, and time at a distance without other individuals in proximity. The Kaiser–Meyer–Olkin measure of validity of the model is 0.877 (maximum of 1). We estimated the factorial values via a regression procedure. For analyses involving measures included in the independence factor, we first worked with the original variables. Because of the intercorrelations, only one such variable could be analyzed at a time, so the analysis could not account for slight interactions between the variables. We usually compared the results of the separate analyses to the results when using the independence factor. If the independence factor explained more of the variance than any one of its components, we used it instead of the original measures as an independent variable in the multiple regression analysis.

Results

Mother–Infant Relationship

Nipple contact

During the first days of life, infants spent most of their time on the nipple. Primiparous mothers were generally very attentive to their infants, but sometimes had difficulty supporting them in gaining nipple access, especially while moving or foraging. This difficulty seemed to result from the infant's small size and weakness rather than the mother's experience. Two of three infants born to primiparous mothers were smaller than average and appeared

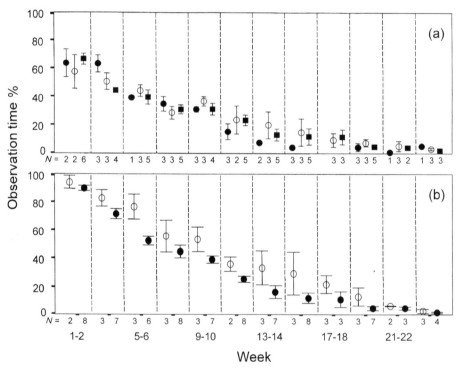

Fig. 1. Average (±S.E.) proportion of time infants spent in (a) nipple contact, and (b) contact to their mother, grouped by infant sex and parity of the mother. Differences in nipple contact (a) were not systematically related to infant sex or maternal parity: full circles, female infants of multiparous mothers; open circles, female infants of primiparous mothers; squares, male infants (multiparous mothers only). Infants of primiparous mothers (open circles) spent more time in contact (b) than infants of multiparous mothers (full circles; $F = 7.9$, $p < 0.05$, repeated measures ANOVA).

to be weak during their first days. Their small size, together with the still short nipples of their mothers, was obviously disadvantageous for gaining nipple access. Infants usually clung to the mothers with their hind-limbs wrapped around her hips, and a small infant could not reach the nipples from this position without maternal support. The one large infant of a primiparous mother had no problem, even from its first days, in gaining or maintaining nipple access while the mother foraged; the same was generally true of the medium to large infants of parous mothers. The two medium-small infants born to parous mothers, however, had low nipple contact rates during the first weeks of life, like the two small infants born to primiparous mothers. These comparisons show that infant size and strength are important in determining the amount of early nipple contact.

The time infants spent on the nipple decreased fairly steadily over the first six months [Fig. 1(a)], reaching $<50\%$ in weeks 5–6 ($\bar{x} = 40.1 \pm 9.4\%$, $n = 9$), and dropping below 5% during the sixth month ($\bar{x} = 2.9 \pm 1.8\%$, $n = 8$). No difference between male and female infants, infants of multiparous vs. primiparous mothers, or infants of differently ranked mothers were revealed by separate repeated measure ANOVAs.

A multiple regression with time spent on nipple as the dependent variable resulted in a different picture. Beside the independence factor ($B = -0.204$, $p < 0.001$, $R^2_{adj} = 0.954$) and time carried by other than mother ($B = -0.478$, $p < 0.001$, $\Delta R^2_{adj} = 0.009$), sex had an effect, with males having slightly lower rates of nipple contact than females did ($B = -0.018$, $p < 0.001$, $\Delta R^2_{adj} = 0.007$). The fact that both the independence factor and the time infants were carried by non-mothers influenced the time an infant was on the nipple is not surprising, since they both increased fairly steadily with the infant's age, but with no significant intercorrelation between the independence factor and the amount of time non-mothers carried infants.

Spatial relationship and distance regulation

Time in contact. We calculated the time infants spent in contact with their mothers by summing up the time they spent clinging, including on nipple, on ventrum, and exploring ventrum, and the time they were off but in physical contact with their mother. Over the first six months of life, infants of primiparous mothers spent more time in contact ($41.7 \pm 1\%$, $n = 3$) than those of multiparous mother [$35.9 \pm 0.7\%$, $n = 9$; Fig. 1(b)].

A multiple regression analysis showed that age in weeks was the most important predictive variable ($B = -0.08$, $p < 0.001$, $R^2_{adj} = 0.8$). Only parity made a small but significant additional contribution to explain differences in contact scores between infants ($B = -0.04$, $p < 0.05$, $\Delta R^2_{adj} = 0.008$), confirming the repeated measures analysis.

Time spent off mother. Infants in their second or third day of life (median: 2.5, range $= 1–5$, $n = 12$) began to explore immediate surroundings by releasing their grip on the mother while maintaining physical contact. The median age at which infants moved out of contact is five days (range = 2–14, $n = 12$). The median age at which they moved more than arm's reach from their mothers is nine days (range = 5–19, $n = 12$).

Again there is a significant difference between infants of primiparous vs. multiparous mothers ($F = 5.5$, $p < 0.05$, repeated measures ANOVA) in time spent off the mother. The average percentage for infants of primiparous mothers ($63.3 \pm 3.7\%$, $n = 3$, months 1–5) is lower than for infants of multiparous mothers ($73.4 \pm 2.3\%$, $n = 9$, months 1–5). However, multiple regression analysis did not confirm this parity effect.

The size of infants at birth significantly influenced the time they spent off their mothers [multiple regression, $B = 0.05$, $p < 0.01$; $\Delta R^2_{adj} = 0.035$; Fig. 2(a)]. Larger infants tended to spend more time off their mothers than smaller infants did, especially during the first four weeks of life. The correlation between size of the infant at birth and overall percentage of time spent off the mother across the first six months ($r = 0.16$, $n = 12$) is not significant however.

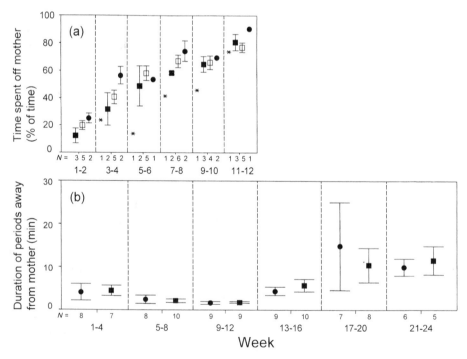

Fig. 2. Time spent off mother (a) and duration of periods away (>1 m) from mother (b), means ($\pm$ S.E.). In (a), asterisk = small, full square = medium-small, open square = medium-big, full circle = big infant. In (b), circles = female and squares = male infants.

Duration of periods away from the mother. The average duration of periods away (>1 m) from the mother did not increase steadily with age [Fig. 2(b)]. Slightly higher values in the first month vs. months two and three were probably caused by the infant's inability to keep up with the mother when she decided to leave. Once an infant began to locomote steadily, it would follow its mother when she left, which led to a decrease in the average length of separation up to the third month. From the third month onward, infants spent ever-longer periods away from their mothers. Average values are probably underestimates, especially for the last two months when infants were often away from their mothers for the entire 75-minutes sample period.

Repeated measures ANOVA revealed no main effect. A multiple regression revealed that aside from age, the rejection rate and the time being carried by others influenced the duration of periods away from the mother ($B = -9.6$, $p < 0.05$, $\Delta R^2_{adj} = 0.047$, and $B = 69.1$, $p < 0.05$, $\Delta R^2_{adj} = 0.040$, for rejection rate and time carried by others, respectively). A higher rejection rate resulted in a decrease in the average duration of periods away from the mother.

Regulation of proximity between mother and infant. There was no overall main effect of sex, parity, rank, size, or time of birth on how proximity between mother and infant was regulated, measured either with the proximity index of Hinde and Atkinson (1970), or as the average number of approaches and leaves. The degree to which the infant was responsible for remaining in proximity to its mother as measured by the proximity index increased more or less steadily with age, whereas especially the number of leaves made by the mother as well as the number of leaves and approaches made by the infant reached a peak between weeks 11 and 14 (Fig. 3).

The overall pattern shows that infants generally were primarily responsible for maintaining proximity to their mothers, as evidenced by the usually higher number of approaches and lower number of leaves by infants vs. mothers and also from the positive proximity index scores from the very beginning of infancy (Fig. 3). Only one infant showed a negative proximity index in the first month. She tried to get away from her mother as often as possible during the fourth week of life, though the mother tried to prevent her escape, and therefore approached her infant more often than the infant approached her.

In the sixth month there was a marked increase in the number of approaches and leaves by male infants. The proportion of leaves by infants is significantly different for males vs. females at 21–22 weeks ($73 \pm 3\%$, $n = 2$, vs. $39 \pm 14\%$, $n = 4$, respectively, $F = 11.1$, $p < 0.05$) and at 23–24 weeks ($70 \pm 5\%$, $n = 3$, vs. $37 \pm 13\%$, $n = 4$, respectively, $F = 16.4$, $p < 0.05$). Significant differences in the same measure (proportion of leaves by the infant) for infants of primiparous vs. multiparous mothers occurred in weeks 5–6 ($6 \pm 10\%$, $n = 3$, vs. $35 \pm 12\%$, $n = 6$, respectively, $F = 13.5$, $p < 0.01$) and 23–24 ($32 \pm 12\%$, $n = 3$, vs. $65 \pm 11\%$, $n = 4$, $F = 14.5$, $p < 0.05$).

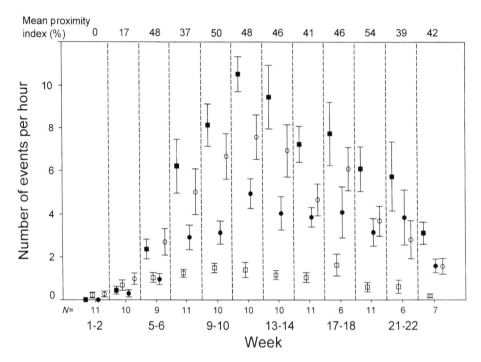

Fig. 3. Proximity regulation between mother and infant, expressed as the average (±S.E.) number of approaches and leaves per hour. Full squares = approaches by infant, open squares = approaches by mother, full circles = leaves by infant, and open circles = leaves by mother. The Hinde and Atkinson (1970) proximity index is the proportion of approaches made by the infant less the proportion of leaves made by the infant, averaged across infants for each period.

The maximum number of transitions occurred in weeks 11–12 for males and weeks 13–14 for females. During this period, mothers would usually leave and infants would try to keep up with them. If an infant had problems catching up, the mother usually waited until it came near again, and then either picked it up or repeated the procedure, i.e., moving ahead and waiting.

Across the entire study period, the number of leaves by mothers is correlated with the proportion of 30-second intervals when infants followed their mothers within a distance of 1 m ($r_s = 0.78$, $p < 0.001$, $n = 97$ 2-week periods, all infants combined), which underscores the predominant responsibility of infants to remain close to their mothers.

Restriction

Mothers restricted their infants from moving out of contact only during the first few weeks. They did so by holding a limb or tail, or by forcing it into ventroventral contact. If others were nearby, mothers frequently shielded clinging infants by embracing them and turning away from other individuals.

By the third to fourth week, restriction was fairly infrequent, occurring in <1% of sampling intervals (Fig. 4). Mothers allowed older infants to move away freely, but would still often pick them up if other individuals approached or if there were other signs of disturbance or danger, e.g., a predator alarm, intergroup encounter, observer, or loud calls by the resident adult male. After the third month, maternal protective behavior was far less frequent, partly because the infants spent much time away from their mothers. When infants were far from their mothers, however, other individuals, especially adult and subadult females, usually offered protection.

Although multiparous mothers tended to restrict their infants far less than primiparous mothers did, and high-ranking mothers were generally least restrictive, the small sample sizes prevent either difference from reaching statistical significance. Mothers of early-born infants, however, showed higher restriction rates during the first three months than mothers of later-born infants (Fig. 4).

Maternal rejection

Rejections occurred infrequently and never exceeded an average of three events per hour for a given infant over a two-week period. Most rejections

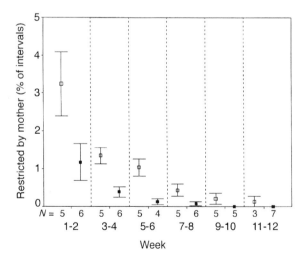

Fig. 4. Average (±S.E.) proportion of half-minute periods in which early- (open squares) and late-born (filled squares) infants were restricted by their mothers from moving out of contact. The difference between early- and late-born infants was significant during the first three months ($F = 17$, $p < 0.01$, repeated measures ANOVA).

were subtle, without overt aggression by the mother. They include pushing or grooming the infant off the nipple, stretching the torso so that the infant lost nipple contact, blocking access to the nipple with the forelimbs or by turning around, and only occasionally hitting or (mock) biting the infant.

Infants sometimes threw tantrums in response to maternal rejections, shrieking loudly while making jerking body movements. After such tantrums mothers usually promptly allowed the infant to suckle. Similar exaggerated reactions occurred even more frequently in response to denied riding opportunities and also resulted in prompt ventroventral contact initiated by the mother. A few prolonged tantrums, to which the mother did not respond positively, resulted in the infant falling out of the tree because of its jerking movements.

The first peak in rejection frequency occurred during weeks 7–8 for mothers with female infants, and about two weeks later for mothers with male infants (Fig. 5). This pattern is independent of parity: both multiparous and primiparous mothers began to prevent their female infants from gaining nipple access earlier than (multiparous) mothers with male infants did. There was, however, considerable variation across individuals, which is masked by mean values: two infants were rejected at an early stage only, two others showed an early peak in rejection frequency as well as a less pronounced peak in the second three months of life, while three others were not rejected at all until the second three months of life. Two infants experienced a rejection rate

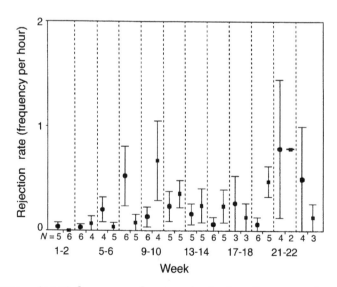

Fig. 5. Average (±S.E.) frequency of maternal rejections from accessing the nipple. Circles = female, squares = male infants.

that increased rather steadily with age, and two infants received relatively low rejection rates throughout the first six months, with no obvious peak.

Neither infant sex or size at birth nor maternal parity or rank had significant influences on the rejection frequency in a factorwise repeated measures ANOVA. A multiple regression model showed the number of leaves mothers made per hour ($B = 0.05$, $p < 0.001$, $R^2_{adj} = 0.13$) as the only useful predictive variable. This measure is highly correlated with the number of approaches per hour made by the infant ($r_s = 0.916$, $p < 0.001$, $n = 112$ two-week time blocks across all infants).

Maternal grooming

For all infants over all ages through the fifth month, neither infant sex nor maternal parity or rank influenced the relative amount of maternal grooming, i.e., grooming time per time spent in contact (repeated measures ANOVA, Fig. 6). In the sixth month, however, mothers groomed their female infants much more than their male infants ($40.7 \pm 22.8\%$ of contact time for females vs. $2.8 \pm 4.8\%$ for males, Fig. 6).

A multiple regression using data from all infants revealed that in addition to the infant's age ($B = 0.017$, $p < 0.05$, $R^2_{adj} = 0.117$), the frequency of rejections ($B = 0.105$, $p < 0.05$, $\Delta R^2_{adj} = 0.035$) and infant sex ($B = -0.035$, $p < 0.05$, $\Delta R^2_{adj} = 0.026$) played a significant role in determining the relative

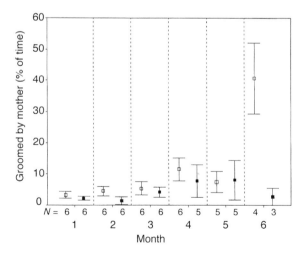

Fig. 6. Average ($\pm$S.E.) proportion of time male and female infants were groomed by their mothers, expressed as a percentage of time spent in contact with the mother. Circles = female, squares = male infants. The sex difference in month six was significant ($F = 7.7$, $p < 0.05$, repeated measures ANOVA).

amount of maternal grooming. Thus at periods with higher rejection frequencies, infants were generally groomed for a higher proportion of time than at periods with lower rejection frequencies, and, accounting for the differences in the previous two variables (age and rejection frequency), female infants were generally groomed for a higher proportion of time than male infants were.

Because sex apparently had some influence on maternal grooming, we carried out separate regressions for male and female infants to see whether these factors were equally important for both sexes. The results for both sexes combined held only for female infants. For males, the mother's rank class was the most important independent variable influencing maternal grooming ($B = 0.08$, $p < 0.05$, $\Delta R^2_{adj} = 0.097$), followed by the independence factor ($B = 0.03$, $p < 0.05$, $\Delta R^2_{adj} = 0.062$), the latter being highly correlated with the infant's age ($r_s = 0.954, p < 0.001, n = 51$ two-week time blocks across all male infants).

Infant Behavior

Mouthing and feeding

Although we observed infants mouthing objects as early as the fourth day of life, we observed them first swallowing food from the fifth week onward (mean: day 44 ± 15, $n = 7$ infants that we monitored sufficiently closely during this period for confident assessment). In the repeated measure analysis, the first month had to be excluded because for some of the first-born infants, feeding behavior was not included in the sampling protocol. Female infants

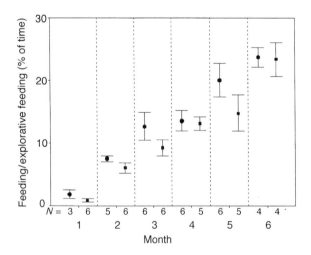

Fig. 7. Average ($\pm$S.E.) proportion of time male and female infants spent manipulating or eating food objects. Females (circles) had significantly higher scores than males (squares; $F = 6.6$, $p < 0.05$, repeated measures ANOVA).

had higher average scores for mouthing and feeding ($14 \pm 1\%$ of observation time, $n = 5$) than males ($11 \pm 1\%$, $n = 5$). This sex difference is consistent over time (Fig. 7) and is supported by multiple regression analysis. Age (in months) and sex accounted for 77.1% of the variance in time spent mouthing or feeding ($B = 0.038$, $p < 0.001$, $R^2_{adj} = 0.718$, and $B = -0.017$, $p < 0.01$, $\Delta R^2_{adj} = 0.044$ respectively).

Independent locomotion and general activity measures

Because the time spent moving independently is significantly correlated with the total time spent off the mother ($r = 0.845$, $p < 0.001$, $n = 66$ four-week time blocks, all infants combined), we used the measure time spent locomoting independently when off the mother for the ANOVA. Strikingly there are no significant interindividual differences in the proportion of time spent in independent locomotion when off the mother and there is no significant change with age in the time off mother allocated to independent locomotion. Similarly, multiple regression analysis failed to identify factors that made significant contributions in explaining the proportion of time spent in independent locomotion when off the mother.

Discussion

Several factors influence the development of mother–infant relationships and infant behavior in blue monkeys. Table III is a summary of the effects of the main group factors—infant sex, maternal parity, maternal rank, birth order, and size of the infant at birth—on five independent variables.

Spatial Relationship between Mother and Infant

The early spatial relationship between infants and their mothers was characterized by interindividual differences in mothering styles. Some of the differences seemed to derive from the mother's prior experience rearing infants and from her rank. Primiparous mothers spent more time than multiparous mothers in contact with and in proximity to their infants throughout most of the first six months. It is not clear whether primiparous mothers or their infants were primarily responsible for this difference relative to multiparous mothers, though primiparous mothers tended to be more responsible for approaches than multiparous mothers were throughout the study period. Infants of primiparous mothers had a significantly lower responsibility for leaves than infants of multiparous mothers did in weeks 5–6 and 23–24.

Similar influences of parity on proximity measures occur in other species. Brown and Dixson (2000) observed that in captive rhesus macaques, infants of

Table III. Summary Table for Effects of Different Group Factors on Variables Representing Development of Infant Behavior and Mother–Infant Relationships in Blue Monkey Infants during Their First Six Months of Life[a]

Variable \ Factor	Sex (M males, F females)	Parity (P primiparous, M multiparous)	Rank	Birth order	Size
Time off mother		Lower for P mothers			Increasing with size in first three months
Explorative feeding/feeding	F more than M, significant after third month				
Restriction		Tendency of higher values for P mothers	Tended to increase with decreasing rank	Higher for early born infants than for later borns	
Rejection	Tended to happen earlier at relatively high levels in F	Tended to be later in infants of P mothers			Tended to be latest in small, and earliest in big infants
Maternal grooming	F were groomed more per contact time in month 6		Tended to increase with rank in M		

[a] Tendencies indicate non-significant but consistent differences over time, or effects seen in the majority of subjects (but where sample size usually prevented the differences from reaching statistical significance). Blanks indicate no discernible effects.

primiparous mothers spent significantly more time in ventroventral contact than infants of multiparous mothers did. Hooley and Simpson (1981) found that primiparous rhesus macaque mothers tended both to approach and to leave their infants more often than multiparous mothers did. These findings together with our results for blue monkeys are consistent with other features of maternal style that characterize primiparous mothers such as their generally greater anxiousness and protectiveness or their lower confidence (Seay, 1966; Mitchell and Stevens, 1969; Ransom and Rowell, 1972, Hooley, 1983; Berman, 1984; Tanaka, 1989; Maestripieri, 1998).

Because the responsibility for maintaining proximity is an important aspect of the mother–infant relationship, Hinde and Atkinson (1970) proposed an index, defined as the percentage of approaches by the infant less the percentage of leaves by the infant, to represent the contribution of each partner to the maintenance of proximity between them. In Kakamega blue monkeys, the proximity index was positive throughout the study period for all but one infant, indicating that from the first weeks of life infants had a greater responsibility to maintain proximity to their mothers than vice versa. This finding contrasts with those for group-living rhesus macaques (Hinde and Spencer-Booth, 1967; Nash, 1978; Berman, 1980b; Johnson and Southwick, 1984) and baboons (Altmann, 1980), in which the proximity index is negative in the first weeks of life, becoming positive only when infants reached ages of five to12 weeks.

Differences among species in the responsibility for proximity maintenance may relate to the dangers that infants face in different environments and social situations. The few cases in which positive proximity indices were recorded from the very beginning of an infant's life involved species or environments wherein intra-group aggression rates were low (Chism, 1986: captive patas monkeys; Hinde and Atkinson, 1970: isolated rhesus macaque mother–infant pairs). In wild blue monkeys, rates of intra-group aggression (Cords, 2000a), especially aggression directed towards infants (Cords, pers. obs.), were also extremely low, whereas most studies that have addressed proximity regulation involve species with relatively high intra-group aggression rates like rhesus macaques or baboons (Chism, 2000). When aggression rates are low, the danger of infant harassment, including potentially fatal kidnapping (Maestripieri, 1994b) by other group members, is reduced; mothers can afford to be less vigilant toward infants and may profit from spending more time foraging or engaging in social interactions that they might otherwise forego in order to maintain vigilance over their offspring.

Protective Actions by the Mother

Maternal restriction of an infant is a way that mothers can directly ensure that their infants stay nearby and safe. Mothers that birthed earliest in a group restricted the movements of their infants more than mothers that birthed later. This seemed to be an immediate result of the higher rates of handling that early-born infants received from other group members, especially adult and large juvenile females that often tried to investigate newborns (Förster and Cords, in prep.). To prevent an infant from being carried too often and for too long, the mother usually restricted its movements as soon as other individuals came close. Mothers that birthed later, however, received much less attention from other group members, and thus the danger of excessive infant handling was lower.

There are also individual differences in the degree of protectiveness, some of which relate to infant size and strength. Mothers with bigger infants generally were less protective and allowed other individuals to hold them earlier than mothers of smaller infants did. Also, big infants were left far behind earlier than small infants were. Other systematic factors that could account for individual differences in mothering behavior include the early experience of the mother when she was an infant herself (Fairbanks, 1989; Berman, 1990) or her rank.

Indeed, in blue monkeys, restrictiveness tended to be related to rank, but the risk of aggression from others did not seem to drive this relationship because aggression was so rare. Probably low-ranking mothers increased restrictiveness to ensure access to their infants. High-ranking mothers had few problems retrieving their infants from caretakers, and reduced restrictiveness seemed advantageous for them. A more relaxed behavioral style by high-ranking mothers could allow them to increase their foraging efficiency, by spending more time and more uninterrupted time. Carrying an infant is energetically costly and can reduce a mother's feeding efficiency (Altmann, 1980; Whitten, 1982 cited in Altmann and Samuels, 1992; Johnson, 1986; Stanford, 1992). The effect of maternal rank on restrictiveness, while weak, is one of the few ways in which rank affects any aspect of non-agonistic social behavior in this species.

Mother–Infant Distance Regulation

The relative role of mother and infant in the development of infant independence is an extensively debated issue (Nicolson, 1987). The proximity index of Hinde and Atkinson (1970) has been used to describe the increasing independence of the infant by relating it to measures such as time spent off or at a distance from the mother (Hinde and Spencer-Booth, 1967; Berman, 1980b). In our study, the proximity index is not significantly related to any behavioral measures, and it did not explain significant amounts of variance in the data set in any of our regression analyses. The absolute number of approaches and leaves, as well as the responsibility for approaches and leaves considered separately, were much more useful to characterize mother–infant relationships.

Maternal Rejection and Its Effects on the Development of Independence

The mother's role in the infant's growing independence is often measured by the rate of rejection infants receive from their mothers when attempting to suckle or to seek contact. There is disagreement about whether maternal rejection promotes infant independence (Nicolson, 1987), but most researchers see rejection as directly responsible for an increase in independence, usually expressed as the time infants spend off or at a distance from their mothers (Hinde and Spencer-Booth, 1967; Altmann, 1980; Berman, 1980b; Rijt-Plooij and Plooij, 1987). In our study, rejection rate correlates with no distance measure. Instead, the rejection frequency is positively related to the number of leaves by mothers. There is no obvious causal connection between these measures. Rejections were seldom followed by leaves, either by mother or infant, but the period of high rejection frequency coincided with a period when it was primarily mothers rather than infants that initiated an increase in distance between them. There is also a correlation between the number of leaves by mothers and the number of approaches by infants. It seems that mothers were trying to promote infant independence by rejecting attempts to get on the nipple and by encouraging independent locomotion with the leave-and-wait game described earlier (Altmann, 1980; Maestripieri, 1995). However, the outcome of this period of maternal rejection was apparently a temporary decrease in infant independence because the higher number of maternal leaves is correlated with a higher number of infant approaches, and not by infants staying away as would be expected if independence were increasing. Indeed, although most infants were already quite independently mobile, as expressed by measures of the mother–infant spatial relationship, when they experienced the highest rates of rejection they seemed to become more attracted to the mother. However, after a period of adjustment to the sudden negative behavior of their mothers, during which infants seemed to learn when and how often they could approach their mothers to suckle, infants shifted their activites away from the mother again, actively increasing their own independence.

Interindividual differences in rejection behaviour were related to maternal parity and infant sex. Two of three primiparous mothers rejected their infants at a later point in life than multiparous mothers did. This difference is consistent with the higher contact and proximity scores for primiparous mothers and their infants, which indicate a closer relationship and a reduced (and delayed) tendency of primiparous mothers to encourage infant independence. There is some evidence, however, that mothers might have simply responded to different abilities of their infants rather than to their own internal motivational states, since the two infants that were rejected latest were the smallest, and the infants that were rejected at high rates earliest in life were the two largest infants. Maestripieri (1995) showed that rhesus macaque mothers adjust their behavior to the competence of the infant.

Like free-living vervet monkeys (Lee, 1984a), the majority of female blue monkey infants were rejected earlier than males were. This sex difference might relate to parental investment theory, which predicts that maternal investment is generally highest in the sex with the most variable reproductive success, i.e., males for most mammals (Trivers, 1972; Altmann, 1980). Because suckling is energetically demanding and suppresses future reproduction, it is generally considered to be a clear form of maternal investment. Rejection, however, is not a direct measurement of the termination of investment, and since there is no sex difference in time spent on the nipple, the existence of a sex difference in actual suckling investment is questionable. Future studies that measure milk consumption could resolve this issue.

Maternal Grooming

In late infancy especially, infant female blue monkeys were groomed much more than the males were. One of the more commonly assumed functions of social grooming is the establishment of long-term cooperative relationships (O'Brien and Robinson, 1993; Cords, 1997). In many cercopithecine species, the relationships are established preferentially with kin (Silk, 1987). For philopatric female blue monkeys, it should be advantageous for both high and low-ranking mothers and their daughters to establish close affiliative and cooperative relationships, which are maintained and manifested by social grooming. These close relationships among female blue monkeys are important for effectively defending home ranges and food resources from neighboring groups (Cords, 2002).

There are very few comparable data on infant grooming and a similar preference for female over male infants as grooming partners of mothers has been reported only for captive patas monkeys (Loy and Loy, 1987), though it was absent in another captive patas monkey group (Rowell and Chism, 1986). A preference for grooming infants of either sex is not apparent in rhesus monkey infants, even though they show a matrifocal social system similar to that of blue monkeys (Hinde and Spencer-Booth, 1967; Brown and Dixson, 2000). Taken together with our results, these studies do not indicate clearly whether or how maternal grooming of infants is used to cement mother–offspring relationships.

Development of Nutritional and Locomotor Independence

There was no major difference among infants in the proportion of time they spent locomoting off their mothers, and there was no major change in such independent locomotion with age. This suggests that the overall activity levels were similar for all infants regardless of sex, size, or mothers' characteristics.

However, some differences occurred in activity budgets, with female infants spending more time than male infants interacting with food objects. Female infants also spent less time playing (Förster and Cords, in prep.), and given similar overall activity levels for both sexes, it would be expected that female infants engaged in other activities more than male infants did. Foraging, i.e., explorative feeding and feeding, was apparently the only activity to which female infants devoted more time than infant males did.

Harrison (1983), Terborgh (1983), van Noordwijk *et al.* (1993), and Nakagawa (2000) reported that sex differences in foraging behavior of juveniles were related to sex differences in foraging behavior of adults, with males usually feeding more and/or using more high-calorie food than females. In adult blue monkeys, males are more frugivorous and less insectivorous than females are (Cords, 1986), but no information is available about sex differences in the time adult blue monkeys spend foraging. Thus a functional explanation for the finding that female infants spent more time feeding than male infants cannot be derived from our data.

Arboreality and Infant Development

The time infants spend in contact with their mothers is a frequently used measure of infant dependence among primates, and it varies both within and among cercopithecine species (Struhsaker, 1971; Altmann, 1980; Chism, 1986; Kirkevold and Crockett, 1987; Maestripieri, 1994a). Chalmers (1972) suggested that arboreality, and especially the risk of falling from trees, might be an important factor influencing the rate at which infants gain independence. However, correlations across species between the degree of terrestriality and the rate at which infants reduce contact with their mothers are not obvious among data from captive and free-living groups (Kirkevold and Crockett, 1987). Regarding only free-ranging groups, Karssemeijer *et al.* (1990) found that arboreal long-tailed macaques developed independence later in life than more terrestrial, similarly-sized species like rhesus macaques or vervet monkeys; however, they used limited comparative data to evaluate the differences among species. Considering data from a larger number of studies in the wild or under natural conditions, we could not discern the contrast they emphasized. For instance, free-living, largely terrestrial patas monkey infants spend more time in contact with the mother at every age during the first six months than infant blue monkeys (Chism, 1986; Loy and Loy, 1987). Our study of wild blue monkey infants does not support the view that infants develop independence more slowly in more arboreal cercopithecines.

Moreover, there is no evidence that the risk of injury from falling from trees is significant. We witnessed infants falling out of trees 20 times from heights of 3–15 m (median: 9.5 m for 16 measured falls), especially between months 2–4. Falls occurred when they tried to keep up with their mothers, while playing with peers, and while throwing tantrums. No falling infant sustained injuries: they either sat motionless until retrieved by their mothers, or climbed up independently. Even after falling from 15 m, infants quickly moved normally.

We suggest that the risks of falling from trees are relatively unimportant in influencing the rate at which primate infants develop independence, even for arboreal species. Infants are also at risk for being prey, or being a victim of aggression or abuse by groupmates. Predation risk may decrease with increasing height above the ground (van Schaik *et al.*, 1983; Boinski *et al.*, 2000), and this effect may counteract the risk of injury from falls, which would increase with greater height, and lead to no discernable relationship between arboreality and the rate at which infants attain independence from their mothers.

Karssemeijer *et al.* (1990) provide data on mother–infant contact in long-tailed macaques, another arboreal species inhabiting forest habitats. Infant long-tailed macaques are considerably slower to break contact with the mother than are infant blue monkeys. The difference between them probably does not reflect differences in the risk of injury from falls, but is likely to result from differences in the risk of predation or infant harrassment by groupmates. We cannot quantitatively compare predation risk at the two sites, but note that predation rates on blue monkeys seem very low. Most infants (*ca.* 80–100%) survive to 12 months, and group members are often very dispersed; group spreads of several hundred meters regularly occur. In addition, the risk of infants receiving conspecific aggression, whether as direct targets or as bystanders, seems very low in blue monkeys, which show very low rates of aggression compared to that in macaques (Cords, 2000). In a low-risk environment, blue monkey mothers may be able to afford relaxed behavior toward their infants, which would promote development of the motor skills and independent movements that are especially important in the highly diverse, three-dimensional environment in which they live.

Summary

We studied 12 blue monkey (*Cercopithecus mitis stuhlmanni*) infants in a wild population during the first six months of their lives, with the goals of (1) identifying factors that explain interindividual differences in the various aspects of the mother–infant relationship and infant behavior, and (2) evaluating the hypothesis that arboreality slows the rate at which infants become independent from their mothers. Infants spent more time in contact with primiparous than multiparous mothers throughout the first six months of life. In addition, primiparous mothers tended to reject their infants at a later stage than multiparous mothers did. A comparison of the time course of development of independence with the timing of maternal rejection suggests that rejection by mothers did not play the major role in the infants' achieving independence. Mothers who gave birth earliest in their groups were more restrictive during the first weeks of the infant's life than mothers who gave birth later. Mothers groomed female infants increasingly more than male infants, suggesting the formation of bonds between mothers and daughters in this matrifocal species at a very early stage in life. The rate at which blue monkey infants attained independence from their mothers resembled that of similar sized terrestrial species, but was faster than the few arboreal cercopithecine species that have been studied to date. We suggest that the development of infant independence in blue monkeys reflects the risks of intra-group aggression and predation more than arboreality.

ACKNOWLEDGMENTS

This research was supported by the National Science Foundation (BCS 98–08273 to MC). We are grateful to the Government of Kenya, Office of the President and Ministry of Education, Science and Technology for permission to conduct research in the Kakamega Forest, to the University of Nairobi's Zoology Department and the Institute for Primate Research (National Museums of Kenya) for local sponsorship, and to the local officers of the Forestry Department and Kenya Wildlife Service for cooperation at the field site. We thank Jim Warfield for sharing his knowledge and his immense literature collection on nonhuman primate social development, and for his comments on parts of the manuscript. Thanks also to Julius A. Omondi, Praxides Akelo, Jasper M. Kirika, Simon Mbugua, and the other monkey watchers of 1999/2000, and to Mary Glenn, Russ Tuttle, and Thelma Rowell for their help in improving the manuscript.

References

Altmann, J. 1980. *Baboon Mothers and Infants*. Harvard University Press, Cambridge.
Altmann, J., and Samuels, A. 1992. Costs of maternal care: infant carrying in baboons. *Behav. Ecol. Sociobiol.* **29**:391–398.
Anderson, C. O., and Mason, W. A. 1974. Early experience and complexity of social organization in groups of young rhesus monkeys (*Macaca mulatta*). *J. Comp. Physiol. Psychol.* **87**:681–690.
Berman, C. M. 1980a. Mother–infant relationships among free-ranging rhesus monkeys on Cayo Santiago: a comparison with captive pairs. *Anim. Behav.* **28**:860–873.

Berman, C. M. 1980b. Early agonistic experience and rank acquisition among free-ranging infant rhesus monkeys. *Int. J. Primatol.* **1**:153–170.

Berman, C. M. 1984. Variation in mother–infant relationships: Traditional and nontraditional factors. In: M. small (ed.), *Female Primates: Studies by Women Primatologists*, pp. 17–36. Alan R. Liss, New York.

Berman, C. M. 1990. Intergenerational transmission of maternal rejection rates among free-ranging rhesus monkeys. *Anim. Behav.* **39**:329–337.

Berman, C. M., Rasmussen, K. L. R., and Suomi, S. J. 1997. Group size, infant development and social networks in free-ranging rhesus monkeys. *Anim. Behav.* **53**:405–421.

Boinski, S., Treves, A., and Chapman, C. A. 2000. A critical evaluation of the influence of predators on primates: effects on group travel. In: S. Boinski, and P. A. Garber (eds.), *On the Move: How and Why Animals Travel in Groups*, pp. 43–72. University of Chicago Press, Chicago.

Bramblett, C. A., and Coelho, A. M. 1987. Development of social behavior in vervet monkeys, Sykes' monkeys, and baboons. In: E. L. Zucker (ed.), *Comparative Behavior of African Monkeys*, pp. 67–79. Alan R. Liss, Inc., New York.

Brown, G. R., and Dixson, A. F. 2000. The development of behavioural sex differences in infant rhesus macaques (*Macaca mulatta*). *Primates* **41**:63–77.

Chalmers, N. R. 1972. Comparative aspects of early infant development in some captive cercopithecines. In: F. E. Poirier (ed.), *Primate Socialization,* pp. 63–82. Random House, New York.

Cheney, D. 1978. Interactions of immature male and female baboons with adult females. *Anim. Behav.* **26**:389–408.

Chism, J. 1986. Development and mother–infant relations among captive patas monkeys. *Int. J. Primatol.* **7**:49–81.

Chism, J. 2000. Allocare patterns among cercopithecines. *Folia Primatol.* **71**:55–66.

Cords, M. 1986. Interspecific and intraspecific variation in diet of two forest guenons, *Cercopithecus ascanius* and *C. mitis*. *J. Anim. Ecol.* **55**:811–827.

Cords, M. 1987a. Forest guenons and patas monkeys: male-male competition in one-male groups. In: B. B. Smuts, D. L. Cheney, R. M. Seyfarth, R. W. Wrangham, and T. T. Struhsaker (eds.), *Primate Societies,* pp. 98–111. University of Chicago Press, Chicago.

Cords, M. 1987b. Mixed-species association of *Cercopithecus* monkeys in the Kakamega Forest, Kenya. *Univ. Calif. Pub. Zool.* **117**:1–109.

Cords, M. 1997. Friendship, alliances, reciprocity and repair. In: A. Whiten, and R. W. Byrne (eds.), *Machiavellian Intelligence II*: *Extensions and Evaluation*, pp. 24–49. Cambridge University Press, Cambridge.

Cords, M. 2000a. Agonistic and affiliative relationships in a blue monkey group. In: P. F. Whitehead and C. J. Jolly (eds.), *Old World Monkeys*, pp. 453–479. Cambridge University Press, Cambridge.

Cords, M. 2000b. The number of males in guenon groups. In: P. Kappeler (ed.), *Primate Males: Causes and Consequences in Group Composition*, pp. 84–96. Cambridge University Press, Cambridge.

Cords, M. 2002. Friendships among blue monkey adult females (*Cercopithecus mitis*). *Behaviour* **139**: 291–314.

Fairbanks, L. A. 1988. Vervet monkey grandmothers: effects on mother–infant relationships. *Behaviour* **104**:176–188.

Fairbanks, L. A. 1989. Early experience and cross-generational continuity of mother–infant contact in vervet monkeys. *Dev. Psychobiol.* **22**: 669–681.

Fairbanks, L. A., and McGuire, M. T. 1985. Relationships of vervet mothers with sons and daughters from one through three years of age. *Anim. Behav.* **33**:40–50.

Fairbanks, L. A., and McGuire, M. T. 1987. Mother–infant relationships in vervet monkeys: response to new adult males. *Int. J. Primatol.* **8**:351–366.

Fashing, P. J., and Cords, M. 2000. Diurnal primate densities and biomass in the Kakamega Forest: an evaluation of census methods and a comparison with other forests. *Am. J. Primatol.* **50**:139–152.

Harlow, H. F., and Harlow, M. K. 1969. Effects of various mother–infant relationships on rhesus monkey behaviors. In: B. M. Foss (ed.), *Determinants of Infant Behavior*, pp. 15–36. Methuen, London.

Harrison, M. J. S. 1983. Age and sex differences in the diet and feeding strategies of the green monkey, *Cercopithecus sabaeus. Anim. Behav.* **31**:969–977.

Hinde, R. A. 1969. Analyzing the roles of the partners in a behavioral interaction—mother–infant relations in rhesus macaques. *Ann. NY Acad. Sci.* **159**:651–667.

Hinde, R. A., and Atkinson, S. 1970. Assessing the roles of social partners in maintaining mutual proximity, as exemplified by other infant relations in rhesus monkeys. *Anim. Behav.* **18**: 169–176.

Hinde, R. A., and Spencer-Booth, Y. 1967. The behaviour of socially living rhesus monkeys in their first two and a half years. *Anim. Behav.* **15**:169–196.

Hinde, R. A., and Spencer-Booth, Y. 1971. Effects of brief separation from mother on rhesus monkeys. *Science* **173**:111–118.

Hooley, J. L. 1983. Primiparous and multiparous mothers and their infants. In: R. A. Hinder (ed.), *Primate Social Relationships*, pp. 142–145. Sinauer Associates Inc., Sunderland, Massachusetts.

Hooley, J. M., and Simpson, M. J. A. 1981. A comparison of primiparous and multiparous mother–infant dyads in *Macaca mulatta. Primates* **22**:379–392.

Johnson, R. L. 1986. Mother–infant contact and maternal maintenance activities among free-ranging rhesus monkeys. *Primates* **27**:191–203.

Johnson, R. L., and Southwick, C. H. 1984. Structural diversity and mother–infant relations among rhesus monkeys in India and Nepal. *Folia Primatol.* **43**:198–215.

Karssemeijer, G. J., Vos, D. R., and van Hooff, J. A. R. A. M. 1990. The effect of some non-social factors on mother–infant contact in long-tailed macaques (*Macaca fascicularis*). *Behaviour* **113**:272–291.

Kirkevold, B. C., and Crockett, C. M. 1987. Behavioral development and proximity patterns in captive DeBrazza's monkeys. In: E. L. Zucker (ed.), *Comparative Behavior of African Monkeys*, pp. 39–65. Alan R. Liss, Inc., New York.

Lee, P. C. 1984a. Early infant development and maternal care in free-ranging vervet monkeys. *Primates* **25**:36–47.

Lee, P. C. 1984b. Ecological constraints on the social development of vervet monkeys. *Behaviour* **91**:245–262.

Loy, K. M., and Loy, J. 1987. Sexual differences in early social development among captive patas monkeys. In: E. L. Zucker (ed.), *Comparative Behavior of African Monkeys*, pp. 23–37. Alan R. Liss, Inc., New York.

Maestripieri, D. 1994a. Mother–infant relationships in three species of macaques (*Macaca mulata, M. nemestrina, M. arctoides*). I. Development of the mother–infant relationship in the first three months. *Behaviour* **131**:75–96.

Maestripieri, D. 1994b. Social structure, infant handling, and mothering styles in group-living Old World monkeys. *Int. J. Primatol.* **15**:531–553.

Maestripieri, D. 1995. First steps in the macaque world: do rhesus mothers encourage their infants' independent locomotion? *Anim. Behav.* **49**:1541–1549.

Maestripieri, D. 1998. Social and demographic influences on mothering style in pigtail macaques. *Ethology* **104**:379–385.

Meancy, M. J., Stewart, J., and Beatty, W. W. 1985. Sex differences in social play: the socialization of sex roles. *Adv. Stud. Behav.* **15**:1–58.

Mitchell, G. 1979. *Behavioral Sex Differences in Nonhuman Primates*, Van Nostrand Reinhold, New York.

Mitchell, G., and Stevens, C. W. 1969. Primiparous and multiparous monkey mothers in a mildly stressful social situation: the first three months. *Dev. Psychobiol.* **1**:280–286.

Nakagawa, N. 2000. Seasonal, sex, and interspecific differences in activity time budgets and diets of patas monkeys (*Erythrocebus patas*) and tantalus monkeys (*Cercopithecus aethiops tantalus*), living sympatrically in northern Cameroon. *Primates* **41**:161–174.

Nash, L., 1978. The development of the mother–infant relationship in wild baboons (*Papio anubis*). *Anim. Behav.* **26**:746–759.

Nicolson, N. A. 1987. Infants, mothers, and other females. In: B. B. Smuts, D. L. Cheney, R. M. Seyfarth, R. W. Wrangham, and T. T. Struhsaker (eds.), *Primate Societies*, pp. 330–342. University of Chicago Press, Chicago.

O'Brien, T. G., and Robinson, J. G. 1993. Stability of social relationships in female wedge-capped capuchin monkeys. In: M. E. Pereira, and L. A. Fairbanks (eds.), *Juvenile Primates: Life History, Development and Behavior*, pp. 197–210. Oxford University Press, Oxford.

Ransom, T. W., and Rowell, T. E. 1972. Early social development of feral baboons. In: F. Poirier (ed.), *Primate Socialization*, pp. 105–144. Random House, Inc., New York.

Rijt-Plooij, H. H. C. v. d., and Plooij, F. X. 1987. Growing independence, conflict and learning in mother–infant relations in free-ranging chimpanzees. *Behaviour* **101**:1–86

Rowell, T. E., and Chism, J. 1986. The ontogeny of sex differences in the behavior of patas monkeys. *Int. J. Primatol.* **7**:83–107.

Sackett, G. P. 1972. Exploratory behavior of rhesus monkeys as a function of rearing experiences and sex. *Dev. Psychol.* **6**:260–270

Seay, B. 1966. Maternal behavior in primiparous and multiparous rhesus monkeys. *Folia Primatol.* **4**:146–168.

Silk, J. B. 1987. Social behavior in evolutionary perspective. In: B. B. Smuts, D. L. Cheney, R. M. Seyfarth, R. W. Wrangham, and T. T. Struhsaker (eds.), *Primate Societies*, pp. 318–329. University of Chicago Press, Chicago.

Simpson, M. J. A. 1985. Effects of early experience on the behaviour of yearling rhesus monkeys (*Macaca mulatta*) in the presence of a strange object: classification and correlation approaches. *Primates* **26**:57–72.

Sokal, R. R., and Rohlf, F. J. 1981. *Biometry*, W. H. Freeman, New York.

Stanford, C. B. 1992. Costs and benefits of allomothering in wild capped langurs (*Presbytis pileata*). *Behav. Ecol. Sociobiol.* **30**:29–34.

Stevenson-Hinde, J., and Simpson, M. J. A. 1981. Mothers' characteristics, interactions, and infants' characteristics. *Child Dev.* **52**:1246–1254.

Struhsaker, T. T. 1971. Social behavior of mother and infant vervet monkeys (*Cercopithecus aethiops*). *Anim. Behav.* **19**:233–250.

Sussman, R. W. 1977. Socialization, social structure and ecology of two sympatric species of lemur. In: S. Chevalier-Skolnikoff, and F. E. Poirier (eds.), *Primate Biosocial Development: Biological, Social, and Ecological Determinants*, pp. 515–528. Garland Publishers, Inc., New York.

Tanaka, I. 1989. Variability in the development of mother–infant relationships among free-ranging Japanese macaques. *Primates*, **30**:477–491.

Terborgh, J. W. 1983. *Five New World Primates: A Study in Comparative Ecology*. Princeton, Princeton University Press.

Trivers, R. L. 1972. Parental investment and sexual selection. In: B. Campbell (ed.), *Sexual Selection and the Descent of Man 1871–1971*, pp. 136–179. Aldine Publishing Company, Chicago.

van Noordwijk, M. A., Hemelrijk, C. K., Herremans, L. A. M., and Sterck, E. H. M. 1993. Spatial position and behavioral sex differences in juvenile long-tailed macaques. In: M. E. Pereira, and L. A. Fairbanks (eds.), *Juvenile Primates*, pp. 77–85. Oxford University Press, New York.

van Schaik, C. P., van Noordwijk, M. A., Boer, R. J., and den Tonkelaas, I. 1983. Party size and early detection of predators in Sumatran forest primates. *Primates* **24**:211–221.

Whitten, P. L. 1982. *Female Reproductive Strategies among Vervet Monkeys*. Ph.D. thesis, Harvard University, Cambridge.

Influence of Foraging   19
Adaptations on Play Activity
in Red-tailed and Blue
Monkeys with Comparisons
to Red Colobus Monkeys

ERIC A. WORCH

Introduction

Although play is considered important for the development of certain skills which help prepare young animals for adult roles (Fagen, 1993; Fairbanks, 1993; Markus and Croft, 1995; Bekoff and Allen, 1998), the amount of play reported for wild primate groups ranges from none to approximately 10% of the daily activity budget (Baldwin and Baldwin, 1976; Fagen, 1981). Many variables may produce within- and between-species differences in play behavior; among those frequently mentioned are diet, food availability, group size, predation pressure, substrate, breeding seasons, and weather. In this study, I focus on the influence of diet and foraging strategy on the amount of time primates play by contrasting the play behavior of two largely frugivorous and insectivorous guenon species with a folivorous colobine.

ERIC A. WORCH • Education Department, University of Michigan, Flint, MI 48502, USA.
The Guenons: Diversity and Adaptation in African Monkeys, edited by Glenn and Cords. Kluwer Academic/Plenum Publishers, New York, 2002.

Dietary quality is especially likely to influence the amount of time devoted to play in primates by affecting their overall energetic and time budgets. The low-quality diets associated with folivory may limit the amount of play in which members of folivorous species engage relative to species with higher-quality diets of fruit or prey. Play quickly drops out of the behavioral repertoire when food resources are marginal (Fagen, 1981) and playfulness, i.e., time spent playing, has been used as an indicator of dietary quality in primates (Sommer and Mendoza-Granados, 1995). A strong positive correlation between time spent playing and food availability indicates that animals are subsisting on marginal food resources, whereas no relationship suggests food availability is not a limiting factor and that other variables are affecting playfulness. On the other hand, for folivorous species, resources are generally more abundant and occur in larger patches. Compared to frugivorous or insectivorous species, folivorous primates therefore tend to spend less time looking for food, travel shorter distances between patches, and spend less time traveling than frugivorous or omnivorous primates (Clutton-Brock and Harvey, 1977). Time for play may therefore be more limited in nonfolivorous groups.

An insectivorous diet may influence play behavior more directly by affecting the proximity of potential play partners. Nonsocial arthropods are likely rather evenly distributed yet relatively rare and elusive food items, and animals that feed extensively on arthropods should space themselves widely to reduce exploitive and interference competition (Miller, 1969; Gaulin, 1979; Waser, 1984), an assumption supported by primate field studies (Homewood, 1976; Struhsaker, 1980; Waser, 1984; Treves, 1998, 1999). Arthropod foraging also requires more consistent movement through the canopy, which requires infants to cling to their mother's ventrums much of the time (Worch, pers. obs.). Playfulness is directly related to the accessibility of potential play partners (Baldwin and Baldwin, 1974; Cheney, 1978; Lee, 1981). In addition to the number of immatures in a group (Baldwin and Baldwin, 1974), the ability to see or make contact with potential play partners should affect playfulness. Therefore, species highly dependent on nonsocial arthropods should play less than those that feed primarily on foods that permit smaller inter-individual distances and greater infant mobility, such as fruit and especially leaves.

Play tends to occur in a relaxed field (Burghardt, 1984), that is, animals need to be in an atmosphere of comfort, low tension, and relative safety before play behavior is manifested. Therefore, group activity and sociality should affect how much primates play. Conditions that enable group members to rest, socialize, and feed in large numbers for extended periods should be conducive to play.

In this chapter, I address the following question: How does a primate's foraging strategy affect the ability of immatures to engage in play? I propose that an arboreal foraging strategy heavily dependent on arthropods provides fewer opportunities to play than one directed primarily toward the consumption of leaves. To test this hypothesis, I examine five variables that are causes or consequences of a species' foraging strategy and that have the potential to affect play in arboreal primates: diet as it relates to foraging effort, day range length, inter-individual spacing, group activity, and ventral clinging by infants.

Methods

Study Site and Subjects

The study took place at the Makerere University Biological Field Station near Kanyawara in the northwest corner of Kibale National Park, Uganda. At an altitude of approximately 1500 m, Kibale is characterized as a medium-altitude, moist evergreen forest (Langdale-Brown *et al.*, 1964). It is the site of numerous studies and has been fully described (Strusaker, 1975; Skorupa, 1986).

From November 1995 through October 1996, two field assistants, John Rusoke and Bruce Aloysius, and I collected data on two groups of red-tailed monkeys (*Cercopithecus ascanius schmidti*), RT1 and RT2, two groups of blue monkeys (*C. mitis stuhlmanni*), BM2 and BM5, and two groups of red colobus monkeys (*Procolobus badius tephrosceles*), RC1 and RC2 (Table I). Both assistants worked on previous primate and phenology studies at Kibale and were familiar with data collection techniques. Study groups were habituated to the presence of human observers.

Observational Methods

We collected data during 16-minute sampling periods beginning on the hour and half hour using focal animal sampling and instantaneous recording at two-minute intervals (Altmann, 1974). The density of the forest canopy and the arboreality of the species rendered individual identification difficult, especially of immatures that had no scars or other deformities. Sexing infant and juvenile red-tailed and blue monkeys was not possible. Therefore, the unit of analysis for this study is age class, to which I assigned individuals based on distinctive physical and behavioral characteristics adapted from Oates (1974), Struhsaker (1975) and the NRC (1981).

Each observer followed a different focal animal for an entire day or for as long as possible. When a focal subject's location was unknown for more than 20 minutes, we selected as a new focal subject the first visible individual from the same age class or, if none was present, the first subject located in a pre-determined age class. To reduce the likelihood of repeatedly sampling a single individual in each age class from day to day, we simultaneously collected data on different individuals of the same age class whenever possible. We followed each species on a rotating basis for six to seven consecutive days per month

Table I. Number of Subjects and 16-Minute Samples per Age Class and Species[a]

Age class[b]	RT1		RT2		BM2		BM5		RC1		RC2	
	Subject	Sample	Subject	Sample	Subject	Sample	Subject	Sample	Subject	Sample	Subject	Sample
Very young infant (VYI)	2	28	0	0	0	0	0	0	1	6	1	10
Young infant (YI)	1	51	2	7	1[c]	23	0	0	4	8	1	101
Old infant (OI)	1	36	3	12	1[c]	94	1	15	2	28	3	140
Unaged infant (UI)	0	0	0	0	0	0	0	0	0	0	4	0
Young juvenile (YJ)	3	63	2	39	2	102	2	35	5	24	4	148
Old juvenile (OJ)	4	70	5	62	2	129	2	22	4	36	4	92
Subadult female (SAF)	3	30	4	14	2	36	2	7	4	9	2	63
Subadult male (SAM)	3	29	1	5	1	28	1	4	3	1	6	14
Adult female (AF)	6	16	9	2	4	36	4	1	10	4	11	40
Adult male (AM)	1	9	1	1	1	15	1	34	7	17	6	48
Unsexed adult/ subadult (UA)	2	0	5	0	0	0	0	0	0	0	0	0
Total	26	332	32	142	13	463	13	118	40	133	42	656

[a] RT1 = red-tailed group 1, RT2 = red-tailed group 2, BM2 = blue monkey group 2, BM5 = blue monkey group 5, RC1 = red colobus group 1, RC2 = red colobus group 2.
[b] Infants cross large gaps clinging to the mother's ventrum. Young infants are distinguished from old ones by bodily size and the color of fur and skin. Juveniles cross large gaps independently. Young juveniles are distinguished from old ones primarily by bodily size. Subadult females are smaller than adults and have button-like nipples. Subadult males are about the size of adult females and have underdeveloped genitalia (Oates, 1974; Struhsaker, 1975; NRC, 1981; Worch, 1998).
[c] Same individual at different times during study.

from 0700 to 1900 h. BM2, RT1, and RC2 were the primary study groups and were each followed ≥ 4 days per month.

At two-minute intervals, we recorded the subject's activity, location, and tree species, and the nearest neighbor's age class, sex, activity, and distance from the focal animal. When the activity was feeding, we recorded the species and food type. When the subject played, we recorded the specific motor activity and the number and age class of play partners.

I defined play as motor patterns resembling those used in serious functional contexts, but used in modified forms in activities that appeared to have no obvious immediate benefits to the player. The motor acts had some or all of the following features: exaggeration of movements and repetition, fragmentation, and disordering of sequences of motor acts relative to how they would be performed in functional contexts (Martin and Caro, 1985). Although phrased somewhat differently, this definition resembles others frequently articulated in the literature (Beckoff and Byers, 1981; Fagen, 1981).

I distinguished play fighting from serious fighting by its lighter blows and bites, longer duration, reversals of aggressive roles, and absence of screams and threat behaviors (Cheney, 1978). It consisted primarily of wrestling, grappling, pushing, and chasing. Locomotor play occurred in solitary and social contexts and consisted mainly of gamboling, leaping, hanging, dropping, bouncing, and running. Object play also occurred in solitary and social contexts and involved manipulation of attached vegetation, detached vegetation, and inanimate objects. I distinguished play from exploration, which is a more cautious, multi-sensory examination of a novel object (Fagen, 1981).

I defined foraging as searching or reaching for, picking, manipulating, or consuming food items. Foraging for arthropods typically involves more searching, manipulating, and moving through the canopy by individuals and groups than foraging for vegetal items, especially foliage. Because I did not count the number of items ingested, my data represent the time and effort individuals spent feeding and foraging on specific food items.

Interobserver Reliability

We ran eight reliability tests, each consisting of four to five independent samples taken simultaneously by all three observers on the same animal for a total of 38 records. For this study, age category and activity are the pertinent variables. Reliability for each variable was expressed as a percentage based on the number of times a field assistant disagreed with my assessment. Measures deviated by 10% or less, with no significant differences among the observers ($\chi^2_{\text{Age}} = 0.889$, df $= 1$, $n = 76$, $p = 0.346$; $\chi^2_{\text{Activity}} = 0.889$, df $= 1$, $n = 76$, $p > 0.346$). Discrepancies within the activity category were associated with problems distinguishing among resting, scanning environment, and watching others. I collapsed these actions into a single activity: rest.

Sampling Biases for Time of Day

Primate activity varies greatly with time of day (Struhsaker, 1975; Oates, 1977). Over the entire study, the average distribution of samples collected per hour is remarkably similar between species (Fig. 1). This suggests that the data are largely unaffected by systematic biases in the sampling schedule.

Food Availability

Chapman and Wrangham (unpub. data) estimated food availability throughout the forest by walking vegetation transects once per month. For each tagged tree, they measured dbh and noted the presence of ripe fruit, unripe fruit, flowers, young leaves, and mature leaves. To estimate the food available on each tree, I multiplied the sum of scores for each food type except mature leaf by the dbh (Chapman *et al.*, 1992). Mature leaf was excluded from the calculation in this study because this food item comprised less than 2% of each study group's diet (Worch, unpub. data). I estimated total food availability by summing the food availability estimates of all tree species on which each primate was known to feed (Struhsaker, 1975; Rudran, 1978; Cords, 1987; Worch, unpub. data). Not all plant species were represented in the transects. For example, red colobus monkeys fed on parts of *Eucalyptus* spp., an exotic tree that grew at the periphery

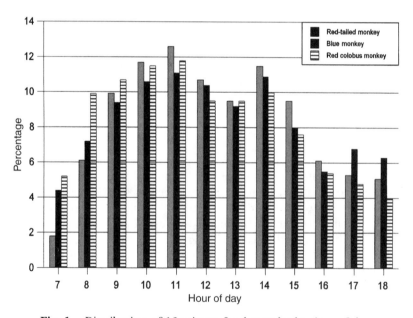

Fig. 1. Distribution of 16-minute focal samples by time of day.

of the forest and outside transect boundaries. Herbaceous plants and lianas also were not measured. I did not estimate arthropod availability.

Statistical Analysis

I grouped data into monthly mean scores for each age class by species to smooth over the possible effects of individual identity, variation in sampling schedule and environmental fluctuations relating to weather and predation events. I performed Kruskal–Wallis, Mann–Whitney and Chi Square tests to compare groups and two-tailed Spearman rank correlations, reported here as r, to calculate associations between variables. Because group size is correlated with the amount of time primates play socially (Baldwin and Baldwin, 1974), I controlled for the number of potential play partners in each group during all analyses involving play. To offset the effects of performing multiple analyses on the data, the criterion for significance is set at 0.025. I performed all analyses using SPSS following Norušis (1993).

Results

Time Spent Playing

Play was a rare behavior in the guenons compared to the red colobus monkeys [Table II, Fig. 2(a, b)]. Typical of the cercopithecine pattern, the amount of time devoted to play declined with age in all three species (Mori, 1974; Owens, 1975; Symons, 1978; Levy, 1980; Koyama, 1985; Pereira, 1985; Fagen, 1993). Play was not observed beyond the juvenile period in any species.

Play represented a similar proportion of the daily activity budget in immature blue and red-tailed monkeys [Fig. 2(a). M–W $U = 61.50$, $n = 24$ months, $p = 0.542$]. The only difference between the species involved infants, who played more often in blue monkeys relative to the number of partners available [Fig. 2(b), $U = 26.00$, $n = 24$ months, $p < 0.025$]. A far more striking difference occurred between the guenons and red colobus, with infant and juvenile red colobus monkeys playing more than ten times as much as their guenon counterparts [Fig. 2(a), K–W $H = 23.25$, df $= 2$, $n = 36$ months, $p < 0.025$], and more than four times when play time was controlled for the number of potential play partners [Fig. 2(b), $H = 19.95$, df $= 2$, $n = 36$ months, $p < 0.025$].

Play, Food Availability, Diet, and Foraging Effort

Food availability data were collected for only the first six months of the study. Spearman rank correlations reveal no relationship between food availability and time spent playing ($r_{BM} = -0.841$, $n = 6$, $p = 0.036$;

Table II. Play, Day Range Length, and Diet of the Study Species

| Species | Play | | Day Range | | Diet | | | | | | |
	(%)[a]	n	(m)	n	Leaf (%)[b]	n	Fruit (%)[b]	Arthropod (%)[b]	Other (%)[b]	n
Red-tailed monkey	2.13 ± 02.30	75	1252 ± 295	48	11.56 ± 11.41		19.60 ± 14.99	66.33 ± 19.29	2.51 ± 2.96	199
Blue monkey	1.83 ± 02.52	77	1060 ± 391	65	23.75 ± 17.56		24.58 ± 20.56	47.92 ± 29.34	3.75 ± 5.60	240
Red colobus monkey	24.38 ± 10.85	1274	670 ± 176	66	93.66 ± 05.20		3.52 ± 03.91	1.06 ± 01.50	1.76 ± 3.68	284

[a] Percent of observation time.
[b] Percent of feeding records.

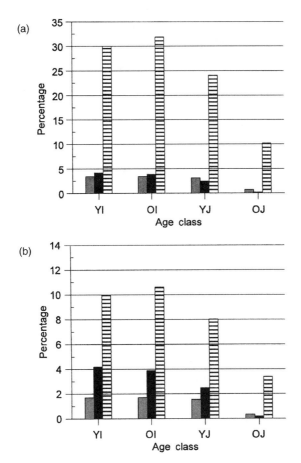

Fig. 2. Percentage of monthly activity budget devoted to play (a) uncontrolled and (b) controlled for the number of potential play partners, by age class (YI = young infant, OI = old infant, YJ = young juvenile, OJ = old juvenile). Shading as in Figure 1.

$r_{RT} = 0.058$, $n = 6$, $p = 0.913$; $r_{RC} = -0.793$, $n = 6$, $p = 0.06$). These results suggest that the fluctuation in food availability was not large enough to influence the amount of time each species played.

Each species showed fluctuations from month to month in the way major items contributed to diet. Within each species, however, correlations across months reveal no relationship between the amount of play and the foraging effort devoted to leaves, fruit, and arthropods (blue monkey play: $r_{Fruit} = 0.071$, $n = 12$, $p = 0.827$; $r_{Arthropod} = -0.207$, $n = 12$, $p = 0.519$; $r_{Leaf} = 0.144$, $n = 12$, $p = 0.654$; red-tailed monkey play: $r_{Fruit} = 0.543$, $n = 12$, $p = 0.068$; $r_{Arthropod} = -0.373$, $n = 12$, $p = 0.232$; $r_{Leaf} = 0.162$, $n = 12$,

$p = 0.615$;   red   colobus   monkey   play: $r_{Fruit} = 0.049$, $n = 12$, $p = 0.879$; $r_{Arthropod} = 0.460$, $n = 12$, $p = 0.132$; $r_{Leaf} = 0.175$, $n = 12$, $p = 0.587$).

Dietary differences between species, however, remained from month to month, with red colobus monkeys spending more time foraging for leaves and less time for fruit and arthropods than the guenons ($H_{Leaf} = 24.62$, df $= 2$, $n = 36$ months, $p < 0.025$; $H_{Fruit} = 8.61$, df $= 2$, $n = 36$ months, $p < 0.025$; $H_{Arthropod} = 20.36$, df $= 2$, $n = 36$ months, $p < 0.025$). With all three species included in the analysis, time spent playing was positively correlated with time foraging for leaves and negatively correlated with time foraging for arthropods ($r_{Leaf} = 0.749$, $n = 36$ months, $p < 0.025$; $r_{Arthropod} = -0.760$, $n = 36$ months, $p < 0.025$). In summary, intensive foraging for arthropods by the guenons and specializing on foliage by red colobus appears to produce highly divergent propensities to play that are not sensitive to monthly fluctuations in the amount of time each group spent foraging on different food items.

Play and Day Range Length

Red-tailed and blue monkeys did not differ in day range length (Table II, $U = 34$, $n = 22$ months, $p = 0.082$); however, both guenon species traveled nearly twice as far each day as red colobus monkeys ($H = 17.29$, df $= 2$, $n = 33$ months, $p < 0.025$). Within species, no significant correlations were found between day range length and the amount of time spent playing ($r_{BM} = 0.169$, $n = 11$ months, $p = 0.619$; $r_{RT} = 0.424$, $n = 11$ months, $p = 0.194$; $r_{RC} = 0.318$, $n = 11$ months, $p = 0.340$), whereas a negative relationship was found when all three species were included in the analysis ($r = -0.407$, $n = 33$ months, $p < 0.025$).

Play and Time Spent Moving and Clinging

Infant red-tailed and blue monkeys did not differ in the combined time they traveled independently and clung to their mother's ventrums ($\bar{x}_{BM} = 45.12\% \pm 12.86$; $\bar{x}_{RT} = 55.03 \pm 17.18$; $U = 38.00$, $n = 22$, $p = 0.140$). Infants in both species spent significantly more time moving and clinging than infant red colobus monkeys ($\bar{x}_{RC} = 36.63\% \pm 12.77$; $H = 7.74$, df $= 2$, $n = 33$ months, $p < 0.025$). Within each species, there was no relationship between time spent playing and time spent moving and clinging ($r_{BM} = 0.164$, $n = 11$ months, $p = 0.631$; $r_{RT} = 0.047$, $n = 11$ months, $p = 0.207$; $r_{RC} = -0.569$, $n = 11$ months, $p = 0.068$); however, with all three species included in the analysis, a negative correlation was found between time spent playing and the total time spent moving and clinging ($r = -0.042, n = 33$ months, $p < 0.025$).

Inter-Individual Spacing

Extensive foraging for arthropods should lead to greater dispersion of group members, whose individual success at insect foraging is likely to be compromised if other monkeys are nearby. It is apparent upon casual observation that individual red-tailed and blue monkeys space themselves farther apart than red colobus monkeys. This pattern is reflected in records of the number of conspecifics within 5 m of the focal subject (Table III, Strusaker, 1980). Red colobus groups were more cohesive than guenon groups ($H = 9.38$, df $= 2, n = 95, p < 0.025$).

Table III. Inter-Individual Spacing Based on the Percentage of Point Samples Having No Neighbors within 5 m of Focal Subjects

Focal animal	Red-tailed monkey		Blue monkey		Red colobus monkey	
	%	n	%	n	%	n
Subadult female	50.3	396	46.3	387	25.0	648
Subadult male	70.0	297	47.9	288	33.3	135
Adult male	76.4	89	51.0	441	36.6	585
Adult female	52.4	143	45.2	261	22.3	287
Adult female w/infant	37.5	16	25.0	72	20.8	106

Play and Group Activity

When play was recorded for the focal animal, the activity of the nearest nonplaying adult neighbor was recorded (Table IV). In all three species, play was observed more often than expected when the nearest adult neighbor was resting or foraging for leaves and fruit and less than expected when it was foraging for arthropods or moving ($\chi^2_{BM} = 17.00$, df $= 1$, $n = 68$, $p < 0.025$; $\chi^2_{RT} = 6.90$, df $= 1$, $n = 58$, $p < 0.025$; $\chi^2_{RC} = 561.47$, df $= 1$, $n = 931$, $p < 0.025$). If the activity of the nearest adult neighbor generally reflects the activity of other group members, play is most likely to occur when groups are more or less stationary in a feeding patch or resting.

Table IV. Activities of the Nearest Adult Neighbors of Focal Subjects (%)

Activity[a]	Red-tailed monkey ($n = 58$)	Blue monkey ($n = 68$)	Red colobus monkey ($n = 931$)
Rest	58.6	33.9	58.7
Forage (F&L)	20.7	41.2	30.2
Forage (A)	5.2	13.1	1.1
Move	15.5	11.8	10.0

[a] Rest includes grooming behavior, F = fruit, L = leaves, A = arthropods.

Discussion

This study supports the hypothesis that an arboreal foraging strategy heavily dependent on arthropods provides fewer opportunities to play than one directed primarily toward the consumption of leaves. Red colobus monkeys played *ca.* ten times more than red-tailed and blue monkeys. Although some aspects of play in these species have been discussed (Galat-Luong, 1975; Starin, 1990), there is scant information regarding the amount of time each species plays. For both red-tailed and red colobus monkeys, Struhsaker (1980) reported values smaller than mine. These discrepancies are most likely due to differences in sampling. My play data are based entirely upon infants and juveniles, the only age classes observed to play, whereas Struhsaker's (1980) values are based on samples made across all age classes. The relative values for day range length and time spent feeding on different food items are consistent with previous work on these species (Struhsaker, 1975; Rudran, 1978).

Within each species, monthly variation in play was not correlated with variations in time spent foraging for different food items or food availability. These findings contrast with observations of primates at other locations. Play behavior is known to decrease in a number of species when food resources decrease (Loy, 1970; Baldwin and Baldwin, 1976; Barrett *et al.*, 1992; Sommer and Mendoza-Granados, 1995). The results of this study suggest play in these groups was not noticeably affected by the fluctuation in food availability.

Play tends to occur when groups engage in activities that permit extended periods of immobility, especially resting and foraging for leaves, and when play partners are readily available. Potential play partners are more difficult to access in red-tailed and blue monkey groups than in red colobus monkey groups. The guenons space themselves 50–100% farther apart than red colobus monkeys (Struhsaker, 1980; Treves, 1998, 1999, unpub. data; Treves and Baguma, 2002). For much of the day, an infant guenon's only companion is its mother, who is constantly on the move searching for arthropods. As a consequence, blue and red-tailed monkey infants spend more time clinging to their mother and more time traveling than red colobus monkey infants, presumably to avoid becoming separated from her. Likewise, juvenile guenons engrossed in play should run a greater risk of becoming separated from their group than red colobus juveniles and this may account for the exceptionally low degree of playfulness in juvenile guenons. In summary, high group dispersion among the guenons and the need to keep up with the group tend to inhibit both social and solitary play.

The lack of association between play and ecological variables within the study groups suggests that playfulness in these species is a function of grade differences. That is, the physical adaptations related to arthropod foraging, such as small body size, high metabolic turnover and rapid gut passage rates (Milton, 1984), and behavioral adaptations, such as wide inter-individual spacing and persistent travel through the canopy, predispose red-tailed and blue monkeys to be relatively unplayful even when opportunities to play are optimal, such as during times of fruit abundance. On the other hand, physical adaptations associated with specializing on leaves, such as large body size, low metabolic rates and slow gut passage rates (Milton, 1984), and behavioral adaptations, such as extended periods of feeding and resting, close inter-individual spacing and shorter day range lengths, facilitate play in red colobus monkeys.

The specific relationships I have described among diet, foraging strategy, and play are not necessarily transferrable to other species. Many factors other than those I have discussed affect a species' playfulness, including sociality, substrate size, and predation pressure. However, it is apparent that a species' foraging adaptations can either facilitate or hinder play in arboreal primates. An understanding of these relationships provides a basis upon which to launch further investigations of play among primates and other animals.

Summary

In this chapter, I addressed the following question: How does a primate's foraging strategy affect the ability of immatures to engage in play? I discussed how diet and foraging strategy affect day range length, group activity, inter-individual spacing, and ventral clinging by infants, which in turn, influence an animal's ability to play. Based on observations of red-tailed, blue, and red colobus monkeys, I provided evidence to support the conclusion that species adapted to diets highly dependent on arthropods engage in behaviors that inhibit both social and solitary play in primates, including persistent movement through the canopy, a long day range, wide inter-individual spacing, and extended periods of infant clinging, whereas, species that specialize on foliage engage in behaviors that facilitate play, including extended periods of adult resting and feeding, a short day range, and close inter-individual spacing.

ACKNOWLEDGMENTS

This study was made possible through the cooperation of the Ugandan President's Office, the National Research Council of Uganda, Makerere University and the Uganda Wildlife Authority, and by the financial support from a Sigma Xi Grant-in-Aid of Research, a David C. Skomp Fellowship in Anthropology, an E. Wayne Gross Fellowship in Science Education, and a Grant-in-Aid of Research and a Collaborative Graduate-Undergraduate Research Fellowship from Indiana University. I am especially grateful to John Rusoke, Bruce Aloysius, and Pascal Baguma for their assistance in the field.

References

Altmann, J. 1974. Observational study of behavior: Sampling methods. *Behaviour* **49**:227–265.

Baldwin, J. D., and Baldwin, J. I. 1974. Exploration and social play in squirrel monkeys (*Saimiri*). *Am. Zool.* **14**:303–315.

Baldwin, J. D., and Baldwin, J. I. 1976. Effects of food ecology on social play: A laboratory simulation. *Z. Tierpsych.* **40**:1–14.

Barrett, L., Dunbar, R. I., and Dunbar, P. 1992. Environmental influences on play behavior in immature gelada baboons. *Anim. Behav.* **44**:111–115.

Beckoff, M. and Allen, C. 1998. Intentional Communication and Social Play: How and why animals negotiate and agree to play. In: M. Bekoff, and J. A. Beyers (eds.), *Animal Play: Evolutionary, Comparative, and Ecological Perspectives*, pp. 97–114. Cambridge University Press, Cambridge.

Beckoff, M., and Byers, J. A. 1981. A critical reanalysis of the ontogeny and phylogeny of mammalian social and locomotor play: An ethological hornet's nest. In: I. Immelmann, G. W. Barlow, L. Petrinovich, and M. Main (eds.), *Behavioral Development: The Bielefeld Interdisciplinary Project*, pp. 269–337. Cambridge University Press, Cambridge.

Burghardt, G. M. 1984. On the origins of play. In: P. K. Smith (ed.), *Play in Animals and Humans*, pp. 5–42. Basil Blackwell, New York.

Chapman, C. A., Chapman, L. J., Wrangham, R. W., Hunt, K. D., Gebo, D. L., and Gardner, L. J. 1992. Estimators of fruit abundance of tropical trees. *Biotropica* **24**:527–531.

Cheney, D. L. 1978. The play partners of immature baboons. *Anim. Behav.* **26**:1038–1050.

Clutton-Brock, T. H., and Harvey, P. H. 1977. Species differences in feeding and ranging behaviour in primates. In: T. H. Clutton-Brock (ed.), *Primate Ecology: Studies of Feeding and Ranging Behaviour in Lemurs, Monkeys and Apes*, pp. 557–584. Academic Press, London.

Cords, M. 1987. Mixed-species associations of Cercopithecus monkeys in the Kakamega forest. *Univ. Calif. Publ. Zool.* **117**:1–109.

Fagen, R. 1981. *Animal Play Behavior*. Oxford University Press, New York.

Fagen, R. 1993. Primate juveniles and primate play. In: M. E. Pereira, and L. A. Fairbanks (eds.), *Juvenile Primates: Life History, Development, and Behavior*, pp. 182–196. Oxford University Press, New York.

Fairbanks, L. A. 1993. Juvenile vervet monkeys: Establishing relationships and practicing skills for the future. In: M. E. Pereira, and L. A. Fairbanks (eds.), *Juvenile Primates: Life History, Development, and Behavior*, pp. 211–227. Oxford University Press, New York.

Galat-Luong, A. 1975. Notes préliminaires sur l'écologie de *Cercopithecus ascanius schmidti* dans les environs de Bangui (R.C.A.). *La Terre et la Vie* **29**:288–297.

Gaulin, S. J. C. 1979. A Jarman/Bell model of primate feeding niches. *Hum. Ecol.* **7**:1–20.

Homewood, K. M. 1976. *Ecology and Behavior of the Tana Mangabey.* Ph.D. Dissertation, University College, London.

Koyama, N. 1985. Playmate relationships among individuals of the Japanese monkey troop in Arashiyama. *Primates* **26**:390–406.

Langdale-Brown, I., Osmaston, H. A., and Wilson, J. G. 1964. *The Vegetation of Uganda and its Bearing on Land-use*, Government of Uganda, Entebbe.

Lee, P. C. 1981. *Ecological and Social Influences on the Development of Vervet Monkeys* (Cercopithecus aethiops). Ph.D. Dissertation, Cambridge University.

Levy, J. 1980. *Play Behavior and its Decline During Development in Rhesus Monkeys* (Macaca mulatta). Ph.D. Dissertation, University of Chicago.

Loy, J. 1970. Behavioural responses of free-ranging vervet monkeys to food shortage. *Am. J. Phys. Anthropol.* **33**:263–272.

Markus, N., and Croft, D. B. 1995. Play behaviour and its effects on social development of common chimpanzees (*Pan troglodytes*). *Primates* **36**:213–225.

Martin, P., and Caro, T. M. 1985. On the functions of play and its role in behavioral development. In: J. S. Rosenblatt, C. Beer, M. Busnel, and P. J. B. Slater (eds.), *Advances in the Study of Behavior* (Vol. 15), pp. 59–103. Academic Press, Orlando.

Miller, R. S. 1969. Competition and species diversity. *Brookhaven Symp. Biol.* **22**:63–70.

Milton, K. 1984. The role of food-processing factors in primate food choice. In: P. S. Rodman, and J. G. H. Cant (eds.), *Adaptations for Foraging in Nonhuman Primates*, pp. 249–279. Columbia University Press, New York.

Mori, U. 1974. The inter-individual relationships observed in social play of the young Japanese monkeys of the natural troop in Koshima Islet. *J. Anthropol. Soc. Jpn.* **82**:303–318.

National Research Council. 1981. *Techniques for the Study of Primate Ecology in the Tropics*, Author, Washington.

Norušis, M. J. 1993. *SPSS for Windows Base System User's Guide, Release 6.0*, SPSS Inc., Chicago.

Oates, J. F. 1974. *The Ecology and Behaviour of the Black-and-white Colobus Monkey* (Colobus guereza Rüppell) *in East Africa*. Ph.D. Dissertation, University of London.

Oates, J. F. 1977. The guereza and its food. In: T. H. Clutton-Brock (ed.), *Primate Ecology: Studies of Feeding and Ranging Behaviour in Lemurs, Monkeys and Apes*, pp. 275–321. Academic Press, London.

Owens, N. W. 1975. Social play behaviour in free-living baboons, *Papio anubis. Anim. Behav.* **23**:387–408.

Pereira, M. E. 1985. *Age Changes and Sex Differences in the Social Behavior of Juvenile Yellow Baboons* (Papio cynocephalus). Ph.D. Dissertation, University of Chicago.

Rudran, R. 1978. Socioecology of the blue monkeys (*Cercopithecus mitis stuhlmanni*) of the Kibale Forest, Uganda. *Smithson. Contrib. Zool.* **249**:1–88.

Skorupa, J. P. 1986. Responses of rainforest primates to selective logging in Kibale Forest, Uganda: A summary report. In: K. Benirschke (ed.), *Primates, the Road to Self-sustaining Populations*, pp. 57–70. Springer-Verlag, New York.

Sommer, V., and Mendoza-Granados, D. 1995. Play as indicator of habitat quality: A field study of langur monkeys (*Presbytis entellus*). *Ethology* **99**:177–192.

Starin, E. D. 1990. Object manipulation by wild red colobus monkeys living in the Abuko Nature Reserve, The Gambia. *Primates* **31**:385–391.

Struhsaker, T. T. 1975. *The Red Colobus Monkey*. The University of Chicago Press, Chicago.

Struhsaker, T. T. 1980. Comparison of the behaviour and ecology of red colobus and red-tailed monkeys in the Kibale Forest, Uganda. *Afr. J. Ecol.* **18**:33–51.

Symons, D. 1978. *Play and Aggression: A Study of Rhesus Monkeys*. Columbia University Press, New York.

Treves, A. 1998. The influence of group size and neighbors on vigilance in two species of arboreal monkeys. *Behaviour* **135**:453–481.

Treves, A. 1999. Vigilance and spatial cohesion in blue monkeys. *Folia Primatol.* **70**:291–294.

Treves, A. and Baguma, P. 2002. Interindividual proximity and surveillance of associates in comparative perspective. In: M. E. Glenn, and M. Cords (eds.), *The Guenons: Diversity and Adaptation in African Monkeys*, pp. 161–172. Kluwer Academic Publishers, New York.

Waser, P. M. 1984. Ecological differences and behavioral contrasts between two mangabey species. In: P. S. Rodman, and J. G. H. Cant (eds.), *Adaptations for Foraging in Nonhuman Primates*, pp. 195–216. Columbia University Press, New York.

Worch, E. A. 1998. *Play in Four Species of East African Monkeys: Implications for Early Childhood and Elementary Education*. Ph.D. Dissertation, Indiana University, Bloomington.

Effects of Natural and Sexual Selection on the Evolution of Guenon Loud Calls

20

KLAUS ZUBERBÜHLER

Introduction

Probably the most striking behavior in male forest guenons is their powerful vocalizations: loud calls or long-distance calls (Gautier and Gautier, 1977). In many species they carry over substantial distances, often greater than the diameter of the home range. Calls are generally low-pitched, which makes them less susceptible to attenuation by the forest vegetation (Wiley and Richards, 1978). Despite their unique and distinct acoustic structure, researchers have long suspected that male guenon loud calls are a functionally heterogeneous class. Whereas some loud calls function as anti-predator calls, some of them clearly have other functions.

In the Taï Forest, four species of guenons can be observed: the Diana monkey (*Cercopithecus diana*), the Campbell's monkey (*C. campbelli*), the lesser spot-nosed monkey (*C. petaurista*), and the putty-nosed monkey (*C. nictitans*). With the exception of the putty-nosed monkey, all species occur at high

KLAUS ZUBERBÜHLER • School of Psychology, University of St Andrews, St Andrews, Fife, KY16 9JU, UK.
The Guenons: Diversity and Adaptation in African Monkeys, edited by Glenn and Cords. Kluwer Academic/Plenum Publishers, New York, 2002.

densities, except where poaching has driven the population levels down. All four species live in groups with one resident adult male and several adult females and their offspring. Females are the philopatric sex and are probably closely related to one another, while males leave the natal group after puberty to live alone or in association with other monkey species, before trying to take over a group of females. Once a male has successfully established himself as the group's resident male, his tenure can be remarkably long: tenures of five years are common (Cords, 1987). Thus, the mating system of forest guenons is best described as female defense polygyny (Van Schaik and Van Hooff, 1983), and this mating system is a notorious target of sexual selection (Anderson, 1994).

Sexual selection, according to Darwin (1859/1993) "... depends, not on the struggle for existence, but on a struggle between the males for possession of the females; the result is not death of the unsuccessful competitor, but few or no offspring." In polygynous mating systems, male contest competition over females is especially high, which typically leads to the evolution of male traits that are useful in assuring preferential access to females or that females find attractive, such as large body size, e.g., elephant seals (Le Boeuf, 1974), weapons, e.g., antlers in red deer (Clutton-Brock, 1982), or conspicuous vocalizations to advertise male quality, e.g., roaring in red deer (Clutton-Brock and Albon, 1979). Hence, guenon loud calls could be the outcome of sexual selection, useful in advertising male quality and resource occupation over long distances, analogous to bird song (Catchpole and Slater, 1995). Although it is generally difficult to identify a trait as the outcome of sexual selection vs. natural selection (Anderson, 1994), a number of findings suggest that sexual selection could have played a key role in the evolution of male guenon loud calls, a hypothesis that has received little attention in the past. First, females do not develop the trait in any of the guenon species studied (Gautier and Gautier, 1977) and dimorphisms are a typical outcome of sexual selection. Second, males do not show the trait before sexual maturity (Gautier, 1978), again a frequently observed phenomenon in sexually selected traits. Third, loud calls possess an acoustic structure that is highly suited for long-distance communication, hence reaching a maximum number of potential male rivals to advertise male presence and dissuade solitary males from trying to approach the group. These findings are consistent with the notion that male loud calls serve to advertise male quality and resource occupation, that is, that they are the result of sexual selection.

At the same time, there are a number of observations that do not fit the notion of male loud calls as a sexually selected trait. For example, during intergroup encounters males often remain silent, while females engage in vigorous fights and vocal battles (Hill, 1994). More importantly, guenon males regularly produce loud calls in response to predators, both as resident males and sometimes also as solitary males living in association with other primate species (Gautier and Gautier, 1977; Zuberbühler, pers. obs.), suggesting that guenon loud calls function as ordinary predator alarm calls. Predator alarm calls, however, are typically treated as a classic example of a behavior evolved in response to natural selection (Maynard-Smith, 1965). Current theory predicts that individuals should produce alarm calls if the calls dissuade the predator (Bergstrom and Lachmann, 2001). This is the case if the predator depends on unprepared prey and the calls act as detection signals: the perception advertisement hypothesis (Flasskamp, 1994). Second, alarm calling is beneficial for the caller if the calls increase the survival chances of closely related nearby recipients: the conspecific warning hypothesis (Maynard-Smith, 1965).

The guenon loud calls, in other words, are an evolutionary puzzle. Their usage as alarm calls suggests an evolutionary pathway through ordinary natural selection. Some of their defining features, however, particularly the sexual dimorphism, the appearance with sexual maturity, and the structural adaptations for long-distance communication, suggest that sexual selection has exerted additional evolutionary effects. In this chapter, I seek to assess the relative importance of sexual and natural selection in the evolution of male guenon loud calls. For this purpose, I first investigate the impact of natural selection by reviewing a number of experimental studies conducted to test the predator-alarm call hypothesis of male loud calls (perception–advertisement and conspecific warning). I then contrast the results of these experiments with a number of observational findings of guenon loud call behavior to assess how sexual selection might have additionally affected the evolution of male guenon loud calls.

Study Site and Methods

I conducted a number of experiments on wild groups of Diana monkeys and Campbell's monkeys in the Taï Forest of western Ivory Coast to assess the relative importance of sexual and natural selection in the evolution of male guenon loud calls (see McGraw and Noë, 1995 for a description of the study site). The loud calls of the male Diana monkeys are best described as roars, consisting of a varying number of syllables. The loud calls of the male Campbell's monkeys are best described as hacks and booms. Taï guenons are faced with four major predators that differ in hunting technique: leopards (*Panthera pardus*), crowned hawk eagles (*Stephanoaetus coronatus*), chimpanzees (*Pan troglodytes*), and human poachers. Predator encounters are exceedingly difficult to study systematically in a dense rain forest habitat. However, previous work showed that primates recognize predators by their vocalizations alone (Hauser and Wrangham, 1990), suggesting that playing back recordings of predator vocalizations is a useful way to study primate antipredator behavior (see Zuberbühler *et al.*, 1997, 1999b; Zuberbühler, 2000c for an overview of the experimental methods).

Results

The Effects of Natural Selection: Loud Calls as Predator Alarm Calls

Loud calls as perception advertisement signals

Alarm calls are a typical outcome of natural selection (Maynard-Smith, 1965). To assess the effects of natural selection on the evolution of male loud calls, I conducted a series of playback experiments. In a first experiment, I played back vocalizations of the four different predators to different groups of Diana monkeys from a concealed speaker. The perception advertisement hypothesis of alarm calls predicts that males call only to predators that rely on unprepared prey such as leopards (Jenny, 1996) and crowned hawk eagles (Mitani *et al.*, 2001), but not to predators that can successfully pursue their prey such as chimpanzees (Boesch and Boesch, 1989) or humans (Martin, 1991).

Crowned hawk eagles and leopards reliably elicited loud calls in male Diana monkeys, while playbacks of chimpanzees and humans did not (Fig. 1). The perception–advertisement hypothesis was additionally supported by the fact that many males combined their calls with approach. For a predator, a calling and approaching male provides a clear and unambiguous signal that the prey is aware of the predator's presence and that hunting by stealth is no longer possible. Since chimpanzees and humans are also highly dangerous monkey predators, it is not plausible that differences in call rates were caused by differences in threat. More likely, the monkeys adjusted their antipredator strategies to the predators' hunting tactics. Males called only to predators that rely on unprepared prey to hunt successfully, but remained

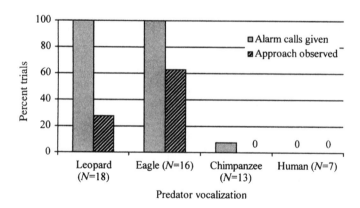

Fig. 1. Male Diana monkey antipredator responses to four different predators (data from Zuberbühler *et al.*, 1997).

quiet in the presence of pursuit hunters. Both chimpanzees and humans search for prey by acoustic cues. Once they have located a group, both are able to reach prey in the canopy, either by group hunting or with weapons. For the monkeys, the only effective strategy against them is to behave cryptically.

A comparative study, involving other primate species in the Taï Forest indicated that the pattern in Diana monkeys is also true for other species. Both leopards and chimpanzees regularly prey on all monkey species in Taï, although the guenons seem to be somewhat more successful in avoiding these two predators compared to the colobines (Boesch and Boesch-Achermann, 2000, Zuberbühler and Jenny, submitted). All species tested called significantly more often to leopards (which give up hunting after detection) than to chimpanzees (which are pursuit hunters; Zuberbühler *et al.*, 1999a). A radio-tracking study revealed that leopards hunt by approaching primate groups and hiding in their vicinity, presumably to attack unwary individuals on the ground (Jenny, 1996). Once the monkeys had detected a hiding leopard, they alarm called at high rates, which typically caused the leopard to move on (Zuberbühler *et al.*, 1999a), suggesting that differences in the alarm call rates are an adaptation to the predators' hunting tactic.

When responding to chimpanzees, the monkeys' choice of antipredator strategy is complicated by the fact that occasionally chimpanzees themselves fall prey to leopards (Boesch, 1991). Chimpanzees give loud and conspicuous alarm screams when they detect a leopard (Goodall, 1986). When I compared the Diana monkeys' responses to a playback of chimpanzee leopard alarm screams to chimpanzee social screams the following pattern emerged. In about half of all Diana monkey groups, both males and females switched from a chimpanzee-specific cryptic response (Fig. 1) to a leopard-specific conspicuous response, suggesting that in some groups, individuals assumed the presence of a leopard when hearing the chimpanzee leopard alarm screams. This was not the case in the control condition, in which I played back acoustically similar chimpanzee social screams to different groups of Diana monkeys (Fig. 2).

The territory of a chimpanzee group covers the home range of several dozen diana monkey groups (Höner *et al.*, 1999; Boesch and Boesch-Achermann, 2000). Chimpanzees do not use their territories evenly, but instead tend to spend most time in a core area (Herbinger *et al.*, 2001). Thus, Diana monkey groups whose home ranges are at the periphery of a chimpanzee group's range will encounter chimpanzees less often than groups whose home ranges are in the core area. A re-analysis of the vocal response of the different Diana monkey groups as a function of the location of their home range revealed that core area Diana monkey groups were significantly more likely to respond with leopard alarm calls to chimpanzee leopard alarm screams than peripheral Diana monkey groups, suggesting inter-group differences in semantic knowledge (Fig. 3).

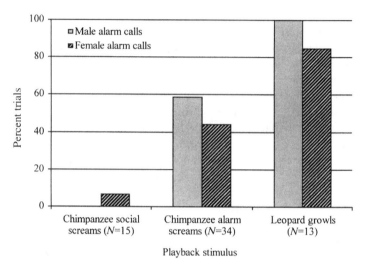

Fig. 2. Alarm call response of different groups of Diana monkeys to chimpanzee leopard alarm screams vs. chimpanzee social screams and leopard growls (data from Zuberbühler, 2000a). Sample sizes refer to the number of different Diana monkey groups tested with each playback condition.

Loud calls as conspecific warning signals

Conspecific warning, the second main function of predator alarm calls, suggests that males call to warn genetic relatives. In particular, this hypothesis predicts that acoustic variation in male calls correlates with the presence of

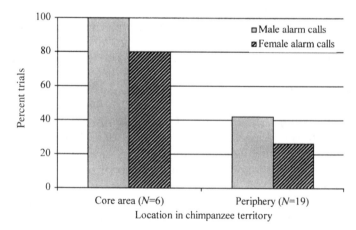

Fig. 3. Relationship between a groups' tendency to respond with leopard alarm calls to chimpanzee leopard alarm screams and their location within the resident chimpanzee territory (data from Zuberbühler, 2000a).

Male Diana monkeys   Female Diana monkeys

Calls to a leopard

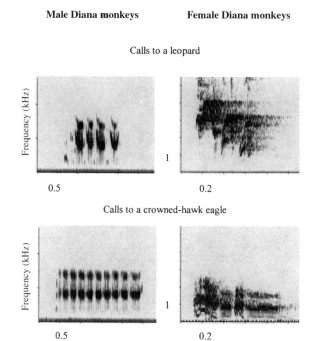

Calls to a crowned-hawk eagle

Time (s)

Fig. 4. Male and female Diana monkey calls to leopards and crowned hawk eagles.

specific predator classes and that conspecifics are sensitive to the variation. Operationally, this can be demonstrated if conspecifics react to a specific call type just as they would to the corresponding predator. For example, monkeys are expected to show the same response to playbacks of leopard vocalizations as to playbacks of male loud calls originally given to a leopard. Seyfarth *et al.* (1980) showed that alarm calls in vervet monkeys (*Cercopithecus aethiops*) inform conspecifics about the presence of particular predators and that the alarm calls alone were sufficient to elicit adaptive antipredator behavior in recipients.

Figure 4 illustrates typical exemplars of male and female calls given in response to leopards and eagles, and shows the striking acoustic dimorphism between the calls of the adult male and female Diana monkeys. Similar predator-dependent acoustic differences also characterize the loud calls of the male Campbell's monkeys (Zuberbühler, 2001). To determine whether male loud calls functioned as labels for the two predator categories, I conducted the following playback experiment. By altering playback stimuli and the position of a concealed speaker, I investigated whether males responded with acoustically different calls depending on a predator's distance (close vs. far),

Table I. Results of Acoustic Analysis of Diana Monkey Loud Calls in Response to Playbacks of Eagle and Leopard Vocalizations Presented from Different Locations (data from Zuberbühler, 2000c)

Acoustic parameter[*]	Predator feature		
	Distance	Elevation	Category
Call unit	–	–	*c*
Pre-syllable unit	–	–	*c*
Syllable unit	*a*	–	–
Inter-syllable unit	–	–	–
Fundamental frequency	–	*a*	–
Dominant frequency onset	–	–	*c*
Dominant frequency middle	–	–	*c*
Dominant frequency end	–	–	*c*
Frequency transition onset	–	–	*c*
Frequency transition offset	–	–	*b*

[a]$p < 0.05$; [b]$p < 0.01$; [c]$p < 0.005$, F-tests, DF = 1,19. [*]See Zuberbühler (2000c) for description of acoustic parameter.

elevation (above vs. below), or identity (eagle vs. leopard). Acoustic analysis of both Diana monkey and Campbell's monkey calls showed that the males consistently responded to predator category regardless of threat or direction of attack (Zuberbühler, 2000c, 2001; Table I).

Another important prediction of the conspecific warning hypothesis concerns the behavior of the recipients. When other monkeys hear male loud calls, are they sensitive to the acoustic variation in the calls and their covariation with predator class? If so, do they simply respond to the acoustic features or are individuals able to process the calls on a more abstract semantic level, possibly involving mental representations of the corresponding predator class? To investigate these possibilities, I played back recordings of loud calls by male Campbell's and Diana monkeys to leopards and eagles to different Diana monkey groups. Then, I compared the female Diana monkey responses to the stimuli to their responses to the actual predators (Zuberbühler *et al.*, 1997, Zuberbühler, 2000b). Figure 5 shows that female Diana monkeys respond as strongly to leopard growls and eagle shrieks as they do to the corresponding male calls.

Some observations indicated that if monkeys repeatedly heard leopard growls (or eagle shrieks) from the same location, then their vocal response decreased, presumably because the costs of calling began to outweigh the benefits. Conspecifics were warned, perception was signaled, and continuous calling might only attract additional predators to the site. I used this behavioral pattern to test whether female Diana monkeys were able to attend to the semantic content when listening to male loud calls. If female Diana monkeys were primed with a conspecific male's loud calls to a leopard (instead of the growls of a leopard) and subsequently tested with the acoustically novel (but

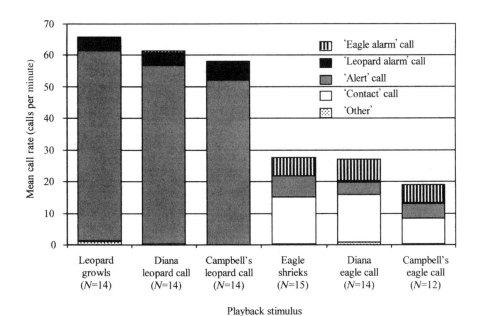

Fig. 5. Female vocal responses of different groups of Diana monkeys to playbacks of Diana monkey male loud calls, Campbell's monkey male loud calls, leopard growls, and eagle shrieks (data from Zuberbühler *et al.*, 1997, Zuberbühler, 2000b).

semantically similar) growls of a leopard, would they also refrain from giving leopard alarm calls to leopard growls, even though the growls normally elicit a strong response? And similarly, if the monkeys were primed with male eagle loud calls and subsequently tested with the acoustically and semantically novel leopard growls, would they show a normally strong response to leopard growls? Figure 6 illustrates the experimental design and the main findings of the study (Zuberbühler *et al.*, 1999b). In the experiment, the monkeys first heard a prime stimulus (a predator vocalization or a series of Diana monkey loud calls) followed by a probe stimulus (the corresponding or non-corresponding predator vocalization) five minutes later. If the monkeys were able to attend to the semantic features of the prime stimuli, then they should ignore semantically similar probe stimuli, regardless of their acoustic features. Results suggested that this was the case. When hearing male calls, Diana monkeys behaved as if they formed a mental representation of the corresponding predator, which caused them to respond weakly when hearing the predator's vocalizations as if they already anticipated its presence (Zuberbühler *et al.*, 1999b).

I used the same paradigm to test interspecific recognition of loud calls. Like the Diana monkeys, the male Campbell's monkeys produce two acoustically different loud calls to crowned hawk eagles and leopards (Zuberbühler, 2001). Diana monkeys often associate with Campbell's monkeys and are sensitive to

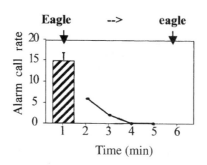

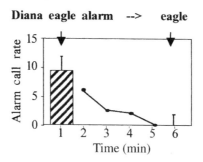

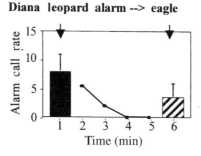

Fig. 6. Results of a prime-probe experiment, in which the acoustic and semantic resemblance between the prime and the probe stimuli were systematically varied. Solid bar: median number of female Diana monkey leopard alarm calls given per minute. Striped bar: median number of female Diana monkey eagle alarm calls given per minute (data from Zuberbühler *et al.*, 1999b).

the acoustic differences because they respond to them as if the corresponding predator were present. To investigate whether Diana monkeys are able to attend to the semantic features when hearing male Campbell's monkey calls to predators, I conducted an analogous experiment, now using male Campbell's monkey calls instead of Diana monkey calls. The acoustic features of the two stimuli always changed between prime and probe, while the semantic features either did or did not change. If monkeys understand the meaning of the loud calls, then they should respond weakly to the predator in cases where they have been primed with the semantically corresponding loud calls. They should, however, respond strongly to the predator in cases where they have been primed with the semantically non-corresponding loud calls because they cannot use the loud calls to predict the predator's presence. As in the previous study (Zuberbühler *et al.*, 1999b), Diana monkeys' responses to the predator were significantly weaker in groups primed with the corresponding Campbell's loud calls than in groups primed with the non-corresponding Campbell's calls, suggesting that Diana monkeys use Campbell's monkey calls as labels for different predator classes (Zuberbühler, 2000b).

Taken together, the experiments show that some guenon loud calls function as classic predator alarm calls. This is because males consistently use them to advertise perception to surprise predators and because the calls show acoustic differences that covary with predator type, which are used by recipients to identify the predator type present.

The Effects of Sexual Selection: Loud Calls to Advertise Male Quality

Male loud calls, however, exhibit several features that are not well explained by the classic notion of an alarm call. First, there is a strong sexual dimorphism between the male and the female calls given to the same predators (Fig. 4). Second, Campbell's monkeys and lesser spot-nosed monkeys exhibit diurnal patterns in their loud call production that are significantly different from random (*Cercopithecus campbelli*, $n = 84$ calls, $p < 0.01$; *C. petaurista*, $n = 56$; $p < 0.01$; *C. diana*, $n = 230$, $p > 0.2$; Kolmogorov–Smirnov tests; $n = 564$ observation hours over 60 days; Zuberbühler, 1993; Fig. 7), similar to other guenon species (Struhsaker, 1970). Although this analysis did not distinguish between different types of male loud calls, diurnal periodicity is not well explained by the hypothesis that loud calls function as predator alarm calls only, suggesting that male guenon loud calls possess an additional communicatory function. Although the males of the different species might have different thresholds for calling, predation-induced calling should lead to more evenly distributed calling frequencies in all three guenon species throughout the day. Instead, particularly the Campbell's monkeys exhibit a biased calling pattern with high calling rates in the evening, typically given during a very narrow time interval. Most other Taï monkeys do not follow this pattern, further suggesting that predation is not the cause of these calls. Moreover, the Campbell's evening calls do not elicit vocal responses by recipients, suggesting that these calls are given without external stimulation and hence do not function as alarm calls. Third, male guenons sometimes call when no predator is present, such as to a falling tree. In Campbell's monkeys, for example, males

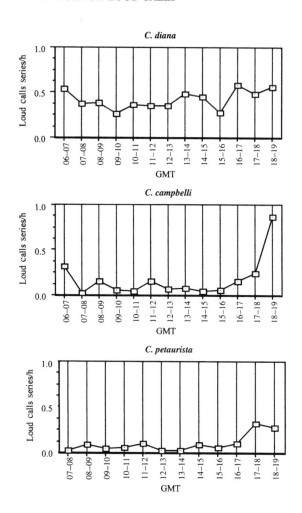

Fig. 7. Diurnal calling patterns of three guenon species in the Taï forest in 1991 (data from Zuberbühler, 1993).

typically respond to a falling tree by producing a series of loud calls that resemble the leopard loud calls, but these calls are preceded by two boom calls (type 1 loud call, according to Gautier and Gautier's 1977 terminology) about half a minute earlier. Experiments showed that adding booms in front of male Campbell's monkey leopard or eagle loud calls no longer elicited alarm calls in Diana monkeys, indicating that the booms had deleted the semantic specificity of the subsequent loud calls. If the booms preceded the loud calls of Diana monkeys, however, they were not effective as semantic modifiers, indicating that they were only meaningful in conjunction with Campbell's loud calls (Zuberbühler, 2002). The dusk-calls of the Campbell's monkey (Fig. 7) are often introduced by two boom calls. These calls are unlikely to contain predator information, but seem to function in male–male competition. Boom calls also occur in other guenon species (*Cercopithecus mitis*, *C. neglectus*, *C. hamlyni*, and the mona superspecies; Gautier, 1988), but few systematic investigations on the function of these calls are currently available (but see Glenn, 1996). Diana monkeys did not evolve boom calls, but they seem to distinguish between predation and non-predation stimuli by varying the number of syllables per call. For example, calls to leopards possess the same acoustic structure as calls to falling trees, but the latter contain more syllables per call than the former (Zuberbühler *et al.*, 1997).

In sum, guenon males in the Taï forest use their loud calls in both predatory and non-predatory contexts, but they use a number of acoustic and syntactic rules to distinguish between the different contexts. Call usage in response to predators suggests that natural selection has acted as the main evolutionary force. The occurrence of sexual dimorphisms and diurnal calling patterns, and the call usage in non-predatory contexts further suggest that sexual selection has acted additionally as a selective factor in the evolution of these calls.

It is interesting to note that during puberty several developmental changes occur in the vocal behavior of male guenons, specifically a drop in pitch and the loss of some of the juvenile vocal repertoire (Gautier and Gautier, 1977; Gautier, 1978; Gautier and Gautier-Hion, 1988). When playing back predator-related stimuli to wild groups of Diana monkeys, in almost all cases only one male responded with loud calls. However, a few times a second individual gave loud vocalizations, but these calls typically sounded imperfect and somewhat immature, probably because they were given by a subadult male. Figure 8 illustrates representative exemplars of the calls of a subadult male vs. calls of an adult female and an adult male, all responding to a leopard. The spectrogram of the subadult male call shows remnants of a female leopard alarm call, but also the first emerging elements of a fully developed male loud call, suggesting that males go through a transition phase when their calls develop from female alarm calls to male loud calls. These observations suggest that some male guenon loud calls have evolved from ordinary alarm calls via sexual selection. Inspection of the acoustic structure of male and female Diana monkey calls (Zuberbühler, 2000c) indicate that adult male calls are the result of a novel call production mechanism, responsible for the sexual dimorphism between male and female call structures (Tobias Riede, pers. com.).

Discussion

A Scenario for the Evolution of Guenon Loud Calls

It seems reasonable to assume that natural selection has been responsible for the evolution of predator-specific alarm calls in ancestral

(a) Adult female

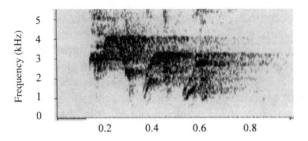

(b) Subadult male

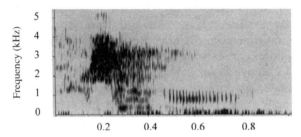

(c) Adult male

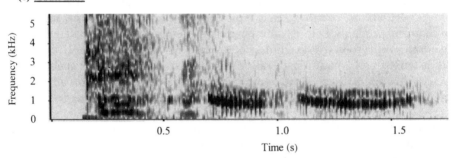

Fig. 8. Diana monkey leopard alarm calls. Comparison between the leopard alarm calls of (a) an adult female, (b) an immature male, and (c) an adult male Diana monkey. In the adult male, the high-pitched noisy call elements are no longer present, while low-pitched call elements are added. The calls of immature males are intermediate and contain both call elements.

guenons, resulting in little or no sexual dimorphism between male and female calls. The vervet monkey could be a good model for such an ancestral guenon. This species typically occupies open habitats where resident males can benefit little by advertising their presence and quality vocally. Male vervet monkeys do not produce loud calls and there is little sexual dimorphism between male and female alarm calls (Struhsaker, 1967). In the visually obstructive forest habitat, however, acoustic advertisement is likely to play a much larger role and most male forest guenons do produce loud calls. According to this scenario, sexual selection has targeted the acoustic structure and usage of the male alarm calls in the forest species to increase conspicuousness and audibility range, turning them into signals of male quality. Crucial in this process was probably the evolution of specialized larynges and vocal annexes in male guenons (Gautier, 1971). Alarm calls to leopards and eagles seem particularly well suited as substrates for sexual selection to act on because they already possessed a conspicuous acoustic structure, probably due to the selective benefits of perception advertisement to these predators (Zuberbühler et al., 1999a).

Assuming that sexual selection was responsible for the transformation of ordinary male alarm calls into highly conspicuous loud calls, a number of questions still remain unanswered. For example, in Diana monkeys it is not clear why sexual selection acted on already existing signals, the leopard and eagle alarm calls, rather than evolving novel ones. In the Taï Forest, male guenons produce predator alarm calls very frequently, often several times per day. Thus, it might have been more economical to modify an existing, reliably occurring, and conspicuous signal rather than to evolve a novel one. The boom calls of the Campbell's monkeys, however, might be a novel invention, unrelated to alarm calls. Second, in many guenon species males play a special role in predation defense, for example, in attacking and charging a nearby crowned hawk eagle or in leading a mobbing party towards a leopard (Gautier-Hion and Tutin, 1988; Zuberbühler et al., 1997). Thus, male calls given to the predators already function as honest indicators of a male's quality in antipredator defense, making these calls ideal substrates for sexual selection (Zahavi, 1975).

As mentioned before, male guenons sometimes call when no predator is present, for example to the noise of a falling tree. It has been suggested that the function of male loud calls in this context is to rally other group members (Gautier and Gautier, 1977), but this hypothesis has never been tested formally. Falling trees and large branches can be quite dangerous and sometimes even fatal for arboreal monkeys, so that it might pay the resident male to advertise to potential rivals that he is still present to defend his resource and that male take over attempts are unlikely to succeed. Although male calls given after a tree fall are acoustically very similar to predator loud calls, they show some distinct structural or syntactic properties as discussed before. Consequently, these calls still possess all the alerting characteristics, conspicuous features, and long distance properties of a predator loud call, but they no longer contain predator-specific information for recipients, allowing a male to advertise his presence to other males without causing anti-predator responses in conspecific group members.

Summary

In the Taï Forest, most guenon loud calls show many properties of classic alarm calls: they advertise perception to predators and warn recipients about the presence of specific predators, suggesting that they have evolved through natural selection. However, several lines of evidence, such as sexual dimorphism in call structure, consistent call usage in non-predatory situations, and ontogenetic evidence suggest that loud calls have been under pressure from sexual selection. Sexual selection seems to have caused the evolutionary transition from regular alarm calls to the structurally distinct loud calls, by selectively affecting the calls' transmission features and by favoring call usage to indicate male quality. In the polygynous mating system of the forest guenons, male competition over females is especially high and this mating system is notorious for leading to the evolution of conspicuous male traits through sexual selection. Male guenon loud calls are another example to be added to the list (Anderson, 1994).

ACKNOWLEDGMENTS

I would like to express my gratitude to Marina Cords, Mary Glenn, Jennifer McClung, and Tobias Riede for discussions, comments, and editorial support. Sonja Wolters, Dana Uster, Winnie Eckardt, Paul Buzzard, Ferdinand Bele, and Bertin Diero have contributed important observations on guenon behavior. The Taï Monkey Project has been funded by the Max-Planck-Institute for Behavioural Physiology and by a grant of the Leakey Foundation. Fieldwork was made possible through research grants of the US National Science Foundation, the University of Pennsylvania, the National Geographic Society, the Swiss National Science Foundation, the University of Zurich, and the Max-Planck-Institute for Evolutionary Anthropology. In Côte d'Ivoire, I thank the ministries of research and agriculture, the Centre Suisse de Recherches Scientifiques, the Centre de Recherche en Ecologie, and the P.A.C.P.N.T. for support and permission to conduct research in the Taï National Park.

References

Anderson, M. 1994. *Sexual Selection*. Princeton University Press, Princeton.
Bergstrom, C. and Lachmann M. 2001. Alarm calls as costly signals of antipredator vigilance: the watchful babbler game. *Anim. Behav.* **61**:535–543.
Boesch, C. 1991. The effects of leopard predation on grouping patterns in forest chimpanzees. *Behaviour* **117**:220–242.

Boesch, C., and Boesch, H. 1989. Hunting behavior of wild chimpanzees in the Taï National Park. *Am. J. Phys. Anthropol.* **78**:547–573.

Boesch, C., and Boesch-Achermann, H. 2000. *The Chimpanzees of the Taï Forest. Behavioural Ecology and Evolution.* Oxford University Press, Oxford.

Catchpole, C. K., and Slater, P. J. B. 1995. *Bird Song: Biological Themes and Variations.* Cambridge University Press, Cambridge.

Clutton-Brock, T. H. 1982. The function of antlers. *Behaviour* **70**:108–125.

Clutton-Brock, T. M., and Albon, S. D. 1979. The roaring of red deer and the evolution of honest advertisement. *Behaviour* **69**:145–170.

Cords, M. 1987. Forest guenons and pata monkeys: male–male competition in one-male groups. In: B. B. Smuts, D. L. Cheney, R. M. Seyfarth, R. W. Wrangham, and T. T. Struhsaker (eds.), *Primate Societies.* University of Chicago Press, Chicago.

Darwin, C. 1859/1993. *The Origin of Species.* Random House, New York.

Flasskamp, A. 1994. The adaptive significance of avian mobbing. An experimental test of the move on hypothesis. *Ethology* **96**:322–333.

Gautier, J.-P. 1971. Étude morphologique et fonctionnelle des annexes extralaryngées des Cercopithecinae; liaison avec les cris d'espacement. *Biol. Gabon* **7**:229–267.

Gautier, J.-P. 1978. Repertoire sonore de *Cercopithecus cephus. Z. Tierpsych.* **46**:113–169.

Gautier, J.-P. 1988. Interspecific affinities among guenons as deduced from their vocalizations. In: A. Gautier-Hion, F. Bourliére, J. P. Gautier, and J. Kingdon (eds.), *A Primate Radiation: Evolutionary Biology of the African Guenons,* pp. 194–226. Cambridge University Press, Cambridge.

Gautier, J.-P., and Gautier, A. 1977. Communication in old world monkeys. In: T. A. Sebeok (ed.), *How Animals Communicate,* pp. 890–964. Indiana University Press, Bloomington.

Gautier, J.-P., and Gautier-Hion, A. 1988. Vocal quavering: a basis for recognition in forest guenons. In: D. Todt *et al.* (eds.), *Primate Vocal Communication,* pp. 15–30, Springer, Berlin.

Gautier-Hion, A., and Tutin, C. E. G. 1988. Simultaneous attack by adult males of a polyspecific troop of monkeys against a crowned hawk eagle. *Folia Primatol.* **51**:149–151.

Glenn M. 1996. *The Natural History of the Mona Monkey* (Cercopithecus mona *Schreber 1774*), *on Grenada, West Indies.* Ph.D. Dissertation, Northwestern University, Evanston, Illinois.

Goodall, J. 1986. *The Chimpanzees of Gombe: Patterns of Behavior.* Harvard University Press, Cambridge.

Hauser, M. D., and Wrangham, R. W. 1990. Recognition of predator and competitor calls in nonhuman primates and birds: a preliminary report. *Ethology* **86**:116–130.

Herbinger, I., Boesch, C., *et al.* 2001. Territory characteristics among three neighbouring chimpanzee communities in the Taï National Park, Cote d'Ivoire. *Int. J. Primatol.* **22**(2): 143–167.

Hill, C. M. 1994. The role of female diana monkeys, *Cercopithcecus diana,* in territorial defence. *Anim. Behav.* **47**:425–431.

Höner, O. P., Leumann, L., and Noë, R. 1997. Dyadic associations of red colobus and diana monkey groups in the Taï National Park, Ivory Coast. *Primates* **38**:281–291.

Jenny, D. 1996. Spatial organization of leopards (*Panthera pardus*) in Taï National Park, Ivory Coast: Is rain forest habitat a tropical haven? *J. Zool.* **240**:427–440.

Le Boeuf, B. J. 1974. Male-male competition and reproductive success in elephant seals. *Am. Zool.* **14**:163–176.

Martin, C. 1991. *The Rainforests of West Africa: Ecology, Threats, Conservation.* Birkhäuser, Basel.

Maynard-Smith, J. 1965. The evolution of alarm calls. *Am. Nat.* **99**:59–63.

McGraw, W. S., and R. Noë 1995. The Tai Forest Monkey Project. *Afr. Primates* **1**:17–19.

Mitani, J., Sanders W. J., and Windfelder, T. L. 2001. Predatory behavior of crowned hawk-eagles (Stephanoaetus coronatus) in Kibale National Park, Uganda. *Behav. Ecol. Sociobiol.* **49**:187–195.

Seyfarth, R. M., Cheney, D. L., and Marler, P. 1980. Vervet monkey alarm calls: Semantic communication in a free-ranging primate. *Anim. Behav.* **28**:1070–1094.

Struhsaker, T. T. 1967. Auditory communication among vervet monkeys (*Cercopithecus aethiops*). In: S. A. Altmann (ed.), *Social Communication Among Primates*, pp. 281–324. University of Chicago Press, Chicago.

Struhsaker, T. T. 1970. Phylogenetic implications of some vocalizations of *Cercopithecus* monkeys. In: J. R. Napier, and P. H. Napier (eds.), *Old World Monkeys: Evolution, Systematics, and Behavior*, pp. 365–444. Academic Press, New York.

Van Schaik, C. P., and Van Hooff, J. A. 1983. On the ultimate causes of primate social systems. *Behaviour* **85**:91–117.

Wiley, R. H., and Richards, D. G. 1978. Physical constraints on acoustic communication in the atmosphere: Implications for the evolution of animal vocalizations. *Behav. Ecol. Sociobiol.* **3**: 69–94.

Zahavi, A. 1975. Mate selection: a selection for a handicap. *J. Theor. Biol.* **53**:205–214.

Zuberbühler, K. 1993. *Acoustic Communication in the Predation Context: the Significance of Vocalizations as Anti-predator Strategies in Two Forest-living Cercopithecines*. M.Sc. Thesis, University of Zürich.

Zuberbühler, K. 2000a. Causal knowledge of predators' behaviour in wild Diana monkeys. *Anim. Behav.* **59**:209–220.

Zuberbühler, K. 2000b. Interspecific semantic communication in two forest monkeys. *Proc. R. Soc. Lond. B* **267**:713–718.

Zuberbühler, K. 2000c. Referential labelling in Diana monkeys. *Anim. Behav.* **59**:917–927.

Zuberbühler, K. 2001. Predator-specific alarm calls in Campbell's guenons. *Behav. Ecol. Sociobiol.* **50**:414–422.

Zuberbühler, K. 2002. A syntactic rule in forest monkey communication. *Anim. Behav.* **63**:293–299.

Zuberbühler, K., Noë, R., and Seyfarth, R. M. 1997. Diana monkey long-distance calls: messages for conspecifics and predators. *Anim. Behav.* **53**:589–604.

Zuberbühler, K., Jenny, D., and Bshary, R. (1999a). The predator deterrence function of primate alarm calls. *Ethology* **105**:477–490.

Zuberbühler, K., Cheney, D.L., and Seyfarth, R.M. (1999b). Conceptual semantics in a nonhuman primate. *J. Comp. Psychol.* **113**:33–42.

Part III

Ecology

Resource Switching and Species Coexistence in Guenons: A Community Analysis of Dietary Flexibility

JOANNA E. LAMBERT

Introduction

Identifying and evaluating the relationships among resource availability, competition, species richness, and guild evolution are central goals in modern ecology (MacArthur, 1972; Pianka, 1976; Diamond, 1978). While several authors (e.g., Connell, 1980; Schoener, 1982; Ben-David *et al.*, 1995; Arlettaz *et al.*, 1997) have questioned the importance of interspecific competition in shaping community interactions, others (e.g., Alatalo *et al.*, 1986; Tokeshi, 1999) have proposed that it can play an important role and that the potential for feeding competition is greatest among closely related, sympatric species occupying similar trophic space. Feeding competition is also generally viewed to be the most intense when resources are either temporally or spatially scarce and/or highly favored as a function of nutritional quality (Richard, 1985; Oates, 1987). Thus, an understanding of the behavioral, anatomical, and physiological solutions by which species cope with limiting resources is critical to understand

JOANNA E. LAMBERT • Department of Anthropology, University of Oregon, Eugene, OR 97403, USA.

The Guenons: Diversity and Adaptation in African Monkeys, edited by Glenn and Cords. Kluwer Academic/Plenum Publishers, New York, 2002.

species coexistence and to interpret guild evolution. When extant species are observed or expected to compete directly for critical resources, one or more of the interacting species are predicted to undergo population decline or character displacement (Gause, 1934). As Tokeshi (1999, p. 221) argued, there is therefore a "dynamic tension between competitive interactions and certain endogenous/exogenous mitigating factors or processes." Behavioral or anatomical features that mitigate competitive interactions over limiting resources should be generally favored and are expected for continued species coexistence (Lambert and Whitham, 2001).

One means by which animals cope with fluctuating food availability is via resource switching, which occurs at several temporal scales including longer-term, seasonal shifts. Many primate species fall back on less favorable or lower quality foods when preferred foods are seasonally scarce (Rudran, 1978; Chapman, 1987; Beeson, 1989; Rogers et al., 1994; Remis, 1997; Kaplin et al., 1998; Wrangham et al., 1998; Lambert et al., 1999; Strier, 2000). Resource switching also occurs on a much smaller time-scale, wherein animals within the course of a particular feeding bout shift to another food as a consequence of being displaced by other animals (Janson, 1988; Whitten, 1988; Barton, 1993). Feeding animals may be displaced to another resource by more dominant members of their own social group, by conspecifics, or by members of other species. Thus, in addition to meeting the challenges of seasonal resource scarcity, a capacity to consume a diversity of food types may also facilitate coexistence of animals with overlapping diets.

Guenons are commonly found in communities with other guenon species, mangabeys and colobines (Lernould, 1988; Tutin, 1999; Chapman et al., 2000a,b). Gautier-Hion (1988) argued that dietary overlap among sympatric guenons is exceptionally high, with a concomitantly high expected level of feeding competition. However, here I suggest that dietary overlap in and of itself need not lead to competition. Indeed, as hinted above "coexistence is considered as a state in which the effects of competition are constantly resisted" (Tokeshi, 1999; p. 221). In this paper, I evaluate whether resource switching is a behavioral mechanism that facilitates coexistence of sympatric *Cercopithecus* species and other Cercopithecoidea with overlapping diets. I do not address seasonal shifts in resource consumption, but instead evaluate resource switching as a mechanism that facilitates shifting foods as a response to the presence of intra- or inter-specific competitors. As many sites in equatorial Africa exhibit high cercopithecoid diversity with sympatric species consuming many of the same foods, explanations for species packing and coexistence are needed (Gautier-Hion, 1978). I argue that it is the cercopithecines' dietary flexibility that facilitates coexistence.

I present data on resource switching and dietary overlap among two guenons (*Cercopithecus mitis stuhlmanni*, Stuhlmann's blue monkey and *C. ascanius schmidti*, Schmidt's red-tailed monkey), and two other sympatric cercopithecoids (*Lophocebus albigena johnstoni*, grey-cheeked mangabey and

Procolobus badius tephrosceles, ashy red colobus) in Kibale National Park, Uganda. My goals are to (1) evaluate dietary overlap among these monkey species; (2) determine whether there are species differences in frequency of resource switching; (3) evaluate whether resource switching is more common under particular feeding circumstances; and (4) assess whether it is more common in animals that may be particularly vulnerable to feeding competition in species-rich primate communities, e.g., smaller animals, such as females, and smaller-bodied species.

I expected digestive strategies to influence a species' capacity to resource switch. Cercopithecines and colobines have divergent gastrointestinal anatomy and digestive strategies (Chivers and Hladik, 1980; Lambert, 1998; Milton, 1998). Colobines have a sacculated stomach, with an alkaline fermenting chamber, while cercopithecines have a simple, acidic stomach and rely on caeco-colic fermentation (Chivers, 1994; Kay and Davies, 1994). Cercopithecines have lengthy digestive retention times (Lambert, 1998, 2002), and my digestive data on guenons are consistent with those collected by other researchers on other cercopithecine species (Clemens and Phillips, 1980; Clemens and Maloiy, 1981; Maisels, 1993). Because cercopithecines have relatively long digestive retention times, but lack the specialized stomachs of colobines, they have the capacity to consume both low-fiber/high-soluble carbohydrate foods when available, as well as lower quality foods that require more fermentation. As digestive flexibility increases, so too does the potential for dietary flexibility and resource switching. As a function of digestive flexibility, the cercopithecines were predicted to exhibit greater dietary diversity and frequency of resource switching than colobines. Among the three cercopithecines, I predicted *Cercopithecus ascanius* would exhibit the greatest frequency of resource switching. They are the smallest Kibale anthropoid and the least likely to succeed in interspecific dominance interactions. Colobines generally do not consume arthropods for dietary protein. While some colobine species consume unripe fruit pulp, an alkaline stomach pH (and the potential for acidosis) precludes colobines from consuming high levels of ripe fruit (Kay and Davies, 1994). Thus, *Procolobus badius* was predicted to exhibit the lowest frequency of resource switching and dietary flexibility—at least in part as a result of their specialized gastrointestinal anatomy and physiology.

Methods

Two full-time assistants and I collected behavioral data between June 1999 and January 2001 on *Cercopithecus mitis stuhlmanni, C. ascanius schmidti, Lophocebus albigena johnstoni* and *Procolobus badius tephrosceles* at the Kanyawara study site in the Kibale National Park, Uganda (for site description see Struhsaker, 1997; Chapman and Lambert, 2000). We followed each species for five continuous days/month and sampled behavior between 0730 h and 1700 h on focal adult subjects via a continuous recording rule in 30 minute periods (Martin and Bateson, 1986). We tried to avoid re-sampling individuals by ensuring that all parts of the group were monitored, by searching for and observing monkeys in both densely vegetated and more open areas, and by monitoring recognizable individuals. All groups included some recognizable individuals, although there were more in the small groups of *Lophocebus albigena* (mean = 13.9, $n = 2$) and *Cercopithecus mitis* (mean = 9.6, $n = 2$) than in the larger groups of *C. ascanius* (mean = 31, $n = 2$) and *Procolobus badius* (mean = 48.5, $n = 2$). Therefore, there may have been an unavoidable bias toward more habituated group members in the latter two species. The density of vegetation, the canopy height at which feeding occurred, and the ease with which the primates (particularly the two *Cercopithecus* species) were alarmed, made it difficult to obtain detailed feeding observations, and the decision to initiate a focal follow was based on both viewing conditions and whether the animal had been sampled previously. In some cases the animal could not be monitored for the full 30 minutes with the required level of accuracy. Here, I include only complete 30-minute follows, for a total of *ca.* 8000 hours of sampling time. We avoided biasing the data toward large fruiting trees, since we followed the animals all day and since we each monitored different areas of the social group, e.g., middle of group spread and periphery of group.

To assess the diversity of food types and food species consumed by a subject during a 30-minute interval, we noted all food consumed, including plant part (mature leaves, immature leaves, leaf buds, petioles, flowers, flower buds, ripe fleshy fruit, unripe fleshy fruit, dehiscent pods, seeds, exudate, and bark/cambium), plant species and other, non-plant foods (soil and invertebrates). We scored resource switching when a focal subject changed between plant and non-plant food, and when it changed the species of plant or the part of the plant consumed. In addition, as there can be substantial nutritional differences according to phenophase (Lambert, 1998), we scored resource switching when subjects changed between ripe and unripe fruit and among leaf buds, young leaves, and mature leaves.

To calculate switching frequency, I included only the first switch to a particular food in each 30-minute sample. For example, if a subject fed on *Ficus exasperata* fruit, then switched to *F. exasperata* leaves, and then continued feeding on *F. exasperata* fruit, I counted only one switch. Switching was also scored in cases in which animals did not switch directly from one food to another, but instead engaged in some other activity such as socializing or locomoting between feeding bouts.

I assessed whether the frequency of switching events increased during conditions of potentially higher feeding competition. As an indirect measure of feeding competition, I recorded three measures of interindividual spacing: nearest neighbor, number of monkeys in the crown of the feeding tree, and the number of monkeys ≤10 m regardless whether they were in the same tree.

I assumed that the potential for feeding competition was greater when nearest neighbors were closer and when more monkeys were within ≤10 m or within the same feeding tree.

Results

Cercopithecus ascanius consumed items in 13 of the 14 food categories, *Lophocebus albigena* ate from 12 of them; *C. mitis* and *Procolobus badius* each ate from 11. We did not observe *Procolobus badius* consuming animal food. Although arthropods were consumed by the three cercopithecines, arthropods were difficult to observe and to identify from a distance. Accordingly, I focus here on the plant diet.

There was considerable overlap in the plant species consumed by the monkeys (Table I). The diet of *Cercopithecus ascanius* included 52 plant species, and the diet of *C. mitis* included 40; 36 plant species were shared by both guenons. *Lophocebus albigena* had a diet with 38 plant species and shared 32 of these plant species with the two *Cercopithecus* species. *Procolobus badius* had a diet that included 40 plant species and shared 37 of these plant species with the three cercopithecines. Of the total 67 plant species used as food by any of the monkeys, 36 (54%) were shared by ≥3 of the monkey species. The diet of *Cercopithecus ascanius* had the greatest percentage of species not shared with other monkeys (13.4%), while *Procolobus badius* had the least unique diet with only 2.9% of the plant species that they ate not consumed by the other species.

In an earlier analysis of primate diets from Kanyawara, Struhsaker (1978) found considerably lower overlap of plant food species among the same monkey species, and concluded that the Kibale monkeys do not typically share their commonest foods. In this study, the two *Cercopithecus* species shared

Table I. Overlap (%) in use of Plant Species by the Four Subject Monkey Species, Based on Data from June 1999–January 2001. Percentage represents the proportion of shared plant species relative to the total number of plant species consumed by the species in that pair. Mean % overlap refers to the average overlap of one monkey species with all other monkey species (e.g., *Cercopithecus ascanius* averages 34.6% dietary overlap with the other three species).

	Procolobus badius	*Cercopithecus ascanius*	*Cercopithecus mitis*	*Lophocebus albigena*
Procolobus badius		32.6	32.6	33.3
Cercopithecus ascanius			39.1	32.2
Cercopithecus mitis				34.6
Mean % overlap	32.8	34.6	35.4	32.8

Table II. Five Most Common Species-Specific Plant Foods Based on an 18-month Study Period

	Percent of plant diet			
Plant Species	*Procolobus badius*	*Cercopithecus ascanius*	*Cercopithecus mitis*	*Lophocebus albigena*
Rothmania longiflora leaves	17.9			
Funtumia africana pods	8.6	12.2	7.3	16.3
Diospyros abyssinica flowers	14.9	7.5		
Diospyros abyssinica fruit			20.2	
Celtis durandii fruit	8.1			
Ficus exasperata fruit	7.3	15.1		
Neoboutonia africana leaves		9.6		
Ficus brachylepis fruit		7.5		
Millettia dura pods			13.2	
Premna angolensis fruit			13.0	
Celtis africana fruit			10.3	
Celtis africana leaves				10.2
Markhamia platycalyx leaves				12.9
Strombosia scheffleri leaves				11.8
Parinari excelsa leaves				6.9

three of the top five most common species-specific foods (Table II). *Lophocebus albigena* overlapped with the guenons in two of five top plant species, and all four monkeys ate *Funtumia africana* extensively. I also found greater dietary overlap than did Struhsaker (1978) in monkey species pairs; in this study, average pair-wise overlap ranged between 32.8 and 35.4%, while Struhsaker (1978) reported a range of 7.75–22.1% in the same primate community.

Several factors may account for differences in our findings. My data were collected by a single research team during the same sample period, while Struhsaker (1978) based his analyses on data collected by multiple researchers employing different methods over several years, and not always simultaneously. However, given the relationship between food availability and diet (Gautier-Hion *et al.*, 1981; Cords, 1986; Tutin and White, 1998), it is more likely that these differences are the result of shifts in food availability, tree diversity, and tree density over several decades. Although it achieved National Park status in 1993, Kibale had a long prior history of logging, and there are documented shifts both in tree dynamics and in feeding behavior of animals as a consequence (Skorupa, 1988; Struhsaker, 1997; Chapman and Lambert, 2000). For example, Lwanga *et al.* (2000) compared data on tree species composition, stem abundance, and basal area in Kibale between 1975 and 1998, and found that both tree species richness and diversity declined, while abundance increased for some species. In comparing his work with earlier work (Waser, 1975, 1977), Olupot (1998) found that the diet of Kanyawara *Lophocebus albigena* included fewer species between 1992 and 1993 than during the 1970s.

Table III. Mean Number of Switching Events/Species/30 minutes, including Standard Error and Sample Size. Calculations based on total time, including 30-minute intervals with non-feeding behaviors.

Species	Mean # switches	Range
Cercopithecus ascanius	0.78 ± 0.018 ($n = 1473$)	1–7
C. mitis	0.71 ± 0.02 ($n = 910$)	1–5
Lophocebus albigena	0.77 ± 0.017 ($n = 1258$)	1–4
Procolobus badius	0.65 ± 0.015 ($n = 1409$)	1–3

The four monkey species ranged in frequency of resource switching (Table III). When both sexes are combined, *Cercopithecus ascanius* and *Lophocebus albigena* switched foods significantly more often than *Procolobus badius* did (ANOVA; F ratio: 11.5; $p < 0.01$). *Cercopithecus mitis* and *Procolobus badius* are indistinguishable in average frequency of resource switching. In an analysis that included only records for which adult sex was positively identified, *Cercopithecus mitis* was the only species to exhibit sexual differences in frequency of switching events (t-test; $p = 0.03$; Table IV). If male *Cercopithecus mitis* are removed from the analysis, *Procolobus badius* are significantly different from the three cercopithecines (ANOVA; F ratio 11.47; $p < 0.01$). Thus, the prediction that cercopithecines would exhibit greater frequency of switching is generally, although not entirely, upheld.

Measures of interindividual spacing are presented in Table V. The only significant difference in distance to nearest neighbor is between *Cercopithecus ascanius* and *Procolobus badius* (ANOVA; F ratio 3.66; $p < 0.01$). *Procolobus badius* had significantly more neighbors ≤ 10 m (ANOVA; F ratio 52.73; $p < 0.01$) and also more neighbors in a tree crown (ANOVA; F ratio 97.59; $p < 0.01$) than the three other monkey species.

I used linear regression analysis to determine whether resource switching was related to feeding circumstances. The various measures of interindividual spacing, averaged for each species, served as independent variables, while the average number of resource switches per 30 minutes, again averaged for each species, was the dependent variable. The relationships between frequency

Table IV. Male vs Female Mean Number of Switching Events/30 minute including Standard Error and Sample Size

Species	Males	Females	p-Value[a]
Cercopithecus ascanius	0.75 ± 0.02 ($n = 537$)	0.81 ± 0.03 ($n = 452$)	$p = 0.25$
C. mitis	0.68 ± 0.03 ($n = 333$)	0.79 ± 0.04 ($n = 281$)	$p = 0.03$[a]
Lophocebus albigena	0.8 ± 0.02 ($n = 551$)	0.8 ± 0.03 ($n = 366$)	$p = 0.93$
Procolobus badius	0.67 ± 0.02 ($n = 560$)	0.64 ± 0.02 ($n = 408$)	$p = 0.32$

[a]t-test comparison of males and females.

Table V. Measure and Descriptive Statistics of Interindividual Spacing, including Standard Error, Sample Size and Range

Species	Mean distance (m) to Nearest Neighbor (NN)	Mean #animals/tree crown	Mean #animals/within 10 m of focal
Cercopithecus ascanius	5.1 ± 0.55 $n = 2246$ range <1–72	3.1 ± 0.04 $n = 2209$ range: 1–20	4.0 ± 0.05 $n = 2275$ range: 0–15
C. mitis	4.0 ± 0.1 $n = 1311$ range: <1–30	3.0 ± 0.05 $n = 1340$ range: 1–10	3.8 ± 0.06 $n = 1303$ range: 0–12
Lophocebus albigena	4.1 ± 0.09 $n = 1747$ range: <1–30	3.2 ± 0.05 $n = 1779$ range: 1–14	3.8 ± 0.05 $n = 1733$ range: 0–12
Procolobus badius	3.6 ± 0.07 $n = 1586$ range: <1–25	4.2 ± 0.07 $n = 2026$ range: 1–25	4.7 ± 0.06 $n = 2008$ range: 1–18

of resource switching and either number of monkeys within 10 m, or number of monkeys sharing a tree crown are weak. The strongest relationship is between nearest neighbor distance and number of switching events, with the best-fit line explaining about 60% of the variance ($y = 0.071x + 0.434$; $r^2 = 0.61$).

Discussion

My study of four sympatric monkeys indicates considerable dietary overlap among species, though *Cercopithecus ascanius* consumed the most plant species, ate from the largest number of dietary categories and exploited the greatest number of unique plant species. The cercopithecines in general switched dietary category/species significantly more than *Procolobus badius* did. The only sexual difference is in *Cercopithecus mitis*: females switched more than males did. Across species, resource switching frequency increased with greater average distance between monkeys.

My results can be interpreted in light of what is known about digestive strategies. Digestive modularity refers to the capacity of an individual (or species) to regulate digestion according to chemical or structural qualities of the diet or both (Karasov and Diamond, 1988; Afik *et al.*, 1997; Lambert, 1998). Research on digestive ecology in a variety of vertebrates indicates considerable morphological and physiological modularity in response to dietary fluctuation. For example, Afik and Karasov (1995) found that warblers switch retention times according to the type of food they are consuming: when eating fruit, they have the shortest retention times, when ingesting insects, they have moderate retention times, and when consuming seeds, they have the longest retention times. Furthermore, the changes in retention times can occur over extremely short periods, e.g., between meals.

We have few data on the effect of food types on digestive retention for most primate species (Milton and Demment, 1988; Caton *et al.*, 1996; Power and Oftedal, 1996). However, cercopithecines readily consume a diversity of high-fiber foods as well as easier to digest foods such as insects and fruits, suggesting a capacity to alternate among foods with differing structural and chemical attributes. In fact, cercopithecines in general are noted for their dietary flexibility and generalist strategy (Rudran, 1978; Struhsaker, 1978; Gautier-Hion, 1988; Beeson, 1989; Richard *et al.*, 1989; Maisels, 1993; Altmann, 1998, Chapman *et al.*, 2002).

Cercopithecus ascanius are noteworthy for several reasons. They are the smallest-bodied among Kibale anthropoids and despite an expectation that they should, accordingly, maintain the highest-quality diet (Bell, 1971; Jarman, 1974; Gaulin, 1979), they actually consume a large percentage of foods high in structural polysaccharides (almost twice as much as larger-bodied mangabeys, e.g., *Cercopithecus ascanius* total foraging effort devoted to leaves = 21%; *Lophocebus albigena*, 12.2%), and more than sympatric chimpanzees (Conklin-Brittain *et al.*, 1998; Wrangham *et al.*, 1998; Lambert, 2002). In addition, of the three cercopithecines, *Cercopithecus ascanius* has the largest group size.

An ability to consume many food types can facilitate larger feeding spheres, which may allow them to avoid the "pushing forward" phenomenon proposed by van Schaik *et al.* (1983) and discussed by Chapman and Chapman (2000). Pushing forward is a response of feeding animals to feeding competitors (Chapman and Chapman, 2000). If the overlap of foraging fields is increased for larger groups, polyspecific associations, or as a result of resource distribution, then per capita encounter rate with food can be decreased (Chapman and Chapman, 2000). However, if a flexible dietary strategy means that more items are food, then resource switching may provide a means to minimize intra- and inter-specific contest in the first place, with switchers able to access a greater diversity of foods. That is, because a given area essentially holds greater food richness for species with extreme digestive flexibility, the likelihood of encroaching on another individual's feeding space is decreased. This is not to imply that animals do not exhibit preference for certain foods and that contest does not happen, but contest may be less likely when animals have greater dietary options as a function of digestive flexibility. Being able to spread out because there are more feeding options may explain why *Cercopithecus ascanius* maintained the greatest mean distance to nearest neighbor (5.1 m) and the greatest maximum distance to nearest neighbor (72 m), why *Procolobus badius* maintained the smallest values (3.6 and 25 m, respectively), and why the strongest positive relationship is between frequency of resource switching and distance to nearest neighbor. I am not arguing that *Procolobus badius* have closer interindividual spacing as a result of the fact that leaves are superabundant. Indeed, the generalization that leaves are non-limiting has been challenged by Chapman and Chapman (1999). Instead, I am attempting to elucidate a specific adaptive mechanism (digestive modularity ⇒ dietary flexibility) that may help to explain the coexistence of closely related cercopithecines in similar trophic space.

What can these patterns tell us about species coexistence and feeding niches? A central question regarding the evolution of a feeding guild concerns the circumstances under which a set of ancient species occupying a particular niche gives rise to or incorporates new species in that trophic space (Tokeshi, 1999). Theoretically, three cases are distinguishable: species can (1) increase overlap between niches, (2) reduce niche width by increasing feeding specialization, and (3) expand the total niche space (Tokeshi, 1999). I suggest that Cercopithecinae, and perhaps *Cercopithecus* spp. particularly, have increased dietary niche overlap (condition #1) which is accommodated by expanded niche space (greater diversity of food types consumed in a given area, condition #3). Conversely, Colobinae appear to illustrate the second condition. Relative to *Cercopithecus* spp. colobines constitute a narrower, specialized guild: that of larger-bodied, arboreal, anatomically specialized folivore. Indeed, while colobine species are variable in degree of dietary diversity (Struhsaker, 1978; Oates, 1987; Bennett and Davies, 1994; Oates, 1994), relative to cercopithecines they consume foods from a smaller spectrum of dietary types. Essentially, this represents a situation wherein the coexistence of generalist and specialist species whose diets largely overlap may be attributed to the generalists being able to use the part of the resource spectrum that is less flexibly exploited by specialists (Tokeshi, 1999). Morris (1996) argued similarly that such a mechanism facilitates the coexistence of a diverse community of rocky mountain rodent species.

Digestive flexibility and dietary diversity need not necessarily be consequences of the "ghost of competition past" (Connell, 1980) among closely related cercopithecine species. Indeed, digestive flexibility and resource switching may either be retained primitive traits, or features that evolved separately in multiple independent lineages. Guenon diversity is almost certainly related to population divergence that took place during the contraction and expansion of habitat over two million years of glacial/interglacial cycles (Struhsaker, 1981; Hamilton, 1988; Kaplin and Moermond, 2000). A flexible dietary strategy may have evolved during these critical periods (Chapman *et al.*, 2002) and may have facilitated coexistence when species later came together (Grant, 1975; Connell, 1980).

In the absence of further information on digestive physiology, my conclusions remain speculative. However, resource switching is almost certainly a behavioral mechanism with gastrointestinal and physiological underpinnings. Clarifying these ecophysiological relationships may help substantially to unravel the complexity of species coexistence and mechanisms by which to mitigate the costs of limiting resources.

Summary

Competition is generally viewed to be an important variable in structuring communities of closely related species occupying similar trophic space and may be most intense over limiting resources. Here, I have evaluated how cercopithecoids with extensive dietary overlap coexist. I examined species patterns of resource switching and expected species differences that relate to differences in digestive strategies between colobines and cercopithecines. Overall, the three cercopithecines had higher frequencies of resource switching than the colobine. *Cercopithecus ascanius* was found to have the most diverse diet and engaged in the most resource switching. I suggest that resource switching is facilitated by digestive flexibility. Such a mechanism may allow species packing and coexistence because it allows animals a means to switch to other dietary resources in the presence of other animal competitors or during times of seasonal scarcity. However, resource switching need not necessarily have evolved as a consequence of past competition. This may be either a retained primitive feature or one that evolved multiple times during population divergence that facilitated coexistence when species later came together.

ACKNOWLEDGMENTS

I thank the Office of the President (Uganda), the Uganda National Council for Science and Technology, the Uganda Wildlife Authority, Gilbert Isabirye Basuta and John Kasenene for granting permission to work in Uganda and in Kibale Forest. I thank Mary Glenn and Marina Cords for inviting me to participate in the IPS Guenon Symposium in Australia, their editorial feedback and for their efforts in putting together such a fine volume. I thank Elise Town for her help with data entry. I thank my anonymous reviewers and Russ Tuttle for critical and editorial input on earlier versions of this manuscript. I thank Colin Chapman for his intellectual and logistical support since my first days in Kibale and I gratefully acknowledge the support of my UO colleague, Professor *Emeritus* Paul E. Simonds, who facilitated my visit to the Australia IPS. Agaba Erimosi and Patrick Kataramu were essential in the data collection. This research was supported by the University of Oregon.

References

Afik, D., and Karasov, W. H. 1995. The trade-offs between digestion rate and efficiency in warblers and their ecological implications. *Ecology* **76**:2247–2257.

Afik, D., Darken, B. W., and Karasov, W. H. 1997. Is diet shifting facilitated by modulation of intestinal nutrient uptake? Test of an adaptational hypothesis in yellow-rumped warblers. *Phys. Zool.* **70**:213–221.

Alatalo, R. V., Gustafsson, L., and Lundberg, A. 1986. Interspecific competition and niche changes in tits (*Parus* spp.): Evaluation of nonexperimental data. *Am. Nat.* **127**:819–834.

Altmann, S. A. 1998. *Foraging for Survival: Yearling Baboons in Africa.* University of Chicago Press, Chicago.

Arlettaz, R., Perrin N., and Hausser, J. 1997. Trophic resource partitioning and competition between the two sibling bat species *Myotis myotis* and *Myotis blythii. J. Anim. Ecol.* **66**: 897–911.

Barton, R. A. 1993. Sociospatial mechanisms of feeding competition in female olive baboons, *Papio anubis. Anim. Behav.* **46**:791–802.

Beeson, M. 1989. Seasonal dietary stress in a forest monkey (*Cercopithecus mitis*). *Oecologia* **78**:565–570.

Bell, R. H. V. 1971. A grazing ecosystem in the Serengeti. *Sci. Am.* **225**:86–93.

Ben-David, M., Bowyer, R. T., and Faro, J. B. 1995. Niche separation by mink and river otters: coexistence in a marine environment. *Oikos* **75**:41–48.

Bennett, E. L., and Davies, A. G. 1994. The ecology of Asian colobines. In: A. G. Davies, and J. F. Oates (eds.), *Colobine Monkeys: Their Ecology, Behavior, and Evolution*, pp. 129–173. Cambridge University Press, Cambridge, UK.

Caton, J. M., Hill, D. M., Hume, I. D., and Crook, G. A. 1996. The digestive strategy of the common marmoset, *Callithrix Jacchus. Comp. Biochem. Physiol.* **114A**:1–8.

Chapman, C. A. 1987. Flexibility in diets of three species of neotropical Costa Rican primates. *Folia Primatol.* **48**:90–115.

Chapman, C. A., and Chapman, L. J. 1999. Implications of small scale variation in ecological conditions for the diet and density of red colobus monkeys. *Primates* **40**:215–232.

Chapman, C. A., and Chapman, L. J. 2000. Constraints on group size in red colobus and red-tailed guenons: Examining the generality of the ecological constraints model. *Int. J. Primatol.* **21**:565–586.

Chapman, C. A., and Lambert, J. E. 2000. Habitat alteration and the conservation of African primates: A case study of the Kibale National Park, Uganda. *Am. J. Primatol.* **50**:169–185.

Chapman, C. A., Balcomb, S. R., Gillespie, T., Skorupa, J., and Struhsaker, T. T. 2000a. Long-term effects of logging on African primate communities: A 28 year comparison from Kibale National Park, Uganda. *Conserv. Biol.* **14**:207–217.

Chapman, C. A., Gautier-Hion, A., Oates, J. F., and Onderdonk, D. A. 2000b. African primate communities: Determinants of structure and threats to survival. In: J. G. Fleagle, C. Janson, and K. E. Reed (eds.), *Primate Communities*, pp. 1–37. Cambridge University Press, Cambridge.

Chapman, C. A., Chapman, L. J., Cords, M., Gathua, J. M., Gautier-Hion, A., Lambert, J. E., Rode, K., Tutin, C. E. G., and White, L. J. T. 2002. Variation in the diets of *Cercopithecus* species: Differences within forests, among forests, and across species. In: M. E. Glenn and M. Cords (eds.), *The Guenons: Diversity and Adaptation in African Monkeys*, pp. 325–350. Kluwer Academic Publishers, New York.

Chivers, D. J. 1994. Functional anatomy of the gastrointestinal tract. In: G. Davies and J. F. Oates (eds.), *Colobine Monkeys: Their Ecology, Behavior and Evolution*, pp. 205–228. Cambridge University Press, Cambridge, UK.

Chivers, D. J., and Hladik, C. M. 1980. Morphology of the gastrointestinal tract in primates: Comparisons with other mammals in relation to diet. *J. Morph.* **116**:337–386.

Clemens, E. T., and Maloiy, G. M. O. 1981. Organic acid concentrations and digesta movement in the gastrointestinal tract of the bushbaby (*Galago crassicaudatus*) and vervet monkey (*Cercopithecus pygerythrus*). *J. Zool. (London)*. **193**:487–497.

Clemens, E. T., and Phillips, B. 1980. Organic acid production and digesta movement in the gastrointestinal tract of the baboon and sykes monkey. *Comp. Biochem. Physiol.* **66**:529–532.

Conklin-Brittain, N. L., Wrangham, R. W., and Hunt, K. D. 1998. Dietary response of chimpanzees and cercopithecines to seasonal variation in fruit abundance. II. Macronutrients. *Int. J. Primatol.* **19**:971–998.

Connell, J. H. 1980. Diversity and the coevolution of competitors, or the ghost of competition past. *Oikos* **35**:131–138.

Cords, M. 1986. Interspecific and intraspecific variation in diet of two forest guenons, *Cercopithecus ascanius* and *C. mitis. J. Animal Ecol.* **55**:811–827.

Diamond, J. M. 1978. Niche shifts and the rediscovery of interspecific competition. *Am. Sci.* **66**:322–331.

Gaulin, S. J. C. 1979. A Jarman/Bell model of primate feeding niches. *Hum. Ecol.* **7**:1–20.

Gause, G. F. 1934. *The Struggle for Existence*. Williams & Williams, Baltimore, Maryland.

Gautier-Hion, A. 1978. Food niches and coexistence in sympatric primates in Gabon. In: D. J. Chivers and J. Herbert (eds.), *Recent Advances in Primatology (Vol 1): Behavior*, pp. 269–286. Academic Press, London.

Gautier-Hion, A. 1988. The diet and dietary habits of forest guenons. In: A. Gautier-Hion, F. Bourlière, J.-P. Gautier, and J. Kingdon (eds.), *A Primate Radiation: Evolutionary Biology of the African Guenons*, pp. 257–283. Cambridge University Press, Cambridge, UK.

Gautier-Hion, A., Gautier, J.-P., and Quris, R. 1981. Forest structure and fruit availability as complementary factors influencing habitat use by a troop of monkeys (*Cercopithecus cephus*). *Terre Vie* **35**:511–536.

Grant, P. R. 1975. The classical case of character displacement. *Evol. Biol.* **8**:237–337.

Hamilton, A. C. 1998. Guenon evolution and forest history. In: A. Gautier-Hion, F. Bourlière, J.-P. Gautier and J. Kingdon (eds.), *A Primate Radiation: Evolutionary Biology of the African Guenons*, pp. 13–34. Cambridge University Press, Cambridge, UK.

Janson, C. H. 1988. Food competition in brown capuchin monkeys (*Cebus apella*): Quantitative effects of group size and tree productivity. *Behaviour* **105**:53–76.

Jarman, P. J. 1974. The social organization of antelope in relation to their ecology. *Behaviour* **58**:215–267.

Kaplin, B. A., and Moermond, T. C. 2000. Foraging ecology of the mountain monkey (*Cercopithecus l'hoesti*): Implications for its evolutionary history and use of disturbed forest. *Am. J. Primatol.* **50**:227–246.

Kaplin, B. A., Munyaligoga, V., and Moermond, T. C. 1998. The influence of temporal changes in fruit availability on diet composition and seed handling in blue monkeys (*Cercopithecus mitis doggetti*). *Biotropica* **30**:56–71.

Karasov, W. H., and Diamond, J. M. 1988. Interplay between physiology and ecology in digestion. *BioScience* **38**:602–611.

Kay, R. N. B., and Davies, A. G. 1994. Digestive physiology. In: A. G. Davies, and J. F. Oates (eds.), *Colobine Monkeys: Their Ecology, Behavior and Evolution*, pp. 229–259. Cambridge University Press, Cambridge, UK.

Lambert, J. E. 1998. Primate digestion: Interactions among anatomy, physiology, and feeding ecology. *Evol. Anth.* **7**:8–20.

Lambert, J. E. (in press). Digestive retention times in forest guenons with reference to chimpanzees. *Int. J. Primatol.* **26** (6).

Lambert, J. E., and Whitham, J. 2001. Cheek pouch use in *Papio cynocephalus. Folia Primatol.* **72**:89–91.

Lambert, J. E., Chapman, C. A., Wrangham R. W., and Conklin-Brittain, N. L. 1999. The hardness of cercopithecine foods: Implications for the critical function of enamel thickness in exploiting fallback foods. *Am. J. Phys. Anthropol* (suppl.) **28**:178.

Lernould, J.-M. 1988. Classification and geographical distribution of guenons: A review. In: A. Gautier-Hion, F. Bourlière, J.-P. Gautier, and J. Kingdon (eds.), *A Primate Radiation: Evolutionary Biology of the African Guenons*, pp. 54–78. Cambridge University Press, Cambridge, UK.

Lwanga, J. S., Butynski, T. M., and Struhsaker, T. T. 2000. Tree population dynamics in Kibale National Park, Uganda 1975–1998. *Afr. J. Ecol.* **38**:238–247.

MacArthur, D. W. 1972. *Geographical Ecology*. Harper and Row, New York.

Maisels, F. 1993. Gut passage rate in guenons and mangabeys: Another indicator of a flexible dietary niche? *Folia Primatol.* **61**:35–37.

Martin, P., and Bateson, P. 1986. *Measuring Behavior: An Introductory Guide*. Cambridge University Press, Cambridge, UK.

Milton, K., and Demment, M. W. 1988. Digestion and passage kinetics of chimpanzees fed high and low fiber diets and comparisons with human data. *J. Nutr.* **118**:1082–1088.

Milton, K. 1998. Physiological ecology of howlers (*Alouatta*): Energetic and digestive considerations and comparison with the Colobinae. *Int. J. Primatol.* **19**:513–548.

Morris, D. W. 1996. Coexistence of specialist and generalist rodents via habitat selection. *Ecology* **77**:2352–2364.

Oates, J. F. 1987. Food distribution and foraging behavior. In: B. B. Smuts, D. L. Cheney, R. M. Seyfarth, R. W. Wrangham, and T. T. Struhsaker (eds.), *Primate Societies*, pp. 197–209. University of Chicago Press, Chicago, IL.

Oates, J. F. 1994. The natural history of African colobines. In: A. G. Davies and J. F. Oates (eds.), *Colobine Monkeys: Their Ecology, Behavior, and Evolution*, pp. 75–128. Cambridge University Press, Cambridge, UK.

Olupot, W. 1998. Long–term variation in mangabey (*Cercocebus albigena johnstoni* Lydekker) feeding in Kibale National Park, Uganda. *Afr. J. Ecol.* **36**:96–101.

Pianka, E. R. 1976. Competition and niche theory. In: R. May (ed.), *Theoretical Ecology*, pp. 114–141. W. B. Saunders Company, Philadelphia, PA.

Power, M. L., and Oftedal, O. T. 1996. Differences among captive callitrichids in the digestive responses to dietary gum. *Am. J. Primatol.* **40**:131–144.

Remis, M. J. 1997. Gorillas as seasonal frugivores: Use of resources that vary. *Am. J. Primatol.* **43**:87–109.

Richard, A. F. 1985. *Primates in Nature*. W. H. Freeman & Company, New York.

Richard, A. F., Goldstein, S. J., and Dewar, R. E. 1989. Weed macaques: The evolutionary implications of macaque feeding ecology. *Int. J. Primatol.* **10**:569–594.

Rogers, E., Tutin, C., Parnell, R., Voysey, B., and Fernandez, M. 1994. Seasonal feeding on bark by gorillas: An unexpected keystone food? *Current Primatology (Vol. I): Ecology and Evolution*, p. 154.

Rudran, R. 1978. Socioecology of the blue monkeys of the Kibale Forest, Uganda. *Smith. Contrib. Zool.* **249**:1–88.

Schaik C. P. van, Noordwijk, M. A. van, Boer, R. J., and Tonkelaar, I. D. 1983. The effect of group size on time budgets and social behaviour in wild long-tailed macaques (*Macaca fascicularis*). *Behav. Ecol. Sociobiol.* **13**:173–181.

Schoener, T. W. 1982. The controversy over interspecific competition. *Am. Sci.* **70**:586–595.

Skorupa, J. P. 1988. *The Effects of Selective Timber Harvesting on Rain-Forest Primates in Kibale Forest, Uganda*. Ph.D. Thesis, University of California, Davis.

Strier, K. B. 2000. Population viabilities and conservation implications for muriquis (*Brachyteles arachnoides*) in Brazil's Atlantic forests. *Biotropica* **32**:903–913.

Struhsaker, T. T. 1978. Food habits of five monkey species in the Kibale Forest, Uganda. In: D. J. Chivers and J. Herbert (eds.), *Recent Advances in Primatology (Vol. 2): Conservation*, pp. 87–94. Academic Press, London, UK.

Struhsaker, T. T. 1981. Forest and primate conservation in East Africa. *Afr. J. Ecol.* **18**:191–216.

Struhsaker, T. T. 1997. *Ecology of an African Rain Forest: Logging in Kibale and the Conflict Between Conservation and Exploitation*. University Press of Florida, Gainesville, FL.

Tokeshi, M. 1999. *Species Coexistence: Ecological and Evolutionary Perspectives*. Blackwell Science, Oxford, UK.

Tutin, C. E. G. 1999. Fragmented living: Behavioural ecology of primates in a forest fragment in the Lopé reserve, Gabon. *Primates* **40**:249–265.

Tutin, C. E. G., White, L. J. T. 1998. Primates, phenology and frugivory: Present, past and future patterns in the Lopé Reserve, Gabon. In: D. M. Newbery, H. H. T. Prins, and N. Brown (eds.), *Dynamics of Tropical Communities*, pp. 309–338. Blackwell Science, Oxford, UK.

Waser, P. M. 1975. Monthly variations in feeding and activity patterns of the mangabey, *Cercocebus albigena* (Lydekker). *E. Afr. Wildl. J.* **13**:249–263.

Waser, P. M. 1977. Feeding, ranging and group size in the mangabey *Cercocebus albigena*. In: T. Clutton-Brock (ed.), *Primate Ecology: Studies of Feeding and Ranging Behavior in Lemurs, Monkeys, and Apes*, pp. 183–222. Academic Press, London.

Whitten, P. L. 1988. Effects of patch quality and feeding subgroup size on feeding success in vervet monkeys (*Cercopithecus aethiops*). *Behaviour* **105**:35–52.

Wrangham, R. W., Conklin-Brittain, N. L., and Hunt, K. D. 1998. Dietary response of chimpanzees and cercopithecines to seasonal variation in fruit abundance. I. Antifeedants. *Int. J. Primatol.* **19**:949–970.

Variation in the Diets of *Cercopithecus* Species: Differences within Forests, among Forests, and across Species

22

COLIN A. CHAPMAN, LAUREN J. CHAPMAN,
MARINA CORDS, JOEL MWANGI GATHUA,
ANNIE GAUTIER-HION, JOANNA E. LAMBERT,
KARYN RODE, CAROLINE E. G. TUTIN,
and LEE J. T. WHITE

COLIN A. CHAPMAN and LAUREN J. CHAPMAN • Department of Zoology, University of Florida, Gainesville, FL 32611, USA.   MARINA CORDS • Department of Ecology, Evolution and Environmental Biology, Columbia University, New York, NY 10027, USA.   JOEL MWANGI GATHUA • Formerly Mammalogy Department, National Museums of Kenya, Nairobi, Kenya.   ANNIE GAUTIER-HION • UMR 6552, CNRS–Université de Rennes I, Station Biologique, 35380 Paimpont, France.   JOANNA E. LAMBERT • Department of Anthropology, University of Oregon, Eugene, OR 97403, USA.   KARYN RODE • Department of Zoology, University of Florida, Gainesville, Florida 32611, USA.   CAROLINE E. G. TUTIN • Centre International de Recherches Médicales de Franceville, Gabon.   LEE J. T. WHITE • Wildlife Conservation Society, 185th Street and Southern Boulevard, Bronx, NY 10460, USA.

The Guenons: Diversity and Adaptation in African Monkeys, edited by Glenn and Cords. Kluwer Academic/Plenum Publishers, New York, 2002.

Introduction

An accumulation of data on the diets of wild primate populations in the last three decades has led to a growing appreciation of the magnitude of dietary variation within species, including differences among populations, among groups in a single population, and within a single group over time (Chapman and Chapman, 1990; Gautier-Hion *et al.*, 1993; Chapman and Chapman, 1999; Davies *et al.*, 1999). Extensive dietary variation challenges the typological conception of species that underlies many comparative analyses and it may obscure large-scale patterns (Nunn and Barton, 2001). Comparative analyses, however, are central to the formulation and testing of hypotheses explaining many aspects of the behavior and ecology of primates (as well as other animals; Milton and May, 1976; Clutton-Brock and Harvey, 1977; Isbell, 1991; Sterck *et al.*, 1997). Dietary variables have figured centrally in such analyses. Typically, feeding data are used to specify the proportions of the diet comprising different food items and thus to classify species according to their main dietary constituents (frugivores, folivores, frugivore/insectivores). For example, frugivorous and folivorous primates have been contrasted with respect to behavioral characters such as day range length, home range size, degree of intra- and intergroup competition, and ability to withstand habitat disturbance (Clutton-Brock and Harvey, 1977; Harvey *et al.*, 1987; Johns and Skorupa, 1987; Isbell, 1991; Grant *et al.*, 1992; Chapman *et al.*, 1999). Aspects of physiology and morphology have also been compared, such as metabolic rate (Ross, 1992), dentition (Kay, 1977; Anapol and Lee, 1994), gastrointestinal tracts (Chivers and Hladik, 1980), and relative brain sizes (Dunbar, 1998).

Most comparative analyses use the species as the fundamental unit, and assume that variation within species is small relative to variation between species or higher-order taxonomic units. The data we present in this chapter challenge this assumption with respect to diet composition, and suggest that dietary flexibility, even at the level of the items included as major constituents, may be considerable for certain species, among them various guenons.

Recent studies among folivorous primates have shown considerable intraspecific dietary variation, both temporal and geographical. For example, Chapman *et al.* (in press) quantified dietary variability in red colobus monkeys (*Procolobus badius tephrosceles*) in Kibale National Park, Uganda and compared eight groups separated by <15 km, neighboring groups with overlapping home ranges, and the same groups over four years. These data were evaluated with respect to dietary variability among *Procolobus badius* populations and all colobine species described from sites across Africa. Within Kibale there were large dietary differences in the use of particular plant parts and species on all spatial and temporal scales. In some cases, these interdemic comparisons revealed greater variation than interspecific comparisons (Chapman *et al.*, 2002). Studies of central and west African colobine populations also demonstrate dietary flexibility: contrary to early descriptions of colobines as strict folivores (Hill, 1964; Napier, 1970), some populations are not obligate leaf-eaters, and seeds can constitute a large proportion of their diets (McKey, 1978; Harrison, 1986; Davies *et al.*, 1988; Davies, 1994; Maisels *et al.*, 1994; Gautier-Hion *et al.*, 1997; Davies *et al.*, 1999).

Unlike colobines, guenons are generally known as being mainly frugivorous. As such, their relationship with plants, especially as seed dispersers, may be different from that of folivorous colobines. Plant requirements for seed dispersal may have led to the evolution of fruits that meet the dietary needs of particular frugivores (Howe and Smallwood, 1982; Gautier-Hion *et al.*, 1985; Chapman, 1995), and because of this mutualistic relationship one might expect a less variable relationship between frugivorous consumers and their plant foods. Despite such theoretical considerations, empirical studies call into question how reliable frugivores are as seed dispersal agents to particular plant species or to plants that share a particular fruit morphology (Gautier-Hion *et al.*, 1980, 1985, 1993; Herrera, 1985, 1998; Fischer and Chapman, 1993; Lambert and Garber, 1998; Chapman and Chapman, 2001).

A comparison of the diets of groups, populations, and species of frugivores is one way to evaluate whether frugivorous diets are relatively invariant. It seems logical to expect that taxa with closer phylogenic proximity or populations with greater spatial proximity will have more similar diets; however, this remains to be verified. We examined the diets of *Cercopithecus* species in three species groups, *cephus*, *nictitans*, and *mona*, and describe dietary variation at four levels: intergroup, interdemic, interpopulational, and interspecific. We also consider variation in diet over time. We primarily consider variation in use of particular plant parts, but also consider use of particular species. The forest guenons are an ideal taxonomic group for such analyses because they are closely related species and their diets are relatively well documented (Gautier-Hion, 1988).

Methods

Kibale National Park, Uganda

Kibale National Park in western Uganda (766 km^2; 0°13'–0°41'N and 30°19'–30°32'E; Struhsaker, 1997; Chapman and Lambert, 2000) has been the site of two studies of *Cercopithecus mitis* (Rudran, 1978a,b; Butynski, 1990). We also report the diets of four groups of *Cercopithecus ascanius* at this site. Mean annual rainfall in the region is 1750 mm (1990–1999). Rainfall is bimodal with two rainy seasons generally occurring from March to May and September to November.

Rudran (1978a,b) studied *Cercopithecus mitis* at the Kanyawara study site within Kibale National Park between February 1973 and January 1974, observing two groups with home ranges separated by *ca.* 500 m. Typically,

Group 1 was observed for five consecutive day periods during the first two weeks of the month, and Group 2 was observed for one 5-day period during the third or last week of the month. Feeding was scored when a particular monkey ingested a particular item from a particular plant species. For plant foods, a new feeding score was recorded for the same combination of monkey, item, and plant only if 30 minutes had elapsed or if the identity of at least one of the three parameters changed. Group 1 was observed for 584 hours (2329 feeding scores), while Group 2 was observed for 558 hours (2268 feeding scores).

Butynski (1990) studied four neighboring groups of *Cercopithecus mitis* at Kanyawara and one group at a site 12 km to the south (Ngogo) between October 1978 and August 1980. Dawn to dusk observations were made on 1, 2, 3, or 5 consecutive days per month for a total of 2673 hours and feeding was scored following Rudran (1978a,b; feeding scores per group at Kanyawara: 1 = 5593, 2 = 1066, 3 = 926, 4 = 1656; Ngogo = 3885).

We also present new observations of dietary variation among *Cercopithecus ascanius* at Kibale, with data from one group at each of four sites each separated by *ca.* 15 km (Chapman *et al.*, 1997; Fig. 1). Within Kibale, there is an elevational gradient from north to south (920–1590 m) that corresponds to an increase in temperature and decrease in rainfall (Struhsaker, 1997). The climatic gradient, other naturally varying abiotic and biotic conditions, and varying histories of human modification to the forest (particularly selective logging) resulted in the four groups of *Cercopithecus ascanius* experiencing different environmental conditions (Chapman and Lambert, 2000). Observations were made from July 1997 to June 1998 at each of the four sites over two consecutive days each month totaling 228 hours (862 feeding scores) at K-30, 152 hours (583 feeding scores) at Dura, 94 hours (415 feeding scores) at Sebatoli, and 94 hours (309 feeding scores) at Mainaro. During each half-hour, an observer took five point samples of different individuals,

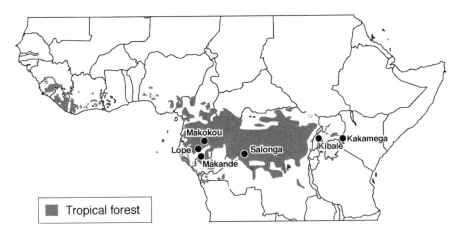

Fig. 1.   A map of central Africa, indicating the location of the study sites.

recording the species and plant part for any feeding monkey (Chapman *et al.*, 1997; Chapman and Chapman, 1999).

Kakamega Forest, Kenya

Kakamega Forest (86 km^2 in main block) in western Kenya (0° 14'N, 34° 52'E, 1580 m), is only 500 km east of Kibale. The two forests are at nearly the same latitude and have similar climates (~2000 mm annual rainfall; Gathua, 2000), physiognomy, and plant species compositions (Cords, 1987, 1990). Kakamega and Kibale were likely part of a continuous forest, until anthropogenic clearing largely converted the forest between them to agricultural land (Hamilton, 1974; Hamilton *et al.*, 1986). They do differ, however, in the richness and biomass of their primate communities (Cords, 1990; Fashing and Cords, 2000), with more primate species and greater biomass at Kibale. Study groups in Kakamega frequented indigenous forest only. Some selective logging and enrichment planting had occurred at the study site *ca.* 30–50 years before the study.

Cords (1986, 1987) reported dietary data collected over 12 months (April 1980–March 1981) from one group each of *Cercopithecus ascanius* ($n = 20$–25 individuals) and *C. mitis* ($n = 40$–45 individuals) at the Isecheno study site. She followed each study group from dawn to dusk for seven to nine days per month (in two sessions) and scored feeding behavior with the same frequency method as Rudran (1978a,b; 9009 scores for *Cercopithecus ascanius*, 10,167 scores for *C. mitis*).

In 1997–98, Gathua (2000), working at the Buyangu study site (*ca.* 11 km north of Isecheno), studied the feeding patterns of two neighboring groups of *Cercopithecus ascanius* over an annual cycle. Each group consisted of approximately 30 individuals, but one was located on the forest edge, whereas the other inhabited only the forest interior. Each group was observed from about 0700–1600 h for about 14 days per month (in two sessions). Diet was evaluated using Rudran's (1978a,b) methods and 13,980 and 13,998 feeding scores were recorded for the two groups.

Lopé Reserve, Central Gabon

The Lopé study area covers about 50 km^2 (0° 10'S, 11° 35'E) in the northern portion of Lopé Reserve (Fig. 1). The forest is heterogeneous in both plant species and structure; however, the most common type is Marantaceae forest (Tutin *et al.*, 1994; White *et al.*, 1995). Parts of the study area were selectively logged during the 1960s and 1970s at an extraction level of approximately 1.5 trees/ha. Mean annual rainfall is 1548 mm (1984–1995) and the climate is characterized by a long dry season between June and September, and a second less well defined dry period from December to February.

To obtain information on diet from many groups in continuous forest, Tutin *et al.* (1997) compiled opportunistically a list of the food consumed by

Cercopithecus species over a ten-year period, and additionally during Ham's (1994) 15-month study of mangabeys. Each food was recorded only once per day per species.

Tutin (1999) also collected dietary data from monkeys inhabiting a nine-hectare forest fragment surrounded by savanna located 90 m from the continuous forest. Between April 1996 and August 1997, she collected data on 127 days (5–12 days per month) on one group of *Cercopithecus cephus* that lived permanently within this forest fragment, using instantaneous scan sampling of all visible individuals at 15-minute intervals.

Forêt des Abeilles, Central Gabon (Makandé)

This forest is part of a large unfragmented forest block in Central Gabon that extends from the left bank of the Ogooué River in the North to the Massif du Chaillu in the South. It is separated from the Lopé Reserve by the Offoué River. The study was carried out at the Makandé site (0°40′39′S; 11°54′35′E) where mean annual rainfall is 1753 mm. There are two rainy seasons and two dry seasons (Jan–Feb and June–Sep). The site is covered with closed canopy forest on hilly topography. Caesalpiniaceae (24% of trees >70 cm dbh) and Burseraceae (13.6%) are the dominant tree families. The study area was selectively logged three years before the study at an extraction level of about 1.5 trees/ha (>70 cm dbh; Lasserre and Gautier-Hion, 1995).

Brugière *et al.* (in press) collected data on the plant diets of *Cercopithecus pogonias* using a frequency method with a half-hour interval. Two thousand eight hundred and seventy three feeding records were collected over an annual cycle (Brugière *et al.*, in press).

Makokou Region, N–E Gabon

Gautier–Hion's (1980) studies on monkey diets were based on stomach contents (*n* = 100 *Cercopithecus nictitans*, *n* = 62 *C. cephus*, *n* = 52 *C. pogonias*) collected over a ten-year period from hunters <50 km around Makokou (0° 34′N, 12° 52′E). In the area, mature seasonal evergreen forest is interspersed with areas of riparian and swamp forest and patches of secondary forest (Quris, 1976; Hladik, 1982). Dry forest types are dominated by Caesalpiniaceae (17.4%), Annonaceae (15.3%), and Euphorbiaceae (16.5%, Maisels and Gautier-Hion, 1994). Mean annual rainfall at Makokou is 1755 mm and its seasonal distribution is similar to Lopé and Makandé.

Salonga National Park, Democratic Republic of Congo

The Salonga National Park is located within the Congo Basin, DRC and covers 36,000 km^2 in two different blocks connected by a 45-km corridor.

Maisels and Gautier-Hion (1994) conducted research on the primate community at Botsima, a site located within a meander of the Lomela River (1°15'S, 22°E). The vegetation is a mosaic of terra firma forest, riverine forest on alluvial levees, and forest with a high degree of seasonal flooding. They collected information on primate diets during eight months (February–September 1991) in an inundated area largely dominated by Caesalpiniaceae (39% of trees). Mean annual rainfall is 1756 mm as measured at Ikela, 120 km from the field site. The area has two rainy seasons and two dry seasons (January–February and June–July). Contrary to the study sites in Gabon, the main dry season occurs in January–February. Maisels and Gautier-Hion (1994) described the diet of *Cercopithecus* through direct observations (3293 feeding records) using a frequency method.

General Considerations

We included in our analyses only populations living under similar environmental conditions. With the exception of the final contrast, all the sample populations inhabited evergreen forests with similar rainfall (range 1548–2000 mm/year) and seasonal patterns, although the timing of the seasons varied. In general, the diurnal primate communities at the sites were similar (Table I) and included two to four *Cercopithecus* species. Arboreal mangabeys (*Lophocebus*) occurred at all sites except Kakamega and the number of colobine species was one or two (Table I).

Comparisons between sites are confounded by the fact that different methods (opportunistic observations, frequency methods, and stomach contents) were used to characterize monkey diets. However, we feel that intersite differences in diet are too large to be explained as methodological artifacts alone. We note where methodological differences may have influenced results. For example, data on stomach contents may be biased by the hunting methods used (e.g., hunters wait at fruiting trees). Opportunistic observations may be biased in many ways, but in particular may underestimate insect foraging. Such methodological differences will influence the evaluation of the diets of the different populations and even slight differences may cause systematic differences (e.g., level of habituation or duration between successive observations). We hope that future studies will investigate how various methods characterize diets differently. It should be noted, however, that studies employing very different methods have previously been used to characterize gross dietary differences and the contrasts they revealed are currently the foundation of a number of theoretical comparisons and the basis of a number of widely held generalizations (Clutton-Brock and Harvey, 1977).

Another methodological difference concerns the categorization of food items: some researchers considered the consumption of both fruit flesh and seeds as fruit eating, while others classed as seed eating both the

Table I. Descriptions of the Monkey Community in Gabon (Lopé, Makandé, Makokou), Uganda (Kibale), Kenya (Kakamega) and DRC (Salonga). Figures indicate biomasses in kg/km^2; – indicates species not present at the site; ? indicates biomass unknown. * *Cercopithecus solatus* is only found south of the Lopé Reserve.

Species	Kibale	Kakamega	Lopé	Makandé	Makokou	Salonga
Cercopithecus						
(*cephus*) *cephus*	–	–	10.2	6	80	–
(*cephus*) *ascanius*	328	220	–	–	–	?
(*nictitans*) *nictitans*	–	–	62.8	41.3	100	–
(*nictitans*) *mitis*	133	645	–	–	–	–
(*mona*) *pogonias*	–	–	10.1	9.0	60	–
(*mona*) *wolfi*	–	–	–	–	–	?
(*lhoesti*) *lhoesti*	13	–	–	–	–	–
(*lhoesti*) *solatus*	–	–	6.9*	73.1	–	–
neglectus	–	–	–	–	110	?
Allenopithecus nigroviridis	–	–	–	–	–	?
Miopithecus ogouensis	–	–	–	–	60	–
Lophocebus albigena	60	–	33.7	13.0	?	–
Lophocebus aterrimus	–	–	–	–	–	?
Cercocebus galeritus	–	–	–	–	?	–
Colobus guereza	317	1035	–	–	rare	–
Colobus angolensis	–	–	–	–	–	?
Colobus satanas	–	–	90.7	56.9	–	–
Procolobus badius	1760	–	–	–	–	?

Kibale: Struhsaker (1975, 1978), Struhsaker and Leland (1979), and Chapman *et al.*, (2000). Kakamega: Fashing and Cords (2000), Fashing unpub. data (for the sites of Cords 1986, 1987, study). Lopé: White (1994a,b, mean of 5 neighboring sites), except *Cercopithecus solatus* biomass: Brugière *et al.* in press. Makandé: Brugière *et al.* in press. Makokou: Waser 1987 — midpoint of population density. Salonga: Maisels *et al.*, 1994.

consumption of seeds from dry fruit and of immature fleshy fruit. Similarly, the amount of time spent eating insects can include only ingestion time or can include the time spent searching for insects.

Results

Intergroup Comparisons

Cercopithecus ascanius *and* C. mitis *in Kibale and Kakamega*

Data from the East African forests allow us to compare neighboring groups of *Cercopithecus mitis* (Kibale) and *C. ascanius* (Kakamega). Rudran (1978a,b) found that two neighboring *Cercopithecus mitis* groups shared only 45% of the plant species used by both groups. The monthly dietary overlap between groups was low to moderate (mean = 40.1%, range 22.6–65.2%; Holmes and Pitelka, 1968, index of dietary overlap). Butynski (1990) studied four groups of *Cercopithecus mitis* at the same site and found that the amount of time feeding on fruit ranged from 22.4 to 35.4% among them. The mean overall dietary overlap of the top 20 most often eaten food items averaged 33.5% among the four neighboring groups (Butynski, 1990). The diets of two neighboring groups of *Cercopithecus ascanius* observed by Gathua (2000) were more similar (Fig. 2). Nine and eight of the top 10 species-specific food items were fruits for the two groups respectively. The top 10 species-specific food items accounted for 60.3 and 60.7% of the two groups' total plant diets, and five of these foods were among the top 10 species-specific food items for both groups (Gathua, 2000). The overlap in monthly diets was considerable, especially for the plant diet. Monthly overlaps of plant species (35–70%) and species-specific foods (33–69%) were lower than those based on items (64–95%), but very similar to each other; these observations suggest that the two groups ate mostly the same plant food items, but used different source plants. One group ate more young leaves for six of the 12 study months, whereas the other group ate more gums and cotyledons (Gathua, 2000). While differences in folivory were not easily explained, the use of some of the other

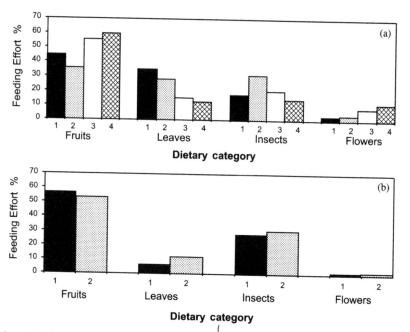

Fig. 2. (a) The percentage of feeding effort that *Cercopithecus ascanius* from four different groups each separated by approximately 15 km in Kibale National Park, Uganda spent eating different plant parts (1 = Sebatoli, 2 = Kanyawara, 3 = Dura River, 4 = Mainaro) and (b) the percentage of feeding effort that neighboring groups of *C. ascanius* from Kakamega, Kenya spent feeding on different plant parts.

minor food items (e.g., gums) likely resulted from availability of the appropriate species. For example, one group fed extensively on the gums of *Prunus africana*, which was common along the forest edge in that group's range. The second group did not feed on gum from this tree because it was not available in the group's range. However, there were also cases in which the same foods were available in both groups' home ranges, but were differentially included in the diet; differential consumption of these items was not explained entirely as a function of differential availability.

Interdemic Comparisons

Cercopithecus mitis *and* C. ascanius *in Kibale and Kakamega*

Butynski (1990) compared the diets of four neighboring *Cercopithecus mitis* groups at one site (Kanyawara) to that of another group *ca.* 12 km away (Ngogo). He found that the amount of time feeding on fruit ranged from 22.4 to 35.4% for the four groups studied at the same site, and was 22.8% at the distant site. Of the top 20 species-specific plant food items that were consumed by the five groups, the Ngogo group shared on average eight with the other four Kanyawara groups. The mean overall dietary overlap of these top 20 food items averaged 33.5% among the four neighboring groups and 18.2% between the Ngogo group and the four Kanyawara groups (Butynski, 1990). In general, differences between the Kanyawara and Ngogo demes exceeded those that occurred between neighboring groups at Kanyawara.

The diet of *Cercopithecus ascanius* varied among four sites in Kibale (Fig. 2). While fruit was the major dietary constituent at all sites over the study period, the amount of fruit varied, ranging from only 36% of feeding records at Kanyawara to 60% at Mainaro. Either leaves or insects were the second most prevalent item in the diet, and the consumption of each of these items varied by a factor of two or more among sites. The degree to which *Cercopithecus ascanius* relied on particular plant species also varied from deme to deme (Table II). Some, but not all, of the variation appeared to be related to the plant density. For example, *Cercopithecus ascanius* ate fruit of *Mimusops bagshawei* at Kanyawara (1.8% of annual diet and 32.0% of the diet in the month when it fruited) and at Dura River (4.5% of total annual diet and 57.2% of the diet in the months it was available), but at Sebatoli they did not eat this fruit even though it was available. At Dura River, *M. bagshawei* is relatively rare (<1.3 trees/ha). It is more common at Kanyawara (3.3 trees/ha) and is most abundant at Sebatoli (6.3 trees/ha). The magnitude and duration of its fruiting were similar at all sites (Chapman *et al.*, 1999).

In contrast to the *Cercopithecus ascanius* at Kibale, the two demes at Kakamega were notably similar in their dietary patterns. On an annual basis, the proportion of fruit (53–61%) and invertebrate feeding (25–31%) varied little across study sites. All of the top 20 food species used by the Isecheno

Table II. The Density (Individual Trees >10 cm DBH) and the Percentage of Time Spent Feeding by *Cercopithecus ascanius* on Trees at Four Sites in Kibale National Park, Uganda. Listed are those species that the *Cercopithecus ascanius* used for >1% of their foraging effort at any site. The plant parts eaten for each tree species are listed in order of importance in the diet: RF = ripe fruit, UF = unripe fruit, YL = young leaves, FL = flower, LB = leaf bud, BA = bark, SD = seed.

Species	Family	Part	Kanyawara %Foraging	Kanyawara Density	Sebatoli %Foraging	Sebatoli Density	Mainaro %Foraging	Mainaro Density	Dura River %Foraging	Dura River Density
Celtis durandii	Ulmaceae	UF/RF/YL/LB	14.84	47.1	4.82	2.50	29.37	33.80	29.79	63.80
Chrysophyllum gorganusanum	Sapotaceae	RF/RL/UF/YL/SD	0.47	2.6	12.53	8.80	15.84	21.20	5.05	47.50
Celtis africana	Ulmaceae	YL/UF/RF/FL	7.66	4.2	4.34	0.00	1.65	1.30	0.00	0.00
Bosqueia phoberos	Moraceae	YL/RF/UF	3.89	50.0	1.93	0.00	0.00	0.00	5.92	22.50
Teclea nobilis	Rutaceae	RF/FL/YL/UF	7.07	17.1	0.00	0.00	1.32	0.00	2.26	0.00
Diospyros abyssinica	Ebenaceae	RF/UF/YL/FL	7.42	40.0	1.93	2.50	0.99	1.30	0.00	0.00
Prunus africana	Rosaceae	YL/RF/LB/BA	2.47	0.0	7.23	2.50	0.00	0.00	0.00	0.00
Warburgia stuhlmanni	Canellaceae	RF/UF/YL/FL	0.00	0.0	0.00	0.00	7.92	0.00	0.52	0.00
Croton sp.	Euphorbiaceae	RF/UF/FL/YL/BA	0.00	0.8	8.19	41.30	0.00	1.30	0.00	0.00
Uvariopsis congensis	Annonaceae	RF/UF	1.06	60.4	0.00	0.00	3.96	43.80	2.79	60.00
Maesa lanofolato	Myrsinaceae	RF	1.65	0.0	6.02	0.00	0.00	0.00	0.00	0.00
Albizia grandbracteata	Leguminosae	YL	0.94	1.3	4.58	0.00	0.00	0.00	1.05	1.30
Bequaertiodendron oblanceolatum	Sapotaceae	YL/RF	0.00	0.0	0.00	0.00	0.00	0.00	6.27	57.50
Mimusops bagshawei	Sapotaceae	RF/UF/YL/FL	1.18	3.3	0.00	6.30	0.33	0.00	4.53	0.00
Ficus exasperata	Moraceae	RF/YL/UF	1.65	3.8	2.17	2.50	1.98	1.30	0.00	0.00
Cynometra alexandri	Leguminosae	FL/YL	0.00	0.0	0.00	0.00	4.29	63.80	0.00	0.00
Celtis mildbraedii	Ulmaceae	RF/YL/UF	0.00	0.0	0.00	0.00	4.29	32.50	0.00	0.00
Ficus natalensis	Moraceae	RF	0.00	0.4	0.00	0.00	0.00	0.00	3.83	0.00

(Cont.)

Species	Family	Part	Kanyawara %Foraging	Density	Sebatoli %Foraging	Density	Mainaro %Foraging	Density	Dura River %Foraging	Density
Linociera johnsonii	Oleaceae	FL/YL	0.00	5.4	0.96	8.80	2.64	0.00	0.00	0.00
Markhamia platycalyx	Bignoniaceae	FL/LP/YL	0.59	50.0	1.45	38.80	0.00	1.30	1.05	8.80
Strychnos mitis	Loganiaceae	RF/TL/FL/UF	0.82	7.5	0.48	0.00	1.32	0.00	0.00	0.00
Olea welwitschii	Oleaceae	YL	2.00	3.3	0.00	0.00	0.00	0.00	0.35	1.30
Bridelia micrantha	Euphorbiaceae	RF/LB	0.82	0.0	1.45	0.00	0.00	0.00	0.00	0.00
Funtumia latifolia	Apocynaceae	RF/FL/YL/UF	0.12	33.8	0.72	2.50	0.00	2.50	1.39	43.80
Ficus sansibarica	Moraceae	RF	1.06	0.0	0.72	0.00	0.00	0.00	0.00	0.00
Monodora myristica	Annonaceae	FL	0.00	0.4	0.00	0.00	0.00	0.00	1.74	3.80
Fagara angolensis	Rutaceae	YL/FL	0.00	0.0	1.69	0.00	0.00	0.00	0.00	0.00
Pseudospondias microcarpa	Anacardiaceae	YL	0.12	1.7	0.48	0.00	0.00	0.00	1.05	3.80
Lovoa swynnertonni	Meliaceae	FL	0.00	0.8	0.00	3.80	0.00	0.00	1.57	3.80
Balanites wilsoniana	Balanitaceae	FL/RF/YL	0.00	1.7	1.20	0.00	0.33	0.00	0.00	1.30
Neutonia buchananii	Leguminosae	YL	0.00	0.0	1.45	26.30	0.00	0.00	0.00	3.80
Chaetacme aristata	Ulmaceae	RF	0.35	17.1	0.00	0.00	0.00	0.00	1.05	3.80
Casearia sp.	Flacourtiaceae	RF/UF	0.00	1.3	0.00	0.00	0.00	0.00	1.39	0.00
Spathodea campanulata	Bignoniaceae	FL	0.00	0.8	0.00	0.00	0.00	0.00	1.22	0.00
Dombeya mukole	Sterculiaceae	YL/FL	1.18	9.2	0.00	0.00	0.00	0.00	0.00	1.30

group (which accounted for 83% of plant feeding records) were also used by the Buyangu groups, and 13 of them also ranked in the top 20 in the plant diets of each study group at Buyangu—where the top 20 species accounted for 89–92% of feeding records. Nevertheless, the relative proportions of species-specific plant food items in the diet of red-tailed monkeys varied across sites, so that overlap (based on shared percentages) between the two sites in the annual diet was 35–42%. There was less overlap between the same species at the two sites than there was between different species (*Cercopithecus ascanius* vs. *C. mitis*) at a single site (70.4%; Cords, 1987). Some differences in the consumption of particular foods present at both sites seemed related to differences between sites in their quantity. For example, at Buyangu, the density of *Strychnos usambarensis* trees was 84–127 times greater and this tree, the 2nd and 4th ranked food tree for the two Buyangu groups, was used for plant feeding 164–248 times more often than at Isecheno (where it was ranked 65th). Other differences in the consumption of particular foods were not so clearly related to availability. For example, *Harungana madagascariensis* was six to eight times less dense at Isecheno, but was eaten as often as Buyangu Group E, and four times more often than Buyangu Group F.

Lopé: Cercopithecus cephus *in continuous forest and a forest fragment*

At Lopé, the diets of *Cercopithecus cephus* in the continuous forest and the forest fragment differed (Table III; Tutin *et al.*, 1997; Tutin, 1999). For example, fruit feeding accounted for 67% of records in continuous forest, but only 49% in the forest fragment. The groups of *Cercopithecus cephus* in the continuous forest ate 85 different plant species/parts, while the group in the forest fragment ate only 52 of them. The majority of the foods eaten by the forest groups (57.6%) were also eaten in the fragment. Most differences were due to variation in the availability of specific foods, as plant species diversity was much reduced in the small forest fragment compared to continuous forest. During months when fruit was scarce in the fragment due to the absence of certain tree species, the monkeys showed extremely high levels of feeding on insects (80% of feeding records) or flowers (30%).

Interpopulational Comparisons

Kibale/Kakamega: Cercopithecus ascanius *and* C. mitis

Because of the geographic proximity and shared history of the Kakamega and Kibale forests, one might expect the diets of the two populations to be similar. At Kibale, the average amount of time that *Cercopithecus ascanius* has been reported to eat fruit, leaves, and insect is 48, 21, and 21% respectively (Table III, averaged across all studies). In contrast, at Kakamega the average

Table III. Diets (Percentage of Dietary Components) of Select Cercopithecines from a Variety of Sites across Africa (leaves include buds, young leaves, petioles; for Makandé and Salonga, seeds include dry seeds and seeds from immature fleshy fruit)

Species	Sites of detailed study	Leaves	Seeds	Fruit (arils)	Flowers	Insects
Cephus Group						
C. ascanius (1)	Kakamega	7	<1	61	2	25
C. ascanius (2)	Kakamega	7	–	57	2	28
C. ascanius (3)	Kakamega	12	–	53	1	31
C. ascanius (4)	Kibale	14	<1	44	15	22
C. ascanius (5)	Kibale	35	–	45	3	18
C. ascanius (6)	Kibale	28	–	36	4	31
C. ascanius (7)	Kibale	15	–	56	8	21
C. ascanius (8)	Kibale	13	–	60	12	15
C. ascanius (9)		42	–	46	5	Not recorded
C. ascanius (10)		74	–	13	9	Not recorded
C. ascanius (11)	Salonga	17	24	44 (32)	16	Not recorded
C. cephus (12)	Makokou	6	–	78	<1	13
C. cephus (13)	Lopé Continuous	11	7	67	6	9
C. cephus (14)	Lopé Fragment	4	5	49	6	35
Mona Group						
C. wolfi (15)	Salonga	30	27	32 (27)	11	Not recorded
C. pogonias (16)	Makokou	1	3	80	<1	16
C. pogonias (17)	Lopé	7	9	69	9	7
C. pogonias (18)	Makandé	13	50	27 (14)	5	Not recorded
Nictitans Group						
C. nictitans (19)		17	–	71	<1	9
C. nictitans (20)		16	11	59	10	3
C. nictitans (21)		17	4	44	9	24
C. nictitans (22)		10	50	36 (13)	4	Not recorded
C. mitis (23)	Kakamega	17	3	55	4	17
C. mitis (24)		–	9	48	6	24
C. mitis (25)		26	–	57	13	6
C. mitis (26)		33	–	54	10	<1
C. mitis (27)		47	–	28	11	1
C. mitis (28)		3	–	91	2	–
C. mitis (29)		6	–	37	20	11
C. mitis (30)		22	2	56	5	9
C. mitis (31)		30	6	45	6	10
C. mitis (32)	Kibale	19	2	43	12	20
C. mitis (33)	Kibale	31	–	26	7	36
C. mitis (34)	Kibale	23	–	30	10	36

(1) Kakamega, Kenya, Cords, 1986; (2) Kakamega, Kenya (Group E) Gathua, 2000; (3) Kakamega, Kenya (Group F), Gathua, 2000; (4) Kibale, Uganda, Struhsaker, 1978, (5) Kibale at Sebatoli, this study; (6) Kibale at Kanyawara, this study; (7) Kibale at Dura River, this study; (8) Kibale at Mainaro, this study; (9) and (10) Budongo Forest, Uganda, Sheppard, 2000; (11) Salonga, DRC, Gautier-Hion, pers. obs.; (12) Makokou, Gabon, Gautier-Hion *et al.* (1980); (13) Lopé (continuous forest), Gabon, Tutin *et al.* (1997a); (14) Lopé (forest fragment), Gabon, Tutin 1999, Tutin *et al.* (1997b); (15) Salonga, Gautier-Hion 1993; (16) Makokou, Gautier-Hion, 1980; (17) Lopé, continuous forest, Tutin *et al.* 1997; (18) Makandé, Gabon, Brugière *et al.* in press (19) Makokou, Gautier-Hion, 1980; (20) Lopé, continuous forest, Tutin *et al.* 1997; (21) Lopé, forest fragment, Tutin *et al.* 1997; (22) Makandé, Brugière *et al.* in press; (23) Kakamega Forest, Kenya, Cords, 1986; (24) Nyungwe Forest, Rwanda, Kaplin and Moermond, 1998; (25) Cape Vidal Dune Forest, South Africa, Lawes, 1991; (26) and (27) Zomba Plateau, Malawi, Beeson *et al.* 1996; (28) Ngoye Forest, South Africa, Lawes *et al.* 1990; (29) Kahuzi-Biega National Park, DRC, Schlichte, 1978; (30) Budongo Forest (logged) and (31) (unlogged), Fairgrieve, 1995; (32) Kibale, (unlogged) Rudran 1978a,b; (33) Kibale (Kanyawara) Butynski, 1990; (34) Kibale (Ngogo), Butynski (1990).

amount of time that this species has been documented to eat fruit, leaves, and insects is 57, 9, and 28% respectively.

To evaluate variability among populations of *Cercopithecus mitis*, we can contrast diets reported by Rudran (1978a,b) and Butynski (1990) to that of Cords (1986, 1987; Table III). This comparison suggests that the blue monkeys at Kibale are less frugivorous (33%), and spend more time eating young leaves (24%) and insects (30%), than the Kakamega population (fruit 54%, leaves 16%, insects 17%). For both *Cercopithecus mitis* and *C. ascanius*, frugivory is higher and folivory is lower in the Kakamega population, while differences in insectivory are inconsistent across species.

Gabon: Cercopithecus pogonias

We examined dietary variation of *Cercopithecus pogonias* in Gabon (Makokou, Gautier-Hion, 1980; Lopé, Tutin *et al.*, 1997; Makandé, Brugière *et al.*, in press) and found that diets differ among three populations (Table III). Fruit eating was well-developed at Makokou (80%) and at Lopé (69%), while seeds were the main food at Makandé (50%). The amount of leaves eaten is lowest at Makokou (1.2%) and greatest at Makandé (12.6%). The number of plant species included in the diets is 121 at Makandé, 63 at Makokou, and 46 at Lope. At Makandé (in 1994), four plant species contributed 50% to the annual plant diet (*Dialium* two spp., *Staudia gabonensis*, and *Xylopia hypolampra*; Gautier-Hion, unpub. data). At Makokou, fruits of the Vitaceae liana *Cissus dinklagei* and of five Apocynaceae lianas provided the commonest fruit items over the year. During the annual period of fruit scarcity in Gabon (June–August), berries of *Polyalthia suaveolens* and arils of *Coelocaryon preussi* and *Pycnanthus angolensis* accounted for 50% of *Cercopithecus pogonias*'s food items at Makokou (Gautier-Hion and Michaloud, 1989). At Lopé, during the same period, the diet of *Cercopithecus pogonias* was dominated by the fruits of *Pseudospondias longifolia* and *Xylopia* (four spp.), as well as arils of *Pycnanthus angolensis* (Tutin *et al.*, 1997). At Makandé, also during the same period, drupes of two species of *Dialium* accounted for 50% of the diet of *Cercopithecus pogonias*.

Interspecific Comparisons

Comparisons within the mona *group*

At Salonga, *Cercopithecus wolfi* ate 32% fruit (4% fleshy fruits and 27% arils), and 27% seeds. Leaves composed 41% of the diet (Table III). Comparisons of *Cercopithecus wolfi* and *C. pogonias* showed that the diet of *C. pogonias* at Makandé, Gabon is more similar with respect to the items eaten to that of *C. wolfi* at Salonga, DR Congo, than to *C. pogonias* at Lopé and Makokou, Gabon. The low consumption of fleshy fruit at Makandé and

Salonga was related to the availability of specific foods: at both sites, the forest is dominated by Caesalpiniaceae which produce dry fruit and species producing fleshy fruit are rare (Maisels and Gautier-Hion, 1994; Brugière *et al.*, in press).

Comparisons within the cephus *group*

There were differences in the use of specific type of foods between closely related *cephus* and *ascanius* species. For example, leaf feeding made up 4% of the foraging effort of the fragment group at Lopé, but up to 35% of foraging effort at Sebatoli (maximum difference; Fig. 3). Comparisons among populations of *Cercopithecus cephus* and of *C. ascanius* showed that the percentage of fruit also varied dramatically, from a minimum for *C. ascanius* at Kibale and Salonga, to a maximum for *C. cephus* at Lopé in the continuous forest and Makokou (Table III). More generally, as in the case of the *mona* superspecies, intraspecific variation among populations was of a similar magnitude or greater than interspecific variation. For example, the use of fruit by *Cercopithecus ascanius* ranged from 13 to 61%, while the three populations of *C. cephus* were documented to rely on fruit for 78, 67, and 49% of their foraging effort.

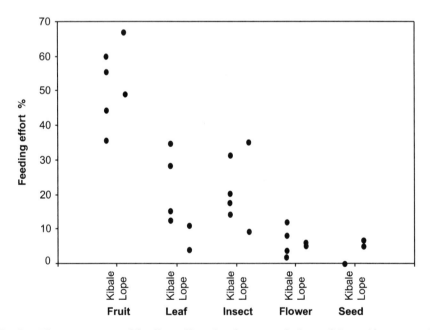

Fig. 3.   The percentage of feeding effort that four populations of *Cercopithecus ascanius* in Kibale National Park, Uganda, and two populations of *Cercopithecus cephus* in Lopé Reserve, Gabon devoted to each food category.

Table IV. **Variability in Diets among Populations of the Same Species (Coefficient of Variation) for Select Cercopithecines from a Variety of Sites across Africa. Studies described in Table III. We first calculated values considering each group as independent, regardless of study site, and second (in parentheses) by averaging the groups at one site before taking means and CVs of these site-specific averages.**

	Leaves	Seeds	Fruit	Flowers	Insects
Cephus Group					
C. ascanius					
Mean	23.93 (26.00)	8.20 (8.20)	46.65 (44.52)	6.93 (8.18)	23.71 (24.57)
CV	83.76 (84.31)	167.93 (167.93)	29.35 (25.83)	78.41 (70.99)	25.80 (19.74)
C. cephus					
Mean	7.13 (6.85)	5.90	64.70 (68.05)	4.07 (3.18)	18.90 (17.33)
CV	53.66 (17.55)	21.57	22.70 (20.89)	76.04 (119.15)	74.35 (38.57)
Mona Group					
C. pogonias					
Mean	6.77	20.47	58.57	4.63	11.30
CV	84.31	125.16	47.79	97.13	60.07
Nictitans Group					
C. nictitans					
Mean	14.98 (14.47)	21.77 (28.88)	52.33 (52.55)	5.95 (4.78)	12.17 (11.48)
CV	20.99 (24.95)	114.30 (104.44)	29.84 (33.33)	71.86 (91.26)	88.10 (25.57)
C. mitis					
Mean	23.21 (20.17)	4.32 (4.48)	47.43 (51.50)	8.79 (8.86)	15.38 (14.04)
CV	53.09 (62.95)	75.23 (73.08)	37.21 (35.13)	56.24 (65.83)	80.62 (74.32)
All species					
Mean	19.68 (17.61)	12.47 (13.92)	51.58 (51.96)	7.15 (7.12)	17.58 (15.17)
CV	75.94 (78.02)	128.64 (124.57)	32.25 (34.32)	68.14 (73.29)	10.74 (56.48)

Comparisons among all arboreal Cercopithecus *species*

We compared the diet of different groups or populations of six *Cercopithecus* species belonging to the *cephus, nictitans*, and *mona* superspecies (Table III) and calculated mean dietary characteristics and coefficient of variation for every species (Table IV). This evaluation was done in two ways. First, we considered each group as a separate unit. This potentially over-represents sites that have a number of studies at different times and locations. Second, we averaged all studies done in one location (e.g., Kibale National Park at any time). This potentially hides interesting within-site variation and it is difficult to determine which studies to average (e.g., should sites that are tens of kilometers apart be averaged).

Overall, fruit dominated the diet of most species (Tables III and IV). However, the variability among groups or populations of the same species in the use of fruit is relatively high (e.g., 13.2–61.3% for *Cercopithecus ascanius* and 26.1–91.1% for *Cercopithecus mitis*). Insects typically made up 10–20% of the diet of most species (Tables III and IV). The variability in insect use was relatively low among populations of *Cercopithecus ascanius* (14.5–31.2%,

Table III) and high among populations of *C. mitis* (0.7–35.9%). Typically, leaves made up just less than 20% of the diet (Tables III and IV). However, great variations occurred among populations especially in *Cercopithecus ascanius* (6.5–73.6%, Table III). The use of seeds varied most between species (Tables III and IV). Seeds were an important dietary component at the Salonga and Makandé sites (49.8 and 50.2% of diets of *Cercopithecus pogonias* and *C. nictitans*, respectively, at Makandé; Table III). Seed eating is very rare in Kibale. However, the absence of seed eating for some populations (not Kibale) may be the result of pooling fruit and seeds as fruit.

Comparisons at Different Temporal Scales

Seasonal variation in diet has been recognized for some time (Clutton–Brock and Harvey, 1977; Gautier–Hion, 1980; Cords, 1986; Chapman, 1987), but interannual dietary variation has received less attention. In Central Africa, periods of food scarcity fall mainly in the long dry season, during which time monkeys rely on fruit, seeds, leaves, or flowers of a few plant species (Gautier–Hion and Michaloud, 1989; Gautier–Hion and Maisels, 1994; Tutin *et al.*, 1997). The identity of the resources during this period varied not only from site to site, but also from year to year. For example, at Makandé, fruit of *Dialium* was the main food of the three arboreal *Cercopithecus* spp. in the 1994 dry season, while in the 1993 dry season, their diets were dominated by flowers of *Pterocarpus soyauxii* (Gautier–Hion, unpub. data). In 1993, few *Dialium* fruited, while *Pterocarpus soyauxii* flowered three to four months early, creating a large difference in foods available to primates in consecutive years.

Interannual variation in diet has rarely been documented for arboreal cercopithecines, because studies typically do not span a number of years. However, a comparison of the diet of the same groups in the same area, as determined by different observers, provides an initial assessment of interannual dietary variation. For example, for *Cercopithecus mitis* at the same site in the Kibale Forest, the percentage of feeding time devoted to fruit ranges from 26.1% (Butynski, 1990) to 49.6% (Wrangham *et al.*, 1998) across studies in different years.

Much of the interannual variation in diet likely reflects interannual differences in food availability. Chapman *et al.* (1999) examined the phenology of 3793 trees from 104 species at two sites over 76 months and found marked variation among years for several species. Of the 14 commonly occurring species considered in detail, six had an irregular fruiting pattern such that they did not predictably fruit during any given time of the year. Similarly, Brugière *et al.* (in press) found that at Makandé most of the Caesalpiniaceae fruited with a supra-annual pattern alternating mast-fruiting years with poor fruit years. Phenological studies at Lopé showed considerable interannual variation in fruiting patterns with only about half of the major primate food species fruiting regularly (Tutin and White, 1998).

Discussion

Interpreting the significance of the dietary variation that we have documented is difficult and will depend on the question being asked. In general, the populations we examined typically ate fruit and young leaves, and augmented their diet with insects (Gathua, 2000). There were, however, exceptions to this general pattern. In several cases, variation in the plant parts used by a single group over time, or between different groups separated by short distances, was greater than that between species. Accordingly, we conclude that dietary flexibility blurs traditional trophic assessment of primate species (Gautier–Hion *et al.*, 1993). Thus, a study of the diet of a single group in a specific habitat at one time may not be representative of the species as a whole.

From a practical perspective, our findings highlight the importance of providing more descriptions of primate diets on a variety of spatial and temporal scales. This will allow more comprehensive and analytical assessments of the level and scale of dietary variability within and among species. Long-term studies of the diet of single primate populations at several different sites will be particularly useful to further understanding the degree to which variation may represent phenotypic versus genotypic plasticity.

The discovery of such variation suggests we should be cautious about continuing to accept conclusions drawn from early comparative studies that proposed links between diet, behavior, and morphology (Milton and May, 1976; Clutton–Brock and Harvey, 1977; Chivers and Hladik, 1980). These early studies were influential and laid the foundation for how we categorize primate diet and adaptations (Gautier–Hion *et al.*, 1993).

Past selective pressures provide insight into dietary flexibility. *Cercopithecus* first appears in the fossil record about 2.9 mya (Leakey, 1988). Their ancestors were probably semiterrestrial frugivores inhabiting woodland and open country habitats. However, once they became rain forest specialists, their diversification was linked to the repeated isolation of populations due to repeated shifts between continuous forests and fragments during glacial/interglacial cycles (Hamilton, 1988). In the last 2.3 million years there have been approximately 20 cycles in which forests were reduced to isolated fragments and subsequently expanded to reconnect fragments into continuous forest (Hamilton, 2001; Maley, 2001; Livingstone, in press). During isolation, populations of *Cercopithecus* inhabited forest fragments that would have differed in edaphic features. This cyclic history would have resulted in strong selective pressures for dietary flexibility (Struhsaker, 1999; Tutin and White, 1999).

The degree of dietary flexibility that we have documented and the evolutionary history proposed to have led to this flexibility suggest that a profitable avenue of future research is the degree of flexibility that different primate lineages have in their digestive strategies (Lambert, 1998; Milton, 1998; Gathua, 2000). In this context, flexibility refers to the capacity to adjust digestive strategy according to the chemical and structural quality of the foods available. To what degree can different individuals, species, or lineages adjust their gastrointestinal capabilities? Within and between populations, is there a single optimal diet to maximize fitness, or can a diversity of diets meet nutritional and energetic requirements? Is the flexibility of a species determined by its intrinsic preference for foods with particular chemical and physical properties (Gathua, 2000)? And, how tight is the relationship among nutritional requirements, diet, and digestive strategies? For example, Conklin–Brittain *et al.* (1998) found remarkably little difference in the macronutrient content of the plant component of diets in *Cercopithecus mitis*, *C. ascanius*, *Lophocebus albigena,* and *Pan troglodytes*. They exhibit a considerable range in their feeding and foraging strategies, suggesting numerous pathways to fulfill dietary requirements.

These pathways may include physiological responses to a fluctuating diet. For example, a variety of omnivorous fish, birds, and mammals alter glucose absorption depending on the concentration of carbohydrates in their diets (Karasov and Diamond, 1988). The pathways may also include morphological responses to a fluctuating diet. Indeed, microtine rodents increase their small intestinal and caecal capacity in response to the quality of their diet (Gross *et al.*, 1985; Lee and Houston, 1993). Given such examples and the evolutionary history of forest-dwelling *Cercopithecus* spp., it would not be surprising to find that they, and perhaps Cercopithecinae generally, can alter their digestive strategies (Lambert, 1998).

A detailed consideration of the flexibility that different primate lineages have in their foraging strategies will have to wait until more studies are conducted using the same methods. At present we know little about how different methodologies influence our impressions of a species' diet. Within Kibale we have documented dietary variation in plant part use by four populations of both *Procolobus badius* and *Cercopithecus ascanius* (Chapman *et al.,* in press) using identical methodologies. This comparison suggests that the colobine species is more flexible than the cercopithecine (maximum difference in use of a plant part—*Procolobus badius* 38%, *Cercopithecus ascanius* 24%). This difference between species accords with what one would expect based on our understanding of fruit/frugivore mutualisms, but more comparisons are needed and alternative explanations need to be explored.

To date, remarkably little research has focused on the physiological ecology of free-ranging primates (Milton, 1998). It is unlikely that we will be able to interpret and explain the dietary flexibility we have identified until more effort is devoted to this area. How different populations, species, and lineages meet their dietary needs and the flexibility in the ways they can do this are exciting directions for future research.

Summary

Dietary data have been used to address numerous theoretical issues, yet we have little understanding of dietary flexibility in primates. Previous comparative research has either explicitly or implicitly assumed that the closer the phylogenetic proximity between two taxa, or the spatial proximity between two populations of the same taxon, the more similar their diets will be. We examine such assumptions by making dietary comparisons among arboreal *Cercopithecus* species at the intergroup, interdemic, interpopulational, and interspecific levels. Our analyses reveal considerable variation and sometimes the magnitude of the variation of particular contrasts is unexpected. We conclude that dietary flexibility blurs our traditional trophic assessment of primate species. Thus, a study of the diet of a single group, in a specific habitat, at one point in time may not be representative of the species as a whole. This flexibility suggests that a profitable avenue of future research is quantifying the degree of flexibility that different primate lineages have in their digestive strategies.

ACKNOWLEDGMENTS

Funding for the research at Kibale was provided by the Wildlife Conservation Society, National Geographic Society, and National Science Foundation (SBR-9617664, SBR-990899). Permission to conduct this research was given by the Office of the President, Uganda, the National Council for Science and Technology, the Uganda Wildlife Authority, and the Ugandan Forest Department. Research at Kakamega was sponsored by the University of Nairobi (Zoology Department) and the National Museums of Kenya under permission from the Office of the President and the Ministry of Education, Science and Technology. Funding sources include the L.S.B. Leakey Foundation (Baldwin Fellowship), Rockefeller Foundation, National Science Foundation (SBER 9726279), British Airways and Columbia University (for Gathua) and Sigma Xi, American Women in Science and the National Science Foundation (for Cords). Special thanks to Nixon Sagita and the late Jackson Gutwa for field assistance. Research at Lopé is possible thanks to the collaboration of the Direction de la Faune and is funded by the Centre International de Recherches Médicales de Franceville and by the Wildlife Conservation Society with additional grants from the Leakey Foundation.

Particular thanks go to Kate Abernethy, Rebecca Ham, Mike Harrison, Richard Parnell and the late Alphonse Mackanga-Missandzou for major contributions to data collection. Funding for researches at Salonga and Makandé were provided by the European programs (EEC, DG8). Particular thanks go to Dr Mankoto Mba Mbaelele, and Dr P. Posso for permission to conduct research in DRC and Gabon and to Fiona Maisels, Jean-Pierre Gautier, and Augustin Moungazi for their contribution to data collection. We thank Tom Gillespie, Mary Glenn, Russ Tuttle, and the anonymous reviewers for comments on the paper.

References

Anapol, F., and Lee, S. 1994. Morphological adaptation to diet in anthropoid primates. *Am. J. Phys. Anthropol.* **94**:239–261.

Beeson, M., Tame, A., Keeming, E., and Lea, S. E. G. 1996. Food habits of guenons (*Cercopithecus* spp.) in Afromontane forest. *Afr. J. Ecol.* **34**:202–210.

Brugière, D., Gautier, J.-P., Moungazi, A., and Gautier-Hion, A. (in press). Primate diet and biomass in relation to vegetation composition and fruiting phenology in a rain forest in Gabon. *Int. J. Primatol.*

Butynski, T. M. 1990. Comparative ecology of blue monkeys (*Cercopithecus mitis*) in high and low density subpopulations. *Ecol. Monogr.* **60**:1–26.

Chapman, C. A. 1987. Flexibility in diets of three species of Costa Rican Primates. *Folia Primatol.* **49**:90–105.

Chapman, C. A. 1995. Primate seed dispersal: Coevolution and conservation implications. *Evol. Anthro.* **4**:74–82.

Chapman, C. A., Balcomb, S. R., Gillespie, T., Skorupa, J. P., and Struhsaker, T. T. 2000. Long-term effects of logging on African primate communities: A 28 year comparison from Kibale National Park, Uganda. *Conserv. Biol.* **14**:207–217.

Chapman, C. A., and Chapman, L. J. 1990. Dietary variability in primate populations. *Primates.* **31**:121–128.

Chapman, C. A., and Chapman, L. J. 1999. Implications of small scale variation in ecological conditions for the diet and density of red colobus monkeys. *Primates.* **40**:215–232.

Chapman, C. A., and Chapman, L. J. in press. Plant-animal coevolution: Is it thwarted by spatial and temporal variation in animal foraging. In: D. Levey, W. R. Silva, and M. Galetti (eds.), *Frugivory and Seed Dispersal: Biodiversity and Conservation Perspectives*, CABI Publishers, New York.

Chapman, C. A., Chapman, L. J., and Gillespie, T. R. (in press). Scale issues in the study of primate foraging: red colobus of Kibale National Park. *Am. J. Phys. Anthropol.*

Chapman, C. A., Chapman, L. J., Wrangham, R., Isabirye-Basuta, G., and Ben-David, K. 1997. Spatial and temporal variability in the structure of a tropical forest. *Afr. J. Ecol.* **35**:287–302.

Chapman, C. A., and Lambert, J. E. 2000. Habitat alteration and the conservation of African primates: a case study of Kibale National Park, Uganda. *Am. J. Primatol.* **50**:169–186.

Chapman, C. A., Wrangham, R. W., Chapman, L. J., Kennard, D. K., and Zanne, A. E. 1999. Fruit and flower phenology at two sites in Kibale National Park, Uganda. *J. Trop. Ecol.* **15**:189–211.

Chivers, D. J., and Hladik, C. M. 1980. Morphology of the gastrointestinal tract in primates: Comparison with other mammals in relation to diet. *J. Morphol.* **116**:337–386.

Clutton-Brock, T. H., and Harvey, P. H. 1977. Primate ecology and social organization. *J. Zool. (London)* **183**:1–39.

Conklin-Brittain, N. L., Wrangham, R. W., and Hunt, K. D. 1998. Dietary responses of chimpanzees and cercopithecines to seasonal variation in fruit abundance. II. Macronutrients. *Int. J. Primatol.* **19**:971–998.

Cords, M. 1986. Interspecific and intraspecific variation in the diet of two forest guenons, *Cercopithecus ascanius* and *Cercopithecus mitis*. *J. Anim. Ecol.* **55**:811–827.

Cords, M. 1987. Mixed-species association of *Cercopithecus* monkeys in the Kakamega Forest, Kenya. *Univ. Calif. Pub. Zool.* **117**:1–109.

Cords, M. 1990. Mixed-species association of East African guenons: general patterns or specific examples? *Am. J. Primatol.* **21**:101–114.

Davies, A. G. 1994. Colobine populations. In: A. G. Davies, and J. F. Oates (eds.), *Colobine Monkeys: Their Ecology, Behavior and Evolution*, pp. 285–310. Cambridge University Press, Cambridge.

Davies, A. G., Bennett, E. L., and Waterman, P. G. 1988. Food selection by two south-east Asian colobine monkeys (*Presbytis rubicunda* and *Presbytis melalophos*) in relation to plant chemistry. *Biol. J. Linn. Soc.* **34**:33–56.

Davies, A. G., Oates, J. F., and Dasilva, G. A. 1999. Patterns of frugivory in three West African colobine monkeys. *Int. J. Primatol.* **20**:327–357.

Dunbar, R. I. M. 1998. The social brain hypothesis. *Evol. Anthropol.* **6**:178–190.

Fairgrieve, C. 1995. *The Comparative Ecology of Blue Monkey* (Cercopithecus mitis stuhlmanni) *in Logged and Unlogged Forest, Budongo Forest Reserve, Uganda: The Effect of Logging on Habitat and Population Density.* Ph.D. Thesis, University of Edinburgh, Edinburgh, Scotland.

Fashing, P. J., and Cords, M. 2000. Diurnal primate densities and biomass in the Kakamega Forest: An evaluation of census methods and comparison among other forests. *Am. J. Primatol.* **50**:139–152.

Fischer, K., and Chapman, C. A. 1993. Frugivores and fruit syndromes: Differences in patterns at the genus and species levels. *Oikos* **66**:472–482.

Gathua, M. 2000. *Intraspecific Variation in Foraging Patterns of Redtail Monkeys* (Cercopithecus ascanius) *in the Kakamega Forest,* Kenya. Ph.D. Thesis, Columbia University, New York.

Gautier-Hion, A. 1980. Seasonal variation of diet related to species and sex in a community of *Cercopithecus* monkeys. *J. Anim. Ecol.* **49**:237–269.

Gautier-Hion, A. 1988. The diet and dietary habits of forest guenons. In: A. Gautier-Hion, F. Bourlière, J.-P. Gautier, and J. Kingdon (eds.), *A Primate Radiation: Evolutionary Biology of the African Guenons*, pp. 257–283. Cambridge University Press, Cambridge.

Gautier-Hion, A., Emmons, L. H., and Dubost, G. 1980. A comparison of the diets of three major groups of primary consumers of Gabon (Primates, Squirrels, and Ruminants). *Oecologia* **45**:182–189.

Gautier-Hion, A., Duplantier, J. M., Quris, R., Feer, C., Sourd, C., Decoux, J. P., Dubost, G., Emmons, L., Erard, C., Hetckestweiler, P., Moungazi, A., Roussilhon, C., and Thiollay, J. M. 1985. Fruit characters as a basis of fruit choice and seed dispersal in a tropical forest vertebrate community. *Oecologia* **65**:324–337.

Gautier-Hion, A., and Michaloud, G. 1989. Are figs always keystone resources for tropical frugivorous vertebrates? A test in Gabon. *Ecology* **70**:1826–1833.

Gautier-Hion, A., Gautier, J.-P., and Maisels, F. 1993. Seed dispersal versus seed predation: An inter-site comparison of two related African monkeys. *Vegetatio* **107/108**:237–244.

Gautier-Hion, A., and Maisels, F. 1994. Mutualism between a leguminous tree and large African monkeys as pollinators. *Behav. Ecol. Sociobiol.* **34**:203–210.

Gautier-Hion, A., Gautier, J.-P., and Moungazi, A. 1997. Do black colobus in mixed-species groups benefit from increased foraging efficiency? *C. R. Acad. Sci., Paris* **320**:67–71.

Grant, J., Chapman, C. A., and Richardson, K. 1992. Defended versus undefended home range size of mammals. *Behav. Ecol. Sociobiol.* **31**:149–161.

Gross, J. E., Wang, Z., and Wunder, B. A. 1985. Effects of food quality and energy needs: Changes in gut morphology and capacity of *Microtus ochrogaster*. *J. Mammal.* **66**:661–667.

Ham, R. M. 1994. *Behaviour and Ecology of Grey-cheeked Mangabeys* (Cercocebus albigena) *in the Lopé Reserve, Gabon*. Ph.D. Thesis, Stirling University, Stirling, Scotland.

Hamilton, A. C. 1974. Distribution patterns of forest trees in Uganda and their historical significance. *Vegetatio* **29**:21–35.

Hamilton, A. C. 1988. Guenon evolution and forest history. In: A. Gautier-Hion, F. Bourlière, J.-P. Gautier, and J. Kingdon (eds.), *A Primate Radiation: Evolutionary Biology of the African Guenons*, pp. 13–34. Cambridge University Press, Cambridge.

Hamilton, A. C. 2001. Hotspots in African forests as Quaternary refugia. In: W. Weber, L. J. T. White, A. Vedder, and L. Naughton-Treves (eds.), *African Rain Forest Ecology and Conservation*, pp. 57–67. Yale University Press, New Haven.

Hamilton, A. C., Taylor, D., and Vogel, J. 1986. Early forest clearance and environmental degradation in south-west Uganda. *Nature* **320**:164–167.

Harrison, M. J. S. 1986. Feeding ecology of black colobus, *Colobus satanas*, in Gabon. In: L. Else, and P. C. Lee (eds.), *Primate Ecology and Conservation*, pp. 31–37. Cambridge University Press, Cambridge.

Harvey, P. H., Martin, R. D., and Clutton-Brock, T. H. 1987. Life history in comparative perspective. In: B. B. Smuts, D. L. Cheney, R. M. Seyfarth, R. W. Wrangham, and T. T. Struhsaker (eds.), *Primate Societies*, pp. 181–196. Chicago University Press, Chicago.

Herrera, C. M. 1985. Determinants of plant-animal coevolution: The case of mutualistic dispersal of seeds by vertebrates. *Oikos* **44**:132–141.

Herrera, C. M. 1998. Long-term dynamics of Mediterranean frugivorous birds and fleshy fruits: a 12-year study. *Ecol. Monogr.* **68**:511–538.

Hill, W. C. O. 1964. The maintenance of langurs (Colobinae) in captivity: Experiences and some suggestions. *Folia Primatol.* **2**:222–231.

Hladik, A. 1982. Dynamique d'une forêt équatoriale africaine: mesures en temps réel et comparaison du potentiel de croissance des différentes espèces. *Acta Oecol., Oecol. Gene* **3**:373–392.

Holmes, R. T., and Pitelka, F. A. 1968. Food overlap among coexisting sandpipers on northern Alaskan tundra. *Syst. Zool.* **17**:305–318.

Howe, H. F. and Smallwood, J. 1982. Ecology of seed dispersal. *Ann. Rev. Ecol. Syst.* **12**:201–228.

Isbell, L. A. 1991. Contest and scramble competition: patterns of female aggression and ranging behaviour among primates. *Behav. Ecol.* **2**:143–155.

Johns A. D., and Skorupa, J. P. 1987. Responses of rain-forest primates to habitat disturbance: A review. *Int. J. Primatol.* **8**:157–191.

Kaplin, B. A., and Moermond, T. C. 1998. Variation in seed handling of two species of forest monkeys in Rwanda. *Am. J. Primatol.* **45**:56–71.

Karasow, W. H., and Diamond, J. M. 1988. Interplay between physiology and ecology in digestion. *BioScience* **38**:602–611.

Kay, R. F. 1977. Molar structure and diet in extant Cercopithecidae. In: K. Joysey, and P. Butter (eds), *Development, Function and Evolution of Teeth*, pp. 309–339. Academic Press, London.

Lambert, J. E. 1998. Primate digestion: Interactions among anatomy, physiology, and feeding ecology. *Evol. Anthro.* **7**:8–20.

Lambert, J. E., and Garber, P. A. 1998. Evolutionary and ecological implications of primate seed dispersal. *Am. J. Primatol.* **45**:9–28.

Lasserre, F., and Gautier-Hion, A. 1995. Impacts environnementaux d'une exploitation sélective en forêt tropicale: l'okoumé en Forêt des Abeilles. Le cas de la société Leroy-Gabon. Rapport.

Lawes, M. J. 1991. Diet of samango monkeys (*Cercopithecus mitis erthrarchus*) in the Cape Vidal dune forest, South Africa. *J. Zool. (London)* **224**:149–173.

Lawes, M. J., Henzi, S. P., and Perrin, M. R. 1990. Diet and feeding behaviour of samango monkeys (*Cercopithecus mitis labiatus*) in Ngoye Forest, South Africa. *Folia Primatol.* **54**:57–69.

Leakey, M. 1988. Fossil evidence for the evolution of the guenons. In: A. Gautier-Hion, F. Bourlière, and J.-P. Gautier (eds.), *A Primate Radiation: Evolutionary Biology of the African Guenons*, pp. 7–12. Cambridge University Press, Cambridge.

Lee, W. B., and Houston, D. C. 1993. The effect of diet quality on gut anatomy in British voles. *J. Comp. Physio.* **163**:337–339.

Livingstone D. A. (in press). An historical view of African inland waters. In: T. L. Crisman, L. J. Chapman, C. A. Chapman, and L. S. Kaufman, (eds.), *Conservation, Ecology, and Management of African Freshwaters*. University Press of Florida, Gainesville, Florida.

Maisels, F., and Gautier-Hion, A. 1994. Why are Caesalpinioideae so important for monkeys in hydromorphic rainforests of the Zaire Basin? In: J. I. Sprent, and D. McKey (eds.), *Advances in Legume Systematics 5: The Nitrogen Factor*, pp. 189–204. Royal Botanic Gardens, Kew.

Maisels, F., Gautier-Hion, A., and Gautier, J.-P. 1994. Diets of two sympatric colobus monkeys in Zaire: More evidence on seed eating in forests on poor soils. *Int. J. Primatol.* **15**:681–701.

Maley, J. 2001. The impact of arid phases on the African rain forest through geological history. In: W. Weber, L. J. T. White, A. Vedder, and L. Naughton-Treves (eds.), *African Rain Forest Ecology and Conservation*, pp. 68–87. Yale University Press, New Haven.

McKey, D. B. 1978. Soils, vegetation, and seed-eating by black colobus monkeys. In: G. G. Montgomery (ed.), *The Ecology of Arboreal Folivores*, pp. 423–437. Smithsonian Institution Press, Washington.

Milton, K. 1998. Physiological ecology of howlers (*Alouatta*): Energetic and digestive considerations and comparison with colobinae. *Int. J. Primatol.* **19**:513–548.

Milton, K., and May, M. L. 1976. Body weight, diet, and home range area in primates. *Nature* **259**:459–462.

Napier, J. R. 1970. Paleoecology and catarrhine evolution. In: J. R. Napier, and P. H. Napier (eds.), *Old World Monkey: Evolution, Systematics, and Behavior*, pp. 53–95. Academic Press, New York.

Nunn, C., and Barton, R. 2001. Comparative methods for studying primate adaptation and allometry. *Evol. Anthropol.* **10**:81–98.

Quris, R. 1976. Données comparatives sur la socio-écologie de huit espéces de cercopithecidae vivant dans use même zone de forest primative périodiquement inondée (Nord-est du Gabon). *Terre Vie* **30**:193–209.

Ross, C. 1992. Basal metabolic rate, body weight and diet in primates: an evaluation of the evidence. *Folia Primatol.* **58**:7–23.

Rudran R. 1978a. Socioecology of the blue monkey (*Cercopithecus mitis stuhlmanni*) of the Kibale Forest, Uganda. *Smith. Contrib. Zool.* **249**:1–88.

Rudran, R. 1978b. Intergroup dietary comparisons and folivorous tendencies of two groups of blue monkeys (*Cercopithecus mitis stuhlmanniii*). In: G. G. Montgomery (ed.), *The Ecology of Arboreal Folivores*, pp. 483–503. Smithsonian Institution Press, Washington.

Schlichte, H. J. 1978. The ecology of two groups of blue monkeys, *Cercopithecus mitis stuhlmanni*, in an isolated habitat of poor vegetation. In: G. G. Montgomery (ed.), *The Ecology of Arboreal Folivores*, pp. 505–517. Smithsonian Institution Press, Washington.

Sheppard, D. J. 2000. *Ecology of the Budongo Forest Redtail: Patterns of Habitat use and Population Density in Primary and Regenerating Forest Sites*. M.Sc. thesis, University of Calgary, Calgary, Alberta, Canada.

Sterck, E. H. M., Watts, D. P., and van Schaik, C. P. 1997. The evolution of female social relationships in nonhuman primates. *Behav. Ecol. Sociobiol.* **41**:291–309.

Struhsaker, T. T. 1975. *The Red Colobus Monkey*. University of Chicago Press, Chicago.

Struhsaker, T. T. 1978. Food habits of five monkey species in the Kibale Forest, Uganda. In: D. J. Chivers, and J. Herbert (eds.), *Recent Advances in Primatology Vol 1. Behavior*, pp. 225–248. Academic Press, New York.

Struhsaker, T. T. 1997. *Ecology of an African Rain Forest: Logging in Kibale and the Conflict between Conservation and Exploitation*. The University Press of Florida, Gainesville, Florida.

Struhsaker, T. T. 1999. Primate communities in Africa: the consequences of long-term evolution or the artifact of recent hunting? In: J. G. Fleagle, C. Janson, and K. E. Reed (eds.), *Primate Communities*, pp. 289–294. Cambridge University Press, Cambridge.

Struhsaker, T. T., and Leland, L. 1979. Socioecology of five sympatric monkey species in the Kibale Forest, Uganda. In: J. Rosenblatt, R. A. Hinde, C. Beer, and M. C. Busnel (eds.), *Advances in the Study of Behavior*, pp. 158–228. Vol. 9, Academic Press, New York.

Tutin, C. E. G. 1999. Fragmented living: Behavioural ecology of primates in a forest fragment in the Lopé Reserve Gabon. *Primates* **40**:249–265.

Tutin, C. E. G., White, L. J. T., Williamson, E. A., Fernandez, M., and McPherson, G. 1994. List of plant species identified in the northern part of the Lopé Reserve, Gabon. *Tropics* **3**:249–276.

Tutin, C. E. G., Ham, R. M., White, L. J. T., and Harrison, M. J. S. 1997. The primate community of the Lopé Reserve, Gabon: Diets, responses to fruit scarcity, and effects on biomass. *Am. J. Primatol.* **42**:1–24.

Tutin, C. E. G., and White, L. J. T. 1998. Primates, phenology and frugivory: Present, past and future patterns in the Lopé Reserve, Gabon. In: D. M. Newbery, H. H. T. Prins, and N. Brown (eds.), *Dynamics of Tropical Communities*, pp. 309–338. Blackwell Science, Oxford.

Tutin, C. E. G., and White, L. J. T. 1999. The recent evolutionary past of primate communities: Likely environmental impacts during the past three millennia. In: J. G. Fleagle, C. Janson, and K. E. Reed (eds.), *Primate Communities*, pp. 220–236. Cambridge University Press, Cambridge.

Waser, P. M. 1987. Interactions among primate species. In: B. B. Smuts, D. L. Cheney, R. M. Seyfarth, R. W. Wrangham, and T. T. Struhsaker (eds.), *Primate Societies*, pp. 210–226. Chicago University Press, Chicago.

White, L. J. T. 1994a. Biomass of rain forest mammals in the Lopé Reserve, Gabon. *J. Anim. Ecol.* **63**:499–512.

White, L. J. T. 1994b. Patterns of fruit-fall phenology in the Lopé Reserve, Gabon. *J. Trop. Ecol.* **10**:289–312.

White, L. J. T., Rogers, M. E., Tutin, C. E. G., Williamson, E. A., and Fernandez, M. 1995. Herbaceous vegetation in different forest types in the Lopé Reserve, Gabon: Implications for keystone food availability. *Afr. J. Ecol.* **33**:124–141.

Wrangham, R. W., Conklin-Brittain, N. L., and Hunt, K. D. 1998. Dietary response of chimpanzees and cercopithecines to seasonal variation in fruit abundance: I. Antifeedants. *Int. J. Primatol.* **19**:949–969.

Diet of the Roloway Monkey, *Cercopithecus diana roloway*, in Bia National Park, Ghana

SHEILA H. CURTIN

Introduction

Roloway monkeys occur in Ghana and southeastern Ivory Coast, east of the Sassandra River, while the Diana monkey lives west of the river as far as Sierra Leone (Oates, 1988) and western Guinea (Barnett *et al.*, 1994). The roloway monkey is alternatively classified as a separate species in the Diana superspecies (*Cercopithecus (diana) roloway*, Groves, 2001; Butynski *et al.*, in prep.), or as a subspecies of *C. diana* (*C. diana roloway*, Grubb *et al.*, 2002), as in this volume. External physical features distinguishing the two include, but are not limited to, length of beard and color of the inner thighs: roloway monkeys have longer beards and yellow rather than red inner thighs. These features were used to distinguish the two forms as early as the 18th century (Schreber, 1774; Erxleben, 1777; Schlegel, 1876; Jentink, 1898; Pocock, 1907; Elliot, 1913). Some specimens from the Sassandra River are intermediate in these characters (Oates, 1988). *Cercopithecus diana* is generally confined to areas of old secondary or primary forest in the Upper Guinea forest region.

The feeding ecology of this species is of interest in view of its peripheral position among forest guenons in dental and cranial anatomy, chromosomal

SHEILA H. CURTIN • 942 Shevlin Drive, El Cerrito, CA 94530, USA.
The Guenons: Diversity and Adaptation in African Monkeys, edited by Glenn and Cords. Kluwer Academic/Plenum Publishers, New York, 2002.

evolution and vocalizations. *Cercopithecus diana* is an "extreme outlier" in dental dimensions relative to other forest guenons (Martin and MacLarnon, 1988), its nearest neighbors being *C. nictitans* and *C. pogonias*. This species is even more peripheral if both cranial and dental measures are combined, its nearest neighbor being *Cercopithecus nictitans* (Martin and MacLarnon, 1988). Of 11 guenon species surveyed for dental/cranial features, only *Allenopithecus nigroviridis* is more isolated. The same picture emerges in chromosomal evolution, where again only *Allenopithecus nigroviridis* is more primitive (Dutrillaux *et al.*, 1982), and regarding vocalizations, where *Cercopithecus diana* occupies a primitive position with only *A. nigroviridis* being closer to the ancestral form (Gautier, 1988). Among forest guenons, *Cercopithecus diana* may be expected to show the greatest similarity in feeding ecology to *C. nictitans*, and indeed Oates (1988) reported that *C. diana* and *C. nictitans* may competitively exclude each other.

I studied feeding and ranging behavior of roloway monkeys at Bia National Park in western Ghana in 1976–1977. At that time, Bia included over 300 km^2 of mostly undisturbed moist evergreen and moist semideciduous tropical forest (Olson, 1986; Martin, 1991). In 1979, more than two-thirds of the park was degazetted and reclassified as Timber Production Reserve, and in the 1980s bush-meat hunting decimated the resident primate population (Oates, 1996; Martin, pers. com.). The three endemic primate subspecies (roloway monkey, Miss Waldron's red colobus, and white-naped mangabey) are now absent from Bia (Oates, 1999), although the forest itself is largely intact. Miss Waldron's red colobus is probably extinct (Oates *et al.*, 2000; McGraw, 2001). Roloway monkeys persist in other forest reserves and parks in Ghana (Magnuson, pers. com.), but are now extremely rare in the Ivory Coast (McGraw, 1998b).

In 1976–1977, no field study had ever been conducted on roloway monkeys and no information on their behavior in the wild was available. To this date this is the only field study carried out on this primate, aside from a six-month population survey undertaken in 2001 (Magnuson, unpub. data).

Description of Study Area

Bia National Park (N 5° 36', E 0° 10') is located at approximately sea level in the western region of Ghana. Much of the park's western boundary is near the international boundary between Ghana and Ivory Coast. During my study, Bia contained populations of nine primate species: roloway monkey, white-thighed black-and-white colobus (*Colobus vellerosus*), Miss Waldron's red colobus (*Procolobus badius waldroni*), olive colobus (*P. verus*), white-naped mangabey (*Cercocebus atys lunulatus*), Lowe's guenon (*Cercopithecus campbelli lowei*), lesser spot-nosed monkey (*C. petaurista*), Prince Demidoff's bushbaby (*Galago demidoff*), and the West African chimpanzee (*Pan troglodytes verus*).

The Bia forest within the study area is representative of moist semideciduous forest of the northwest subtype (Hall and Swaine, 1981). The area is marked by low relief, with elevation changes ≤20 m, and the continuity of the mature high forest is broken in several locations by seasonal swamps dominated by the raphia palm, *Raphia hookeri*. Characteristic emergent tree species of the undisturbed forest, which is the tallest of all Ghanaian forest classifications (Hall and Swaine, 1981), include *Entandrophragma utile*, *Guibourtia ehie*, and *Khaya anthotheca*.

Methods

I observed roloway monkeys for eight months between November 20, 1976 and July 24, 1977 for a total of 2222 hours. From December to July, there were an average number of 28 contact days (range 23–31 days, SD = 2.9) and 274 contact hours (range 164–328 hours, SD = 58.89) per month.

Study Subjects

The focal study group initially numbered 14 individuals: two adult males, one subadult male, one subadult female, six adult females, two juveniles and two infants. One adult male was expelled in January, 1977, and at the end of the study, the group contained three juveniles and three infants, or a total of 15 monkeys. Thanks to the assistance of game rangers assigned to this project by the Ghana Department of Game and Wildlife (now called the Wildlife Division of the Forestry Commission), contact could be maintained almost continuously with the main study group once they were habituated.

Feeding Data Collection

I used two different measures to describe the feeding behavior of the Bia roloway monkeys: food item visits and scans. In a food item visit (FIV), ≥1 monkey entered a tree and began to feed on a discrete food item, e.g., immature seeds of *Pycnanthus angolensis*. If ≥ 1 individual then began to feed on a second food item, e.g., insects on new leaves, in the same tree, this event was a new FIV. I scored the same food item and same tree a second time only if all feeding ceased and resumed, with or without exit from that tree and re-entry. I recorded scans every five minutes; if any subject was in view, I noted its age and sex and identity (if known), nearest neighbor, its feeding behavior, forest strata level, and location, including numbered tree. Since dietary proportions obtained by the two methods were similar (e.g., insects account for 25% total FIV and 23.8% total scans), FIV have been used here as the principle data.

I described feeding in terms of previously numbered and mapped trees and climbers, or if new trees were fed upon, I described and located these trees relative to transect lines or numbered trees and then tagged, mapped, and measured them and recorded their phenological state. Measurements include height (as determined by rangefinder and clinometer readings), dbh, and climber investment (on a scale of 0–4). Phenological information includes a description of crown (degree of leaf loss, percentage of new and mature leaves), size of new leaves, data on flowering (beginning flowering, flowering, and finishing flowering), and fruiting (some, few, many of immature, mature, and senescent fruit). During this eight-month study, I tagged, mapped, and identified specifically a total of 1250 feeding trees.

I distinguished the following plant food item categories: new leaves (buds, buds plus small new leaves, less than full size, full size, and petiole), mature leaves (entirety, base including petiole, and epiphyte bulbs), fruit pulp (of immature fruit, of mature fruit, of senescent fruit, of mixed immature and mature fruit, entirety, i.e., pulp plus seeds as in *Ficus* species), seeds (aril, seeds of mature fruit, of immature fruit, and of senescent fruit), and flowers (buds, mature flowers, ovaries of mature flowers, and ovaries of senescent flowers). Faunal food items included: spiders and spider webs, contents of avian nests (including insects, possibly eggs, and small nestlings) moths, galls, small grubs, ants and termites, cocoons, orthopterans, millipedes, caterpillars, and small immobile insects in terminal branches (especially important in *Piptadeniastrum africanum*). Since the roloway monkeys consumed spiders and millipedes infrequently and I could not positively confirm feeding on eggs or nestlings, I use the term insects, rather than arthropods to denote animal prey. I had to observe actual ingestion of an insect food item, not just insect-hunting or insect-foraging, to score an insect FIV.

Phenology Data Collection

I described the phenological states of 106 trees of nine species per month for five consecutive months, March–July. I ranked each tree 0–4 on a scale of abundance for new leaves, mature leaves, flowers, and fruits. I distinguished within each category sub-sets, e.g., for new leaves: leaf buds, new leaves less than half-size, and new leaves full-size. I recorded leaf loss on a scale of 0–4.

Enumerations

I sampled composition and structure of the vegetation in the 189-hectare territory of the focal study group in the following manner. I chose a stratified random sample of ten plots, each 20 m × 20 m. I mapped all trees in each plot

≥50 cm diameter at breast height, measured the dbh, and identified species. I did not sample lianas. I enumerated a total of 444 trees ≥50 cm dbh in the 10 plots covering 0.4 ha, 80% of which were identified.

Results

Roloway monkeys made 2296 FIV to trees, 888 visits to climbers, and 40 visits to epiphytes, for a total of 3224 recorded FIV. They fed on food items (including insects) from 101 tree species, 42 climber species, and eight epiphytes.

The distribution of FIV among the different food item categories for trees, climbers, and epiphytes is shown in Table I. This table provides an overview of the roloway monkey diet at Bia during this study.

Top Ranked Food Species

The 20 top ranked food species are presented in Table II. These food species account for 2312 FIV, representing 71.6% of all FIV. Several of the most important species are described below.

Pycnanthus angolensis is well-represented in the moist semideciduous forest at Bia and constitutes a significant portion of the upper canopy (Importance Value (IV) = 5; IV is the sum of relative density, relative

Table I. Number of Food Visits in Different Plant Types and for Different Food Items

Food item	Trees (101 spp.)		Climbers (42 spp.)		Epiphytes (8 spp.)	
	n	%	n	%	n	%
Insects	775	33.8	38	4.3	1	2.5
Mature leaves	2	0.1	5	0.6	13	32.5
New leaves, buds	147	6.4	89	10.0	3	7.5
Flowers, buds	131	5.7	56	6.3	2	5.0
Pulp immature fruit	46	2.0	118	13.3	0	0.0
Pulp mature fruit	434	18.9	561	63.2	3	7.5
Seeds immature fruit	85	3.7	1	0.1	0	0.0
Seeds mature fruit	676	29.4	20	2.3	0	0.0
Root nodes	0	0.0	0	0.0	9	22.5
Bulbs	0	0.0	0	0.0	9	22.5
Total	2,296	100.0	888	100.1	40	100.0

Table II. 20 Top Ranked Food Species

Scientific name	Rank	n^a	Family	Common name	Food item[b]	Seasonality[c]	Description[d]
Pycnanthus angolensis	1	548	Myristicaceae	Otie	smfr, ins, simfr, nl, fb	Deciduous	Large tree
Piptadeniastrum africanum	2	216	Mimosaceae	Dahoma	ins, nl, simfr, smfr	Deciduous	Large tree
(c) —	3	209	—	(c) rope 13	pimfr, smfr, pimfr, simfr, ins, nl		Lg.woody climber
(c) Santaloides afzelii	4	167	Connaraceae	(c) rope 12	pimfr, pimfr, nl, fb, ins	Evergreen	Lg.woody climber
(c) Landolphia hirsuta	5	165	Apocynaceae	(c) Aman	pimfr, pimfr, ins	Evergreen	Lg.woody climber
Funtumia elastica	6	139	Apocynaceae	Funtum	smfr, simfr, fb	Evergreen	Medium tree
Parkia bicolor	7	134	Mimosaceae	Asoma	pimfr, ins, flow, pimfr	Evergreen	Large tree
(c) Salacia howesii	8	106	Celastraceae	(c) rope 11	pimfr, ins, pimfr	Evergreen	Med. woody climber
Sterculia oblonga	9	70	Sterculiaceae	Oha	ins, nl, smfr	Deciduous	Large tree
Chlorophora excelsa	10	68	Moraceae	Odum	pimfr, ins, pimfr	Briefly decid.	Large tree
Strombosia glaucescens	11	67	Olacaceae	Afena	ins, pimfr, simfr, pimfr, nl	Evergreen	Med.-lg. tree
Klainedoxa gabonensis	12	66	Irvingiaceae	Kroma	ins, smfr, pmfr, nl	Not clear	Large tree
Khaya spp.[e]	13	48	Meliaceae	Mahogany	ins	Briefly decid.	Large tree
Parinari excelsa	14	46	Chrysobalanaceae	Kotosima	pmfr, ins	Evergreen	Large tree
Antiaris welwitschii	15.5	45	Moraceae	Kyenkyen	pmfr, pimfr	Deciduous	Large tree
Bussea occidentalis	15.5	45	Caesalpineaceae	Atewa	ins, fb, simfr	Not dist. seas.	Med.-lg. tree
Cordia platythyrsa	17.5	44	Boraginaceae	Tweneboa	pimfr, ins, pimfr, fb, nl		Large tree
Pachypodanthium staudtii	17.5	44	Annonaceae	Nwo-no-kyene	pimfr, pimfr, ins, nl	Evergreen	Med.-lg. tree
Albizia zygia	19	43	Mimosaceae	Okuro	ins, nl, fb	Deciduous	Large tree
Aningeria robusta	20	42	Sapotaceae	Asanfena	pimfr, fb, ins, simfr, smfr, pimfr, nl	Evergreen	Large tree
Total		2312 = 71.6% total FIV					

(c) = Climber.

[a] n = food item visits.

[b] In order of importance. ins = insects; nl = new leaves; fb = flower buds; flow = flowers; pimfr = pulp immature fruit; pmfr = pulp mature fruit; simfr = seed immature fruit; smfr = seed mature fruit.

[c] Hall and Swaine, 1981.

[d] Hall and Swaine, 1981. Large tree = 30 m maximum height, Medium tree = 10–30 m.

[e] K. anthotheca and K. ivorensis not distinguished.

frequency, and relative productivity). The fruit is a single-seeded capsule that dries and splits to reveal a large black seed with a red aril. Bia roloway monkeys preferentially ate the mature seed, which is rich in oil, though they also fed on the seeds of immature fruit when no mature fruit were available. Seeds were broken open and consumed (destroyed, Lambert and Garber, 1998). They did not consume the aril though Diana monkeys in Sierra Leone do so (Whitesides, 1991). Fruiting by individual trees of this species stretched over months. Only a few mature fruit were available in a tree at one time, and fruiting of different trees was asynchronous. *Pycnanthus angolensis* thus provided a source of high energy nutrition (kernels are 54% solid fat; Irvine, 1961) over a long period. The distribution of feeding visits to *Pycnanthus angolensis* during the study is presented in Figure 1. Feeding on seeds of mature fruit peaked in the late dry season (late March/early April), then fell sharply (Fig. 2).

Piptadeniastrum africanum is also a common tree at Bia (Importance value = 9). Trees of this species occur scattered throughout the forest, but in some areas they are so numerous they virtually dominate the canopy. Roloway monkeys feed primarily on small immobile insects in the terminal branches (211 of 216 FIV). They probably spend more time feeding in *Piptadeniastrum africanum* than in any other tree species, since insect-foraging sessions may last ≥ 1 hour, and on some days the monkeys moved from one *P. africanum* to another throughout the day. Feeding peaked in April, at the end of the dry season/beginning of the wet season (Fig. 1).

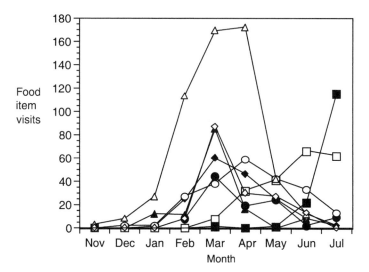

Fig. 1. FIV to top-ranked food species by month. Δ = *Pycnanthus angolensis* ($n = 548$), $\blacksquare$ = *Funtumia elastica* ($n = 139$), $\diamondsuit$ = climber *Landolphia hirsuta* ($n = 165$), $\square$ = climber. 13 ($n = 209$), $\blacklozenge$ = climber *Santaloides afzelii* ($n = 167$), $\bigcirc$ = *Piptadeniastrum africanum* ($n = 216$), $\bullet$ = climber *Salacia howesii* ($n = 106$), and $\blacktriangle$ = *Parkia bicolor* ($n = 134$).

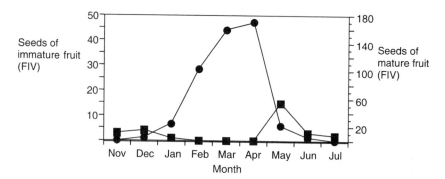

Fig. 2. *Pycnanthus angolensis* seed-eating. Based on FIV. ■ = seeds of immature fruit (*n* = 28), and ● = seeds of mature fruit (*n* = 485).

Parkia bicolor is one of several species, including *Khaya* spp. and *Klainedoxa gabonensis*, whose over-mature fruit were targeted by roloway monkeys for intensive insect-hunting, with the prey being grubs or larvae (or perhaps wasps; Tutin, 1999). They also ate the showy red flowers and mature fruit pulp. The distribution of feeding records on *Parkia bicolor* over the course of the study is in Figure 1. The pronounced peak in March represents feeding on mature fruit pulp.

Antiaris welwitschii was important to the roloway monkeys not only for its sweet fruit pulp but as a crucial source of moisture in the latter part of the dry season. The trees are emergents with large crowns, and they bore huge crops of juicy fruit during January and February in this study. Day ranges centered around these trees while they were in fruit.

Four of the top eight food species are large woody climbers (Table II). Three of the four provided large crops of juicy fruit in the dry season and early wet season, when moisture was at a premium; the fourth was a staple food item in the late wet season (Fig. 3). The predominance of mature fruit pulp in feeding visits to climbers is presented in Figure 4. In contrast to feeding visits to trees, where pulp represented only 18.9% of total visits, mature fruit pulp accounted for 63.2% of total visits to climbers. The only climbing species in which pulp of immature fruit represented a significant fraction of total FIV was *Santaloides afzelii* (Fig. 5). The roloway monkeys ate immature and mature fruit pulp on 76.5 % of FIV to climbers but on only 20.9% of FIV to trees (Table I).

Seasonal Variation

Rainfall at Bia National Park was highly correlated with food availability. In general, the fruits of trees and climbers were more abundant in the dry

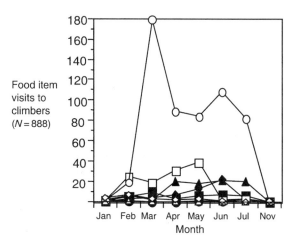

Fig. 3. Food item choice in climbers by month. ■ = insects, ● = mature leaves, ▲ = new leaves, leaf buds, ◆ = flower buds, flowers, □ = pulp immature fruit, ○ = pulp mature fruit, △ = seed immature fruit, and ◇ = seed mature fruit.

season, while young leaves, flowers, and arthropods were more abundant in the wet season. The year of this study was a very dry year, with only 938 mm of rainfall, as compared to 1441 mm in 1976 (Olson, pers. com.). In January–March 1977 (the main dry season) only 29 mm of rain fell compared to 440 mm in January–March 1976. The period of greatest food scarcity was the end of the dry season, in late March–early April.

This analysis of seasonal variation in the roloway monkey diet is based on 25 full-day follows (0600–1830 h) of the main study group in February and

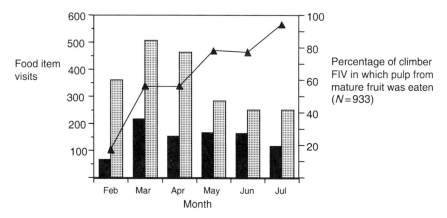

Fig. 4. FIV to climbers. ■ = FIV to climbers ($n = 876$), ▦ all other FIV ($n = 2121$), and ▲ = % all pulp from mature fruit ($n = 933$).

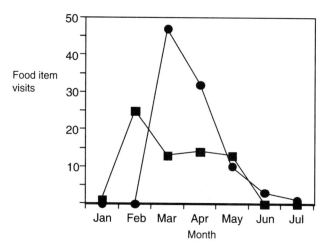

Fig. 5.   Climber *Santaloides afzelii* FIV. ● = pulp mature fruit (*n* = 93), and ■ = pulp immature fruit (*n* = 66).

March (dry season) and 25 full-day follows in May and June (wet season). Aspects of the diet include: food species (tree, climber, and epiphyte); food items (botanical parts and insects); FIV; and number of feeding trees per day. The ten top-ranked food species for each seasonal sample are shown in Table III. The first-ranked species overall, *Pycnanthus angolensisis* (Table II), also ranked first in the dry season, and third in the wet season. The second-ranked food species overall, *Piptadeniastrum africanum*, ranked fifth in the dry season, and second in the wet season. The third-ranked species, climber 13,

Table III.   Top Ranked Food Species[a] in Dry and Wet Seasons

Dry season (*n* = 63 spp.)	% Total FIV (*n* = 637)	Wet season (*n* = 80 spp.)	% total FIV (*n* = 518)
1. *Pycnanthus angolensis*	21.5	1. (c) *"rope 13"*	12.4
2. (c) *Landolphia hirsuta*	8.3	2. *Piptadeniastrum africanum*	7.1
3.5. (c) *Santaloides afzelii*	8.0	3. *Pycnanthus angolensis*	5.5
3.5. *Parkia bicolor*	8.0	4. *Diospyros manii*	5.0
5. *Piptadeniastrum africanum*	7.9	5.5. (c) *Santaloides afzelii*	4.3
6. (c) *Salacia howesii*	5.7	5.5. *Sterculia oblonga*	4.3
7. *Klainedoxa gabonensis*	3.0	7. (c) *Strychnos soubrensis*	3.9
8. *Antiaris welwitschii*	2.8	8. *Cordia platythyrsa*	3.5
9.5. *Parinari excelsa*	2.4	9. (c) *Dichapetalum pallidum*	3.3
9.5. *Diospyros heudelotii*	2.4	10. (c) *Landolphia hirsuta*	2.9
Total	69.9		52.0

[a]Based on FIV. *n* = 25 days in each season.
(c) = climber.

Table IV. Seasonal Variation in Feeding[a]

	Number of feeding trees/day[b]				Number of feeding species/day[c]			
Season	$\bar{x}$	SD	Median	Range	$\bar{x}$	SD	Median	Range
Dry	19.60	4.68	18	10–32	11.12	3.22	11	6–17
Wet	16.92	4.55	16	8–27	12.16	2.34	12	8–16
	Number of food items/day[c]				Number of feeding visits/day[d]			
	$\bar{x}$	SD	Median	Range	$\bar{x}$	SD	Median	Range
Dry	12.88	3.85	12	7–22	25.60	6.78	25	14–39
Wet	13.96	3.26	14	9–21	20.64	5.31	19	12–33

[a] Based on FIV. $n = 25$ days in each series.
[b] Significant seasonal difference, 2-sample t-test, $0.1 < p < 0.05$.
[c] No significant seasonal difference, $p > 0.05$.
[d] Significant seasonal difference, $p < 0.01$.

was absent from the dry-season rankings but ranked first in the wet season. The prominence of climbers in the diet again is evident in the rankings for the seasons: three of the top ten species in the dry season are climbers, and five of the top ten species in the wet season are climbers. Nine of the ten species in both samples are classed as large trees (>30 m, Hall and Swaine, 1981) or woody climbers; only *Diospyros heudelotii* in the dry and *D. mannii* in the wet season are medium-sized trees (8–30 m, Hall and Swaine, 1981).

A comparison of dry and wet season samples on different measures is presented in Table IV. The higher numbers of feeding trees/day and feeding visits/day in the dry season correlate with longer day ranges in dry (mean = 2280 m) vs. wet (mean = 1505 m) seasons (Curtin and Olson, 1982). Roloway monkeys are traveling farther, feeding in more individual trees and making more feeding visits per day overall in the season of greater food and moisture scarcity.

The major components of the diet measured by FIV in the dry season are the pulp of mature fruit of woody climbers and trees, seeds of mature fruit, mainly *Pycnanthus angolensis*, and insects. In the wet season, insects become even more important, the pulp of mature fruit remains a significant food item, seeds drop out, and new leaves and flower buds produced by the rains assume greater importance (Table V).

Insectivory

In the six-month period from February–July, a total of 730 FIV involving insectivory occurred in 86 different botanical species. An additional 11 FIV occurred in tree/climber species that could not be identified. During these 6 months, the tree species most favored by roloway monkeys during insect-hunting is *Piptadeniastrum africanum* (205 records, 23.1% of total insect FIV).

Table V. Dry and Wet Season Feeding Visits by Food Item Category

Food item	Dry ($n = 25$ days)		Wet ($n = 25$ days)	
	n	%	n	%
Insects	147	23.0	149	28.8
New leaves, buds	33	5.2	64	12.4
Flowers, buds	13	2.0	72	13.9
Pulp immature fruit	24	3.8	39	7.5
Pulp mature fruit	246	38.5	145	28.0
Seeds immature fruit	14	2.2	13	2.5
Seeds mature fruit	159	24.9	24	4.6
Other	3	0.5	12	2.3
Total	639	100.1	518	100.0

Most of the visits involved feeding on small immobile insect prey in new and mature leaf-clumps in the terminal branches. Other tree species in descending order of use are *Klainedoxa gabonensis* (50 records), in which the typical target were saltatory insects located in terminal twigs that were either deciduous or in leaf-flush; *Khaya* spp. (46 records), in which the target insect was a larva that feeds on multiple compressed samaras in the dehiscent fruit pod; *Bussea occidentalis* (40 records), no particular pattern of insect food-item choice; *Sterculia oblonga* (40 records), *Strombosia glaucescens* (30 records), and *Albizia zygia* (20 records). These seven tree species accounted for 59% of all insect predation during six months (431 of 730 records). I observed few actual insect captures in vine-trees although insect foraging in vine-trees was common. Trees heavily invested with vines offer poor observation conditions. Insect predation by roloway monkeys within the cover of vine-trees must be greatly underestimated in my data.

The following motor patterns indicated insectivory:

1. Shifting positions every few moments, visually scanning nearby leaves, examining them manually, and plucking objects from apices of leaf-rakes. In emergents, this activity occurred mainly in terminal branches at the outer edge of the canopy.

2. Sudden leaps out or down in the canopy, snatching an item from a branch or foliage (sometimes clapping hands together), and bringing it to mouth. In downward leaps of >6 m the monkey may end up hanging upside down by its feet from a branch while it secures the active prey with one hand (McGraw, 1998a).

3. Systematic movement through the open crown of an emergent or within vines of a heavily clad tree, examining, turning over, and opening dead leaves, presumably for cocoons.

4. Systematic movement as in (3), picking up and rolling an object between hands or with one hand on the surface of a branch, usually while seated. The food item is thought to be caterpillars with urticating hairs. Rolling may continue for ≤2 min for very large (15 cm) caterpillars.

5. Systematic movement as in (3) and (4) but very rapid; the entire group (*n*=14 animals) fed on aggregations of caterpillars in *Celtis zenkeri*. Tree foliage was extremely insect-damaged. The ground beneath the tree was covered with caterpillar droppings and fallen caterpillars dislodged by the monkeys' rapid movements.

6. Breaking chunks of rotten bark from boles or primary branches, eating from the branch or the piece of bark itself, and licking the underside of the bark. Food items appeared to be small grubs or insect detritus.

7. Moving head down along the bark of primary branches, apparently mouthing insects from a trail (ants, termites?); the monkey may suspend itself beneath the branch and move along the trail upside-down.

8. In a movement pattern that mimics feeding on fruit, the monkeys dispersed throughout the crowns of giant *Khaya* and extracted small grubs from old dehiscent seed pods. Pods showed evidence of insect damage. Food item was a 19 mm larva, black and white with a brown-red head.

In many of the insect-feeding situations, roloway monkeys display an almost feline agility (McGraw, 1998a,b) as in moving among the small flexible supports in the terminal branches of *Piptadeniastrum africanum* (1), making abrupt acrobatic leaps in pursuit of orthopterans (2), searching through dead leaves in trees heavily-invested with vines where supports are small, limber, and unsteady (3), and following ant trails under, on, and around primary branches (7).

Table VI. Insect Food Items and Insect FIV in Dry and Wet Seasons

Season	Daily number of insect food items[a]				Daily number of insect FIV[a]			
	$\bar{x}$	SD	Median	Range	$\bar{x}$	SD	Median	Range
Dry	4.20	2.25	4	1–10	5.88	3.73	6	1–13
Wet	4.48	2.68	4	0–11	5.96	4.22	5	0–16
Season	Insect food items/total food items[a]				Insect FIV/total FIV[a]			
	$\bar{x}$ (%)	SD (%)	Median (%)	Range (%)	$\bar{x}$ (%)	SD (%)	Median (%)	Range
Dry	32.0	12.4	32.6	6.7–55.6	21.9	10.9	22.0	5–46.2
Wet	30.2	13.9	30.8	0–64.7	27.6	15.6	25.0	0–70%

[a] Seasonal differences were not significant, 2-sample *t*-test, $p > 0.05$.

Table VII. Insect Prey Location Sites

Site	Food item	Comments
Apex of compound new leaf rakes	?	Rubbed on branch
Apex of compound new leaf rakes	Small immobile insect	Plucked from leaf surface
Stems of new leaf rakes	Scale insects?	
New leaves and flowers	Moths?	"Clouds" of insects
New leaves/ mature leaves	Lg. saltatory insects (grasshopper)	Caught with sudden leaps in terminal branches
New leaves and bark	Caterpillars (small, brown/black)	Ground under trees littered with caterpillar droppings
Underside of mature leaves	Cocoon?	Mature leaves insect-damaged
Apex of mature leaf rakes	Caterpillars? pupae?	Senescent leaves insect-damaged
Mature and senescent leaves	Galls	Piece of leaf torn off and eaten
Dead leaves	Cocoons	Turn over and open lvs; senescent lvs insect-damaged
Twigs	Scale-insects? galls?	Plucked from twig surface
Twigs and small branches at axils	Galls	
Apices of semi-deciduous twigs	? Ants	Twigs stripped with hand, hand licked
Deciduous terminal twigs	Cobwebs (and insects within)	Gathered with swiping motions
Rotten bark	Insects and insect detritus	Leaves insect-damaged; bark cracked and peeling
Large bark flake from bole	?	Licked from underside
Hollow, bell-shaped flowers of climber	Beetles	
Surface of primary branch	Ants	Trail followed mouth-to-bark
Underside of primary branches	Small caterpillars?	Plucked from bark surface
Mature fruit	Larvae in fruit pulp	
Senescent seeds	Larvae in seeds	Seeds (invested with larvae) eaten whole
Seeds of dehiscent mature and senescent fruit	Larvae	Sift thru many seeds before feeding
Bird's nest	Insects? eggs? fledglings?	Nest pulled apart

A comparison of insect-feeding in the two seasons is presented in Table VI. Insect food items and insect FIV are not significantly different between the two seasons. A similar picture is shown by the scan data (Table VII): both insect-foraging and insect-feeding scans are broadly similar in the dry and wet seasons.

Table VII represents a list of substrates insects occupied when captured and eaten by roloway monkeys. A few types of substrates, such as new leaf rakes of *Piptadeniastrum africanum* and senescent seeds of *Parkia bicolor* and *Khaya* spp., accounted for most of the insect-feeding records. The diversity of substrates on which the monkeys pursued insect prey, ranging from the stems of semi-deciduous twigs to birds' nests and rotten bark, is remarkable. Roloway monkeys appear to be opportunistic insect predators, despite the fact that most of their predation is focused on a few prey items in specific contexts.

Discussion

The diet of Bia roloway monkeys consisted of 31% pulp of mature fruit, 25% insects, 22% seeds of mature fruit, 7% leaf buds and new leaves, 6% flower buds and flowers, 5% pulp of immature fruit, 3% seeds of immature fruit, and <1% each of mature leaves, root nodes, and bulbs, as measured by food item visits. Pulp of mature fruit, particularly of climbing species, was the most important food category in the dry season, followed by seeds of mature fruit and insects. In the wet season, pulp of mature fruit and insects were equally important, followed by flower buds and flowers and leaf buds and new leaves. Pulp of mature fruit was the preferred food item in 63% of feeding visits to climbers but in only 19% of feeding visits to trees. The two most important food items over the course of the study were seeds of mature fruit of *Pycnanthus angolensis* and small immobile insects on the new and mature foliage of *Piptadeniastrum africanum*.

The distinction between tree species in which insect-hunting appeared to be prey-specific and trees in which hunting appeared more random may be an artifact of different observation conditions. In other words, it may be that roloway monkeys usually focus on a class of target insects, e.g., saltatory insects in terminal branches of *Klainedoxa gabonensis*, or on certain foraging situations, e.g., dead leaves in vines, but that this is not evident because of the difficulties of observing prey capture when visibility is limited.

Table VIII is a comparison of the roloway monkey diet at Bia with three studies of *Cercopithecus diana diana*: two at Tiwai Island, Sierra Leone (Whitesides *et al.*, 1988; Whitesides, 1989, 1991; Oates and Whitesides, 1990) and by Hill (1991, 1994), and one in the Taï National Park, Ivory Coast (Wachter *et al.*, 1997). Other publications on the Taï population are by Galat and Galat–Luong (1985), Holenweg *et al.* (1996), Bshary (1997), Honer *et al.*

Table VIII. Summary of *Cercopithecus diana* Studies: Dietary Proportions

Food item	*C. d. roloway*[a] Bia N.P.	*C. d. diana*[b] Tiwai Island	*C. d. diana*[c] Tiwai Island	*C. d. diana*[d] Tiwai Island	*C. d. diana*[e] Tiwai Island	*C. d. diana*[f] Taï N.P.
Insects	25.2	30.8	24.5	11.8	8.0	37.8
Mature leaves	0.6	–	–	16.4	19.2	0.1
New leaves, leaf buds	7.4	–	–	21.4	6.8	1.6
New and mature leaves	–	11.0	14.1	–	–	1.8
Flower buds, flowers	5.9	16.7	15.5	9.1	20.7	5.1
Pulp immature fruit	5.1	–	–	11.4	14.8	2.7
Pulp of mature fruit	31.0	–	–	11.4	10.1	42.1
Pulp of fruit	–	21.3	31.7	–	–	7.2
Seeds of immature fruit	2.7	–	–	–	–	–
Seeds of mature fruit	21.6	–	–	–	–	–
Seeds	–	6.9	5.1	8.0	11.6	–
Pod exudate	–	12.2	8.3	–	–	–
Unidentified	–	–	–	9.0	7.4	–
Other	0.6	1.1	0.8	1.6	1.4	1.6
Total	100.1	100.0	100.0	100.1	100.0	100.0

[a] Curtin, this study. n = 3224 FIV.
[b] Whitesides (1988, 1989, 1991). Oates and Whitesides (1990). Group W, n = 2482 scan samples.
[c] Whitesides (1988, 1989, 1991). Group E, n = 1080 scan samples.
[d] Hill (1991, 1994). Group W, n = 1571 scan samples.
[e] Hill (1991, 1994). Group E, n = 1103 scan samples.
[f] Wachter et al. (1997). n = 4415 scan samples.

(1997), Noë and Bshary (1997), Zuberbühler *et al.* (1997), McGraw (1998b,c), and Zuberbühler (2000a,b). In Sierra Leone, Whitesides and Hill studied the same two groups of diana monkeys, and data for the two groups are presented separately. There are large differences in dietary proportions (a) between conspecific groups in closely adjacent territories, (b) between the same group in different years, and (c) between conspecific groups at different study sites, which complements the dietary flexibility of other arboreal *Cercopithecus* spp. (Gautier-Hion *et al.*, 1993; Chapman *et al.*, 2002).

Bia roloway monkeys differ from Diana monkeys in Sierra Leone and Ivory Coast in the high percentage of seeds in their diet (24.3% new and mature seeds combined, vs. 0–11.6% in the other three studies). According to Chapman *et al.* (2002), seeds were the most variable of dietary categories in studies of six arboreal *Cercopithecus* spp. (excluding *C. diana*) but were always a minor component, except for *C. nictitans* in the continuous forest at Lopé, Gabon (11.1%, Tutin, 1999). Gautier-Hion *et al.* (1993) reported a figure of 27.3% (legume and other seeds combined) for *Cercopithecus wolfi* in Salonga National Park, Zaire. There are even higher percentages as measured by stomach contents for *Cercopithecus pogonias* (49.8%) and *C. nictitans* (50.2%) at Makandé, Gabon (Gautier-Hion, unpub. data, cited by Chapman *et al.*, 2002). Bia roloway monkeys primarily ate the mature seeds of two species: *Pycnanthus angolensis* (Myristicaceae) and *Funtumia elastica* (Apocynaceae) (77% of all seed FIV).

All African guenons consume a significant proportion of mature fruit pulp (Gautier-Hion, 1988), and the roloway monkeys are no exception. Chapman *et al.* (2002) cite a mean of 54.3% fruit consumption (immature and mature fruit lumped) for six arboreal species; the corresponding figure for the roloway monkeys at Bia is 36.1%. They ate the fruit pulp of 35 species of trees and 22 species of climbers. Most FIV to climbers were for mature fruit pulp; the fruit of four climbing species and two tree species appeared to provide crucial moisture during an unusually dry year. Very large woody climbers such as *Landolphia hirsuta* and *Santaloides afzelii*, investing $\geqslant$2 adjacent emergent trees, appeared to be key resources for these animals.

The proportion of insects in the diet of the Bia roloway population during my study (25.2%) falls well within the range reported for the five Diana monkey groups (8–37.8%, Table VIII). For arboreal *Cercopithecus* spp. (Chapman *et al.*, 2002), this figure ranged from <1% (*C. mitis*, Beeson, 1989) to $\leqslant$80% of monthly feeding scores for *C. cephus* in a nine-hectare forest fragment when fruit was scarce (Tutin, 1999; the overall figure for the fragment group is 3.5%, and for *C. cephus* in continuous forest, 9.1%). The single most important prey for roloway monkeys is a small immobile insect on new and mature foliage of *Piptadeniastrum africanum* (25.9% of all insect FIV).

The Bia roloway monkey appears to be a highly opportunistic insectivore, a seed-eater, and a frugivore preferring mature juicy fruit, especially of climbers. The high canopy habits of this species have been noted by many observers, as have its confinement to areas of old secondary and tall primary forest (Booth, 1956; Oates, 1988; Whitesides, 1991). Much of the diet of *Cercopithecus diana roloway* at Bia occurs in the terminal branch niche of large emergent trees (arthropods, young leaves, fruit, and seeds) and in the large woody climbers that invest them (juicy climber fruit, insects). Both terminal branches and climbers are characterized by an abundance of twigs or other small, unsteady supports, and McGraw (1998a) has noted that diana monkeys use twigs more frequently and branches less frequently than other arboreal monkeys, despite their relatively large body size for a guenon.

Cercopithecus diana has been described as an "extreme outlier" in cercopithecine dental and cranial dimensions (Martin and MacLarnon, 1988), with its closest neighbor being *C. nictitans*. Among all African guenons, *Cercopithecus diana roloway* may be most similar to *C. nictitans* in feeding ecology (particularly the high percentage of seeds in the diet) and stratum use, and the two species may competitively exclude other (Oates, 1988).

Summary

The roloway monkey, *Cercopithecus diana roloway*, occupies a peripheral position among African guenons regarding dental and cranial anatomy, and exhibits primitive features in its chromosomes and details of the vocalization system. I studied feeding and ranging behavior of roloway monkeys at Bia National Park in western Ghana in 1976–1977 with the aim of better understanding the dietary strategy of this species relative to other guenons. At that time, Bia included over 300 km^2 of mostly undisturbed moist evergreen and moist semideciduous tropical forest. I mapped, measured, and identified 1250 feeding trees and recorded >3000 feeding visits to these trees. I studied feeding behavior using two methods, food item visits (FIV) and scans. The roloway monkey diet was found to be highly diverse, including plant parts from >130 species of trees, climbers, and epiphytes. Major components of the diet were mature fruit pulp, arthropods, oil-rich seeds, and young leaves. Most food items for this monkey were located among twigs and small supports in the terminal branches of emergent trees and within large woody climbers. Among other forest guenons, the roloway monkey's dietary strategy may be most similar to that of *Cercopithecus nictitans*, which also consumes a high proportion of seeds, and the two species may competitively exclude each other.

ACKNOWLEDGMENTS

I wish to thank Patrick Curtin, Dana Olson, Herschel Franks, Chris Davies, Game Warden Claude Martin, Head of the Department of Game and Wildlife E.O.A. Asibey, and Bia game rangers: Philip E. Mensah, Daniel K. Sarfo, Alfred A. Awuah, Gabriel K. Asante, A. A. Ampofo, John K. Frimpong, and Isaac Y. Nyameckye.

References

Barnett, A., Prangley, M., Hayman, P. V., Diawara, D., and Koman, J. 1994. A preliminary survey of Kounounkan Forest, Guinea, West Africa. *Oryx* **28**:269–275.

Beeson, M. 1989. Seasonal dietary stress in a forest monkey (*Cercopithecus mitis*). *Oecologia* **78**: 565–570.

Booth, A. H. 1956. The cercopithecidae of the Gold and Ivory Coasts: geographic and systematic observations. *Ann. Mag. Nat. Hist.* **9**:476–480.

Bshary, R. 1997. Red colobus and Diana monkeys provide mutual protection against predators. *Anim. Behav.* **54**:1461–1474.

Butynski, T. M., Kingdon, J., and Happold, D. C. D. (eds.) (in prep.). *The Mammals of Africa*, Volume 1. Academic Press, London.

Chapman, C. A., Chapman, L. J., Cords, M., Gathua, J. M., Gautier-Hion, A., Lambert, J. E., Rode, K., Tutin, C. E. G., and White, L. J. T. 2002. Variation in the diets of *Cercopithecus* species: differences within forests, among forests, and across species. In: M. E. Glenn, and M. Cords (eds.), *The Guenons: Diversity and Adaptation in African Monkeys*, pp. 325–350. Kluwer Academic Publishers, New York.

Curtin, S. H., and Olson, D. K. 1982. Ranging patterns of black-and-white colobus and Diana monkeys in Bia National Park, Ghana. Paper presented at the 53rd annual meeting of the American Association of Physical Anthropologists.

Dutrillaux, B., Couturier, J., Muleris, M., Lombard, M., and Chauvier, G. 1982. Chromosomal phylogeny of forty-two species or sub-species of cercopithecoids (Primates, Catarrhini). *Ann. Génét.* **25**:96–109.

Elliot, D. G. 1913. A review of the primates. *Am. Mus. Mongr.* **2**:1–382.

Erxleben, J. C. P. 1777. *Systema regni animalis. Classis I, Mammalia*: pp. 30, 42.

Galat, G., and Galat-Luong, A. 1985. La communauté de primates diurnes de la forêt de Tai, Côte d'Ivoire. *Terre Vie* **30**:3–30.

Gautier, J.-P. 1988. Interspecific affinities among guenons as deduced from vocalizations. In: A. Gautier-Hion, F. Bourlière, J.-P. Gautier, and J. Kingdon (eds.), *A Primate Radiation: Evolutionary Biology of the African Guenons*, pp. 194–226. Cambridge University Press, Cambridge.

Gautier-Hion, A., Bourlière, F., Gautier, J.-P., and Kingdon, J. (eds.). 1988. *A Primate Radiation: Evolutionary Biology of the African Guenons*. Cambridge University Press, Cambridge.

Gautier-Hion, A., Gautier, J.-P., and Maisels, F. 1993. Seed dispersal versus seed predation: an inter-site comparison of two related African monkeys. *Vegetatio* **107/108**: 237–244.

Groves, C. P. 2001. *Primate Taxonomy*. Smithsonian Series in Comparative Evolutionary Biology, Smithsonian Institution Press, Washington, DC.

Grubb, P., Butynski, T. M., Oates, J. F., Bearder, S. K., Disotell, T. R., Groves, C. P., and Struhsaker, T. T. 2002. An assessment of the diversity of African primates. Unpublished report. IUCN/SSC Primate Specialist Group, Washington, DC.

Hall, J. B., and Swaine, M. D. 1981. *Distribution and Ecology of Vascular Plants in a Tropical Rain Forest; Forest Vegetation in Ghana*. W Junk, The Hague.

Hill, C. M. 1991. *A Study of Territoriality in* Cercopithecus diana. *Do Females Take an Active Part in Territorial Defense?* Ph.D. Thesis, University College London, London.

Hill, C. M. 1994. The role of female Diana monkeys, *Cercopithecus diana*, in territorial defense. *Anim. Behav.* **47**:425–431.

Holenweg, A. K., Noë, R., and Schabel, M. 1996. Waser's gas model applied to associations between red colobus and Diana monkeys in the Tai National Park, Ivory Coast. *Folia Primatol.* **67**:125–136.

Honer, O. P., Leumann, L., and Noë, R. 1997. Dyadic associations of red colobus and Diana monkey groups in the Tai National Park, Ivory Coast. *Primates* **38**:281–291.

Irvine, F. R. 1961. *Woody Plants of Ghana, with Special Reference to Their Uses.* Oxford University Press, London.

Jentink, F. A. 1898. On the "Diana" and the "Roloway". *Notes Leyden Mus.* **20**:233–239.

Lambert, J. E., and Garber, P. A. 1998. Evolutionary and ecological implications of primate seed dispersal. *Am. J. Primatol.* **45**:9–28.

Martin, C. 1991. *The Rainforests of West Africa: Ecology–Threats–Conservation.* Birkhäuser, Basel.

Martin, R. D., and MacLarnon, A. M. 1988. Quantitative comparisons of the skull and teeth in guenons. In: A. Gautier-Hion, F. Bourlière, J.-P. Gautier, and J. Kingdon (eds.), *A Primate Radiation: Evolutionary Biology of the African Guenons*, pp. 160–183. Cambridge University Press, Cambridge.

McGraw, W. S. 1998a. Posture and support use of Old World monkeys (Cercopithecidae): the influence of foraging strategies, activity patterns, and the spatial distribution of preferred food items. *Am. J. Primatol.* **46**:229–250.

McGraw, W. S. 1998b. Three monkeys nearing extinction in the forest reserves of eastern Côte d'Ivoire. *Oryx* **32**:233–236.

McGraw, W. S. 1998c. Comparative locomotion and habitat use of six monkeys in the Tai forest, Ivory Coast. *Am. J. Phys. Anthropol.* **105**:493–510.

McGraw, W. S. 2001. A survey of forest primates near the Ivory Coast's Ehi Lagoon: further evidence for the extinction of *Colobus badius waldroni. Am. J. Phys. Anthropol.* **32**:106–107.

Noë, R., and Bshary, R. 1997. The formation of red colobus – Diana monkey associations under predation pressure from chimpanzees. *Proc. R. Soc. Lond. B* **264**:253–259.

Oates, J. F. 1988. The distribution of *Cercopithecus* monkeys in West African forests. In: A. Gautier-Hion, F. Bourlière, J.-P. Gautier, and J. Kingdon (eds.), *A Primate Radiation: Evolutionary Biology of the African Guenons*, pp. 79–103. Cambridge University Press, Cambridge.

Oates, J. F. 1996. *African Primates: Status Survey and Conservation Action Plan.* IUCN/SSC action plans for the conservation of biological diversity, Gland, Switzerland.

Oates, J. F. 1999. *Myth and Reality in the Rain Forest: How Conservation Strategies Are Failing in West Africa.* University of California Press, Berkeley.

Oates, J. F., and Whitesides, G. H. 1990. Association between olive colobus (*Procolobus verus*), Diana guenons (*Cercopithecus diana*), and other forest monkeys in Sierra Leone. *Am. J. Primatol.* **21**:129–146.

Oates, J. F., Abedi-Lartey, M., McGraw, W. S., Struhsaker, T. T., and Whitesides, G. H. 2000. Extinction of a West African red colobus monkey. *Conserv. Biol.* **14**:1526–1532.

Olson, D. K. 1986. Determining range size for arboreal monkeys: methods, assumptions, and accuracy. In: D. M. Taub, and F. A. King (eds.), *Current Perspectives in Primate Social Dynamics*, pp. 212–227. Van Nostrand Reinhold, New York.

Pocock, R. I. 1907. A monographic revision of the monkeys of the genus *Cercopithecus. Proc. Zool. Soc. Lond.* **1907**:677–746.

Schlegel, H. 1876. Les singes. Simiae. *Mus. Hist. Nat. Pays-bas.* **7**:92–93.

Schreber, J. V. S. 1774. *Die Saugthiere in Abbildungen nach der Natur.* Wolfgang Walther.

Tutin, C. E. G. 1999. Fragmented living: behavioral ecology of primates in a forest fragment in the Lope Reserve, Gabon. *Primates* **40**:249–265.

Wachter, B., Schabel, M., and Noë, R. 1997. Diet overlap and polyspecific associations of red colobus and Diana monkeys in the Tai National Park, Ivory Coast. *Ethology* **103**:514–526.

Whitesides, G. H. 1989. Interspecific associations of Diana monkeys, *Cercopithecus diana*, in Sierra Leone, West Africa: biological significance or chance? *Anim. Behav.* **37**:760–776.

Whitesides, G. H. 1991. *Patterns of Foraging, Ranging, and Interspecific Associations of Diana Monkeys* (Cercopithecus diana) *in Sierra Leone, West Africa*. Ph.D. Thesis, University of Miami, Miami.

Whitesides, G. H., Oates, J. F., Green, S. M., and Kluberdanz, R. P. 1988. Estimating primate densities from transects in a West African rain forest: a comparison of techniques. *J. Anim. Ecol.* **57**:345–367.

Zuberbühler, K. 2000a. Referential labeling in Diana monkeys. *Anim. Behav.* **59**:917–927.

Zuberbühler, K. 2000b. Interspecies semantic communication in two forest primates. *Proc. R. Soc. Lond. B* **267**:713–718.

Zuberbühler, K., Noë, R., and Seyfarth, R. M. 1997. Diana monkey long-distance calls: messages for conspecifics and predators. *Anim. Behav.* **53**:589–604.

Part IV

Conservation

Conservation of Fragmented Populations of *Cercopithecus mitis* in South Africa: the Role of Reintroduction, Corridors and Metapopulation Ecology

24

MICHAEL J. LAWES

Introduction

Blue monkeys belong to the polytypic *Cercopithecus mitis/albogularis* subgroup of the *C. nictitans* supergroup. They are arguably the most widely distributed guenon species in Africa. Their range extends from the forests of southern Sudan to Eastern Cape Province in South Africa. *Cercopithecus mitis* is also one of the most recently divergent species among the Cercopithecini (Dutrillaux *et al.*, 1988; Leakey, 1988; Ruvolo, 1988). Their wide distribution and polytypic pattern is generally attributed to a recent radiation in response to vegetation adjustments following climatic changes during the late Quaternary (Lawes, 1990). According to this view, blue monkeys are thought to be more tolerant than most other guenons of changes in habitat quality and area (Lawes, 1992).

MICHAEL J. LAWES • School of Botany and Zoology, Forest Biodiversity Programme, University of Natal, Pietermaritzburg, South Africa.
The Guenons: Diversity and Adaptation in African Monkeys, edited by Glenn and Cords. Kluwer Academic/Plenum Publishers, New York, 2002.

After all, they occur in a wide variety of forested habitats (Lawes, 1990; Lawes, 1992) and much of their current forest range was naturally fragmented during the last 100,000 years (Moreau, 1966; White, 1981; Hamilton, 1988; Eeley et al., 1999). In many respects the *Cercopithecus mitis* subgroup appears to be well-adapted to cope with modern fragmentation effects. I examine this premise here.

In South Africa, blue or samango monkeys (*Cercopithecus mitis labiatus*) inhabit the highly fragmented Afromontane forests of the midlands of KwaZulu–Natal Province. However, not all forests are occupied and the expectation is that, in the face of further fragmentation, they will persist in meta-populations (Lawes et al., 2000a). A metapopulation is a set of local populations within some larger area, where typically migration from one local population to at least some other populations is possible (Hanski and Simberloff, 1997). Clearly, the influence of connectivity between forest patches on metapopulation persistence will be considerable. By reviewing data from earlier work by myself and coworkers, I examine (1) whether samango monkey populations in small forests are demographically vulnerable, (2) the likelihood that samango monkeys will persist as metapopulations, and (3) the effect of habitat connectivity and quality on monkey survival.

Demographic Viability of Low-Density or Small Populations

In 14 forests in KwaZulu–Natal Province, South Africa, samango monkey density ranged from 0.19 to 2.02 individuals ha^{-1} ($\bar{x} \pm 1$ SE $= 0.59 \pm 0.13$ ind ha^{-1}, $n = 14$) (Lawes, 1992). Although the correlation between forest size and samango population density, measured either as troops km^{-2} ($r_s = 0.48$, d.f. $= 14$, $p < 0.08$) or individuals ha^{-1} ($r_s = 0.49$, d.f. $= 14$, $p < 0.08$), is not significant, these data show a trend toward lower population densities in smaller forests.

The size of forest blocks occupied by samango monkeys in KwaZulu–Natal varies considerably (68–3500 ha). Nearly half (42%, $n = 22$) of the known populations occur in forests smaller than 500 ha (median $= 470$ ha, $n = 54$). Based on the estimate of mean density (0.59 ind ha^{-1}; Lawes, 1992) this suggests that a large number of populations contain < 150 reproducing individuals. The viability of these populations in the long-term must be questioned, particularly should any further reduction in forest area or density occur.

Using an Allee-effect density-dependent (Allee, 1931; Courchamp et al., 1999; Stephens and Sutherland, 1999) and age-structured demographic model, we examined the demographic viability of low-density samango monkey populations under conditions of environmental change and population fluctuation (Swart et al., 1993). We estimated the threshold density (relative to carrying capacity) below which populations are likely to collapse demographically.

The model did not consider genetic effects on small population viability, such as inbreeding effects, because in dynamic landscapes, shifts in habitat area and quality occur so rapidly that populations probably become demographically vulnerable and go extinct before detrimental genetic effects can operate (Swart and Lawes, 1996).

We derived estimates of the dynamics, or parameter values, for the birth-and-death branching processes typical of populations of *Cercopithecus mitis*, from captive studies and free-ranging populations (Swart *et al.*, 1993). Because of the uncertainty in parameter values and the exact shape of the density and fecundity functions, it was not possible to give an exact value of the density or size below which a population would collapse demographically. Nevertheless, the model showed that free-ranging populations could not tolerate $>60\%$ decrease across all age–sex cohorts from the equilibrium-density troop structure. Should further major perturbation of the system follow closely on such a decrease or if the periodicity of moderate disturbance was frequent, the population would proceed to extinction very rapidly.

The absence of monkeys from *ca.* 30% of potentially available forest, those forests large enough to hold viable populations, could not be explained by demographic response to catastrophic disturbance or environmental variation. On the whole, the monkeys showed a moderate ability to recover from considerable declines in population density. However, the critical factor in the above analysis, and an important factor upon which all predictions of recovery potential depended, is that carrying capacity potential was not altered by the event, such as habitat loss or decline in habitat quality, causing the original population decline. A population was unlikely to go extinct if population size was simply depressed within reasonable limits. Any explanation of samango monkey persistence and patch occupancy would have to encompass issues such as patch area, quality, and connectivity.

Patch Occupancy and Potential Metapopulation Dynamics

A comparison of the frequency distribution of forest sizes (in 50 ha increments) that samango monkeys used with ones that are potentially available shows that it is mostly the small forests (<300 ha) that are not used (Fig. 1). For forests <500 ha, there is a significant positive correlation between the proportion of forests used and the forest size class ($r_s = 0.99$, d.f. $= 8$, $p<0.001$). For forests $\geqslant 500$ ha in area there is no correlation ($r_s = 0.08$, d.f. $= 8$, $p = 0.82$). Collectively the data show that samango monkeys will use forest wherever it is available, but the likelihood of occupancy decreases as forest area becomes small. The latter suggests a potential causal relationship between the reduction in area of a forest and the loss from the system of the monkeys. This conclusion is supported by subsequent analysis, described

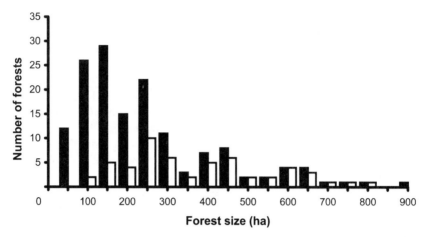

Fig. 1. A comparison of the frequency distributions of forest areas < 1000 ha available to (solid bars), and occupied by (open bars), samango monkeys in KwaZulu–Natal, South Africa (reprinted from Lawes 1992, *Biological Conservation*, **60**(3):197–210, Fig. 5; with permission from Elsevier Science).

below, of the patch occupancy of samango monkeys in 199 forest patches, ranging from 0.035 to 1732 ha ($\bar{x} \pm 1$ SE = 25.8 ± 10.34 ha, $n = 199$), in the Karkloof/Balgowan Afromontane forest archipelago (Fig. 2).

Samango monkeys are rare mammals in South Africa (Taylor, 1998). In the Afromontane forests of the Karkloof hills, this rarity is ascribed to

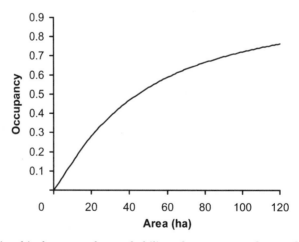

Fig. 2. Relationship between the probability of occupancy of a patch and the area of a patch as derived from a generalised linear model (reprinted from Lawes *et al.* 2000, *Conservation Biology* **14**(4):1089–1099, Fig. 3c; with permission from Blackwell Science).

the fragmentation and destruction of forest habitat (Lawes, 1992; Lawes *et al.*, 2000b). As a consequence of heavy logging activities from 1860 to 1940 the Karkloof forests are about 40% of their original size (Lawes, 1992). Although the forests are now under threat from agriculture and commercial forestry plantations (Lawes, 1992; Macfarlane, 2000), the major reduction in area and fragmentation of the forests occurred >50 years ago (Fourcade, 1889; Rycroft, 1944; Macfarlane, 2000). This is sufficiently long ago for most forest mammal populations to have responded to these anthropogenic effects.

In a test of vulnerability to fragmentation, we examined the pattern of patch occupancy of samango monkeys in forests of the Karkloof hills. Only 14 of the 199 forests surveyed were occupied by samango monkeys, and the occupied forests were on average 12 times the size ($\bar{x} \pm 1\ SE = 316.8 \pm 126.6$ ha, $n = 14$) of the mean forest patch area (25.8 ha) in the archipelago (Lawes *et al.*, 2000b). In addition, the occupied forests were on average further (1.12 ± 0.04 km, $n = 14$) from suitable patches compared to the mean isolation distance for all patches (0.49 ± 0.05 km, $n = 199$), but the difference is not significant (Welch's approximate $t = 9.84$, d.f. $= 73$, $p < 0.001$).

Although the theoretical effects of habitat isolation and patch size are well known (Hanski, 1994a; Hanski and Gilpin, 1997), detailed studies of their influence on individual species are rare (Harrison *et al.*, 1988; Thomas *et al.*, 1992). Part of the difficulty of assessing the persistence and pattern of distribution of a species in a patchy landscape is that data are frequently limited to presence–absence records, which do not easily reveal the processes responsible for the observed patterns.

Using observational data on a large set of habitat patches, it is possible to devise incidence functions that describe the probability of a species being present in a patch as a function of its area and isolation from a mainland source patch (Hanski, 1994b). This useful approach is based on the metapopulation theory of Hanski (1991, 1994a) in which the separate patch populations may have a finite lifetime, and stochastic fluctuations within the populations cause local extinctions that generate empty habitat patches that are then available for recolonization.

By formulating Hanski's (1994a) mainland–island incidence function model as a generalized linear model (GLM), we were able to use the presence–absence data from the single census of 199 patches described above, to show how patterns of patch occupancy are related to patch area, patch isolation, and various environmental factors (Lawes *et al.*, 2000b). The final model, which included only the effect of patch area and in which neither the environmental factors nor patch isolation distance entered the model, was a good fit (Lawes *et al.*, 2000b; $p > 0.99$). Thus, current land use in the matrix had no discernible effect on patch occupancy by samango monkeys. Nor was patch occupancy affected by wood removal or burning of the forest margins, livestock disturbance or general disturbance such as past logging activities.

The incidence probability of samango monkeys in forest patches was invariant with increasing isolation so that the isolation factor did not enter the model, and they occupied a few, large patches (Fig. 3). Many suitable large patches were unoccupied and there were very few patches of small area occupied unless they were very close to the mainland. It appears that samango monkeys are unable to cross open areas between patches.

The monkeys are very sensitive to area-dependent extinction effects. The area exponent, the x-factor in $1/A^x$ of Hanski (1992), is 1.2 and the estimate of minimum critical patch area is thus 44.39 ha. This is commensurate with the estimates of actual home range area for the Karkloof range (63.3 ha, Lawes, 1992).

Clearly, the mainland–island metapopulation model does not adequately describe the patterns of patch occupancy by samango monkeys. Creation of a functional metapopulation depends to a larger degree on environmental conditions suitable for encouraging monkeys to move among patches. Current conditions in the matrix between forests are such that this is unlikely to happen unless a well coordinated management plan is put in place. To manage samango monkey populations an understanding of the influence of greater connectivity between patches on metapopulation persistence is essential.

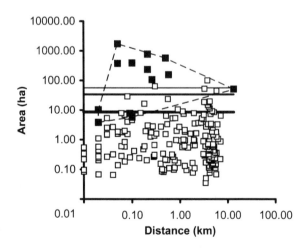

Fig. 3. The log-log relationship between area and isolation distance of occupied (solid) and vacant (open) forest patches for samango monkeys in the Karkloof/Balgowan archipelago. Solid lines give fitted percent occupancies from logistic model (——90% contour; —50% contour; ▬10% contour). The dashed line is a convex polygon enclosing the occupied patches and vacant patches with the potential for occupation (reprinted from Lawes *et al.* 2000, *Conservation Biology* **14**(4):1089–1099, Fig. 2c; with permission from Blackwell Science).

The Effect of Connectivity, Corridors and Translocation Events on Metapopulation Persistence

Using a stochastic, spatial population model in which we allowed for a variable number of patches in a metapopulation, we examined the effects of different levels of connectivity between forest patches on potential metapopulation persistence (Swart and Lawes, 1996). To accurately reflect the sociobiology of samango monkeys we devised a troop-based model in which one could follow the life history of each individual. We focused on a typical landscape of fixed spatial extent comprising a number of forests, usually eight in our analysis and always ≤10, that may be linked by corridors along which dispersal can occur, thus creating a simple metapopulation. Each forest has a fixed outer boundary and is partitioned into a number of elastic compartments simulating home ranges, each of which can accommodate a troop. Some compartments are core areas and others are dominated by edge habitat. In other words, we modeled the responses of core and edge troops separately. At low population density troop home range could expand. As population density increased and approached carrying capacity, more troops were formed and compartments approached a minimal viable size.

Important events in the life of samango monkeys occur during the short breeding season once per year, when fission and subsequent dispersal are likely to occur. Accordingly, we used a discrete model with a time step of one year. A stochastic model was also required to deal with the small troop numbers and to investigate uncertainty. Furthermore, to model both stable and dynamic growth modes typified by high- and low-density populations, respectively, the model had to be density dependent. We proceeded in the following fashion (Swart and Lawes, 1996). First, we defined an initial metapopulation by specifying the number of forests in the metapopulation, the number of compartments in each forest, the number of troops in each forest, and the age and sex structure of each troop. Then we updated the model from one year to the next by (1) determining stochastically if each monkey survived, (2) determining stochastically if each adult female produced an offspring, using birth intervals that ranged between 14 and 29 months (Butynski, 1990; Swart *et al.*, 1993) and whether it was male or female, and (3) modelling social interactions such as eviction of young males, possible changes in male residency, possible infanticide, group fission and dispersal. Dispersal was possible along any corridor to a linked forest in the metapopulation and to any vacancy in that patch. To add further realism to the model we included the effects of the major types of nonspatial variability that might affect the populations: (1) demographic stochasticity, (2) environmental stochasticity arising from changes in the parameter due to normal changes in weather patterns such as rainfall, and (3) natural disasters, such as extreme winters or diseases. These events could be applied to some or all of the troops within a given metapopulation. For natural disasters, a member of the affected population had a probability of death that was independent of death from other causes.

Finally, we investigated metapopulation persistence under the following scenarios: (1) no corridors between forests and dispersal between forests is not possible; (2) dispersal occurs along corridors to nearest neighboring forests only; (3) random selection and dispersal to any linked forest patch; (4) intelligent search in which dispersal is to the linked forest with the lowest population density; (5) reintroduction of troops in low-density areas where no corridors for dispersal exist. We modeled a realistic policy of reintroduction, given the difficulties of capturing and translocating whole groups, as a one time addition of an extra troop to a metapopulation; and (6) different numbers of forests in a metapopulation (one, three, and eight forests).

Using an initial occupancy for each forest in the metapopulation of 62.5%, the survival profiles under the various connectivity assumptions showed that compared with the totally fragmented, no corridor situation, nearest-neighbour dispersal dramatically improved the likelihood of survival in the medium to long term (>500 years) (Fig. 4). In the model, survival was further improved by providing more connectivity with intelligent searching as an upper bound. The corresponding 200-year hazard profiles clearly illustrate the substantial

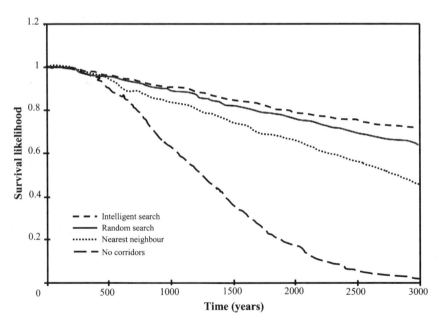

Fig. 4. The effect of corridors and different levels of connectivity on survival profiles corresponding to an 8-forest metapopulation configuration with initial occupancy of 62.5% (reprinted from Swart and Lawes 1996, *Ecological Modelling* **93**(1): 57–74, Fig. 2; with permission from Elsevier Science).

reduction in future hazard as a result of at least some connectivity between the patches (Fig. 5). In the case of full connectivity, the probability of the metapopulation becoming extinct within the next 200 years remained <5% for all time. Nearest-neighbour linkage yielded a corresponding probability of extinction of <10%, whereas in the case of unconnected patches the 200 year hazard profile had a mean level >35% likelihood of extinction in the longer term (Fig. 5).

Furthermore, the survival likelihood to 500 years, under the above assumptions on connectivity, over a range of frequency of disaster likelihoods, from a disaster every six years to one every three years, showed that increased levels of connectivity lead to increased survival likelihoods (Fig. 6). In addition, increasing the number of forest patches in the metapopulation resulted in higher survival probabilities or longer time frames of persistence.

Which management scenario is appropriate—reintroduction or creation of corridors? The reintroduction of troops, modeled here as higher initial occupancy levels, benefited the metapopulation in the short-term and then only if the population density was low (Fig. 7, compare trends with those in Fig. 5). In general, neither management option had any meaningful effect on short-term (<200 years) survival. In the long term, however, corridors

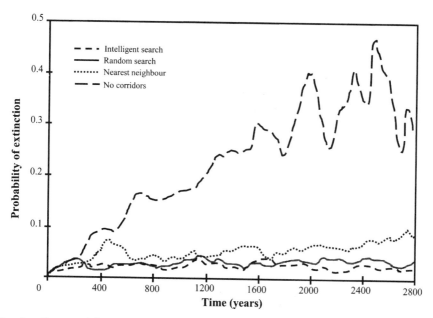

Fig. 5. Future risk associated with different levels of connectivity in an 8-forest metapopulation configuration with initial occupancy of 62.5% (reprinted from Swart and Lawes 1996, *Ecological Modelling* **93**(1):57–74, Fig. 3; with permission from Elsevier Science).

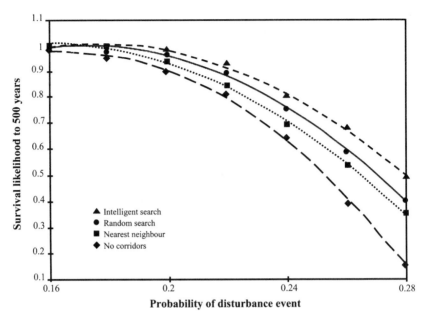

Fig. 6. The effect of corridors and levels of connectivity on survival likelihood to 500 years under variable frequencies of disaster. An 8-forest metapopulation configuration with an initial occupancy of 62.5% is illustrated (reprinted from Swart and Lawes 1996, *Ecological Modelling* **93**(1):57–74, Fig. 4; with permission from Elsevier Science).

significantly improved metapopulation persistence. In the absence of interventive management to improve connectivity, samango monkey metapopulations comprised of eight forest patches, show poor persistence probabilities, <0.75 persistence probability to 800 years and <0.34 to 1500 years. In contrast the survival probabilities are 0.9–800 years and 0.8–1500 years when connectivity is provided. The creation of corridors is of the utmost importance for the long-term survival of fragmented populations of *Cercopithecus mitis*.

Discussion

In developing conservation strategies for *Cercopithecus mitis* in South Africa we have been guided by the principles of metapopulation theory (Hanski and Gilpin, 1991; Fahrig and Merriam, 1994; Hanski, 1994a; Thomas, 1994a,b; Hanski and Gilpin, 1997). However, the interpretation of the model outcomes is mitigated by some of their limiting assumptions. For instance, the models do not take into account the quality of dispersal routes, and the likelihood that dispersers survive routes of low quality, such as plantation monocultures of *Pinus* spp. that may act as sinks (*sensu* Pulliam, 1988a), and lower regional

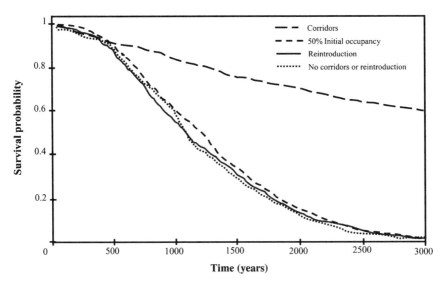

Fig. 7. Survival profiles for a low density, 8-forest metapopulation configuration, corresponding to the management options of reintroduction or connectivity. The initial occupancy in each forest is 12.5 % (reprinted from Swart and Lawes 1996, *Ecological Modelling* **93**(1):57–74, Fig. 5; with permission from Elsevier Science).

population abundance (Henien and Merriam, 1990). Furthermore, unnatural corridors that provide suitable cover, such as plantations, are probably of dubious conservation value because they create problems for other taxa, for example by selective or differential filtering of species (Noss, 1991; Beier and Noss, 1998). Corridors may also promote the spread of contagious disease (Simberloff and Cox, 1987; Hess, 1994). We did not explicitly model the possible spread of contagious disease among a metapopulation.

A significant further limiting assumption of our metapopulation analysis is that the incidence models implicitly assume habitat stability (Hanski and Gilpin, 1991; Hanski, 1991, 1994a,b; Thomas, 1994b; Sjögren-Gulve and Ray, 1996; Wiens, 1997). In other words, patches that become vacant through local (stochastic) extinction remain suitable for subsequent recolonization. However, faunal extirpation can be a deterministic population response to habitat degradation that leaves patches unsuitable for recolonization (Thomas, 1994a). If the processes that create empty habitat are not all stochastic in nature, then the basis for a stable equilibrium between processes of extinction and colonization at the metapopulation level breaks down (Thomas, 1994a). This affects patch occupancy and extrapolated metapopulation dynamics.

The processes that create empty habitat in the case of samango monkeys appear to be mostly highly deterministic area-dependent extirpation effects. Small samango monkey populations in small forest patches are demographically vulnerable (Swart *et al.*, 1993), and small patches are seldom colonized

(Lawes *et al.*, 2000b). While habitat degradation may also affect local extinction, the effect of habitat quality (disturbance) on patch occupancy was not implicated in the incidence function model. Yet in an earlier study samango monkey group density was lower in disturbed forest than undisturbed forest of a particular type (Lawes, 1992). Furthermore, of seven primate species examined in the Kibale forest, *Cercopithecus mitis stuhlmanni* was the most affected by logging intensity and habitat degradation (Chapman *et al.*, 2000; but see Skorupa, 1986). In general, it would be unwise to dismiss the effect of habitat quality on patch occupancy because: (1) the incidence function model is based on presence/absence data and cannot be used to interrogate differences in monkey density among forest patches of differing quality; (2) also there were 11 patches that were suitable in terms of size and isolation but were not occupied by monkeys (Fig. 3, Lawes *et al.*, 2000b), suggesting a role for habitat quality in determining patch occupancy trends, and also hints at constraints on dispersal among patches; and (3) primate carrying capacity of a forest may be more severely affected by disturbance, especially logging, in a more seasonal forest habitat, such as the sub-tropical forests of KwaZulu–Natal (Lawes, 1992).

There is no doubt that the persistence of a metapopulation depends to a large extent on whether local populations are able to track the shifting mosaic of suitable habitat patches (Thomas, 1994b). This is an ability that is clearly dependent on both the dispersal capability of the animal and the extent of connectivity in the landscape between suitably-sized patches (Taylor *et al.*, 1993; Wiens, 1997). Although *Cercopithecus mitis* is apparently a highly mobile species and potentially a good disperser, as evidenced by its extensive distribution, very few patches (7%, $n = 199$) in the Karkloof range were occupied by samango monkeys. In fact, the final incidence function model did not include an isolation effect showing that any isolation distance between patches was too much for samango monkeys to cross. Thus, against expectation, the patch occupancy data strongly support a limited ability or unwillingness by samango monkeys to disperse across a highly fragmented landscape. Low patch occupancy levels among *Cercopithecus mitis* are not confined to samango monkeys, suggesting that the group as a whole consists of reluctant dispersers and are thus susceptible to fragmentation effects. For example, blue monkeys (*Cercopithecus mitis stuhlmanni*) do not move between or use forest fragments or corridors near Kibale forest, though red-tailed monkeys (*Cercopithecus ascanius*) that have a similar diet and social organization, frequently move between forest fragments via forest corridors and crossing agricultural areas (Chapman and Onderdonk, 1998; Onderdonk and Chapman, 2000; Chapman *et al.*, in press).

Taken with their apparently poor dispersal ability and obvious area-dependent extirpation, samango monkey metapopulations, if they exist at all, are probably of the transient, nonequilibrium (declining) metapopulation type (Harrison and Taylor, 1997), which is characterized by local extinctions that occur in the course of a species' decline to regional extinction (cf. Brown, 1971), with recolonization occurring infrequently or not at all (Harrison, 1991; Harrison and Taylor, 1997). The current pattern of occupancy of patches appears to describe remnant populations that persist because of their relatively large size, with the ongoing extirpation of samango monkeys from very small patches (Fig. 3). By identifying large forest patches for careful protection and management, the survival of monkey populations can be prolonged. However, since no functional metapopulation may exist for them, this is an emergency measure. In addition, in the absence of evidence to suggest that samango monkeys continuously disperse to patches where the population growth rate is negative, i.e., a sink, it is unlikely that a source–sink system (Pulliam, 1988b) is operating.

Why do samango monkeys, indeed blue monkeys, display such low levels of occupancy and potentially poor dispersal ability? The apparent reluctance of *Cercopithecus mitis* to disperse among patches in fragmented landscapes may be due to a combination of the constraining influence of life-history traits, especially their social structure and their considerable dietary flexibility, and the nature of the matrix. Group dynamics provide a strong centripetal force that may inhibit coordinated group dispersal across hostile habitat (Recher *et al.*, 1987; Laurance, 1990; Swart and Lawes, 1996). This effect has been recorded among social arboreal marsupials in tropical rainforest patches on the Atherton Table-land (Laurance, 1990). As previously continuous forest is fragmented, the local range of *Cercopithecus mitis* contracts to the larger remaining fragments. Populations in small fragments and even some large fragments may go extinct, but the fragments are not recolonized because of social constraints on group dispersal. However, social constraints alone are insufficient to explain the limited patch occupancy of *Cercopithecus mitis* since other highly social primates disperse freely among forest fragments near Kibale (Chapman *et al.*, in press).

Dietary flexibility may provide further explanation for why the likelihood of occupancy of a patch by *Cercopithecus mitis* is very area-dependent (Lawes, 1992; Swart *et al.*, 1993; Swart and Lawes, 1996). The gut of *Cercopithecus mitis erythrarchus* is well-adapted to cope with a folivorous, low quality diet (Bruorton and Perrin, 1988; Bruorton *et al.*, 1991) and all populations of *C. mitis* show an ecological shift in the diet toward foliage (Lawes, 1991). The competitive advantage that folivory confers on an essentially frugivorous primate, has allowed the *Cercopithecus mitis* species group to become almost exempt from the effects of climatic seasonality and food restriction (e.g., Beeson, 1989, 52% leaf in the diet; Beeson *et al.*, 1996), causing substantial separation of their feeding niches from those of other cercopthicines (Rudran, 1978; Struhsaker, 1978; Cords, 1987; Lawes, 1991).

Dietary plasticity, particularly folivory, not only enabled *Cercopithecus mitis* to coexist with other guenons and persist in forest patches, but also to radiate further south than other arboreal guenons have (Lawes, 1990). Thus, conservative patch occupancy in remnant but large forest patches by *Cercopithecus mitis* populations is somewhat enigmatic. I suggest low patch occupancy arises because *Cercopithecus mitis* can tolerate variation in food abundance better than most other guenons can and they need not disperse to, or move frequently among, fragmented forest patches, which ensures that the species is remarkably sedentary within its range. *Cercopithecus mitis* does not respond to fragmentation like other species whose need to seek out alternative feeding sites creates strong incentive for coordinated group movement among forest patches.

It is difficult to test conclusively for the effect of dietary plasticity and sociality on dispersal and patch occupancy. However, it is clear that the conservation of *Cercopithecus mitis* entails not only understanding the interplay of their ecology with their radiation history, but also identifying deterministic factors that cause population declines and extinction. The *mitis*-species group has achieved wide distribution in the past by virtue of dietary flexibility, but they are now unable to cope with the rate of forest fragmentation within their ranges and persist mostly in large remnant, and hence very slowly declining, populations that are likely to be virtually extinct within 500 years (Swart and Lawes, 1996). Populations of *Cercopithecus mitis* in both the Karkloof hills (Lawes *et al.*, 2000a,b) and Kibale forest (Chapman *et al.*, in press) do not persist as metapopulations in the way that other guenons and social primates do. Within South Africa we are dealing with at best transient, non-equilibrium (declining) metapopulations and a proactive management approach is required to halt further population extirpation and declines. Samango monkey life-history traits appear to require a large minimum viable forest patch area. Identifying large habitat patches that are close to one another and minimizing disturbance in and on the edges of the patches, and linking them in such a way that individuals and groups can track changes in the landscape mosaic, is the most prudent management strategy for the species.

Summary

In spite of a distribution comprising different forest habitats, blue monkeys (*Cercopithecus mitis* subgroup) do not tolerate fragmentation or isolation as well as less widely distributed social primate species. Small populations are demographically vulnerable and this monkey is generally absent from small forests (< 150 ha). They are reluctant to disperse over open ground and exist in transient, non-equilibrium or declining metapopulations in which local extinctions are caused by the reduction in forest area and declining habitat quality. Dietary flexibility, particularly the use of low quality folivorous foods may have conferred a competitive advantage on this species group and exemption from the effects of seasonal variation in food supply and quality. This has in turn reduced the need to move among forest fragments, unlike other species whose need to seek alternative feeding sites creates a strong incentive for coordinated group movement among forest patches. Nevertheless, the survival of populations of blue monkeys depends on their persistence in metapopulations. Corridors significantly improve blue monkey metapopulation persistence in the long term. To maintain populations of blue monkeys, large, adjacent and occupied forest patches must be linked in a way that individuals and groups can track changes in the landscape. Translocations and reintroductions of monkey groups is a short-term solution that is not recommended.

ACKNOWLEDGMENTS

I thank Harriet Eeley for editing this manuscript and assisting with the preparation of the figures. I am grateful to the South African National Research Foundation together with the University of Natal Research Foundation for their financial backing during the period when the research reported here was conducted. The Mazda Wildlife Fund provided logistical support.

References

Allee, W. C. 1931. *Animal Aggregations. A Study in General Socioecology.* University of Chicago Press, Chicago.

Beeson, M. 1989. Seasonal dietary stress in a forest monkey (*Cercopithecus mitis*). *Oecologia* **78**: 565–570.

Beeson, M., Tame, S., Keeming, E., and Lea, S. E. G. 1996. Food-habits of guenons (*Cercopithecus* spp.) in Afro–Montane Forest. *Afr. J. Ecol.* **34**:202–210

Beier, P., and Noss, R. F. 1998. Do habitat corridors provide connectivity? *Conserv. Biol.* **12**: 1241–1252.

Brown, J. H. 1971. Mammals on mountain tops, nonequilibrium insular biogeography. *Am. Nat.* **105**:467–478.

Bruorton, M. R., Davis, C. L., and Perrin, M. R. 1991. Gut microflora of vervet and samango monkeys in relation to diet. *Appl. Env. Microb.* **57**:573–578.

Bruorton, M. R., and Perrin, M. R. 1988. The anatomy of the stomach and caecum of the samango monkey, *Cercopithecus mitis erythrarchus* Peters, 1852. *Z. Saugetierkunde* **53**:210–224.

Butynski, T. M. 1990. Comparative ecology of blue monkeys (*Cercopithecus mitis*) in high- and low-density subpopulations. *Ecol. Monogr.* **60**:1–26.

Chapman, C. A., Lawes, M. J., Naughton-Treves, L., and Gillespie, T. R. (in press). Primate survival in community-owned forest fragments: are metapopulation models useful amidst intensive use? In: L. Marsh (ed.), *Primates in Fragments: Ecology and Conservation*, pp. 63–78, Kluwer Academic/Plenum Publishers, New York.

Chapman, C. A., and Onderdonk, D. A. 1998. Forests without primates: primate/plant codependency. *Am. J. Primatol.* **45**:127–141.

Chapman, C. A., Balcomb, S. R., Gillespie, T. R., Skorupa, J. P., and Struhsaker, T. T. 2000. Long-term effects of logging on African primate communities: a 28-year comparison from Kibale National Park, Uganda. *Conserv. Biol.* **14**:207–217.

Cords, M. 1987. *Mixed-species Associations of* Cercopithecus *Monkeys in the Kakamega Forest, Kenya.* University of California Press, Berkeley.

Courchamp, F., Clutton-Brock, T., and Grenfell, B. 1999. Inverse density dependence and the Allee effect. *Trend. Ecol. Evolut.* **14**:405–410.

Dutrillaux, B., Muleris, M., and Couturier, J. 1988. Chromosomal evolution of Cercopithecinae. In: A. Gautier-Hion, F. Bourlière, J.-P. Gautier, and J. Kingdon (eds.), *A Primate Radiation: Evolutionary Biology of the African Guenons,* pp. 150–159. Cambridge University Press, Cambridge.

Eeley, H. A. C., Lawes, M. J., and Piper, S. E. P. 1999. The influence of climate change on the distribution of indigenous forest in KwaZulu–Natal, South Africa. *J. Biogeog.* **26**: 595–617.

Fahrig, L., and Merriam, G. 1994. Conservation of fragmented populations. *Conserv. Biol.* **8**:50–59.

Fourcade, H. G. 1889. *Report on the Natal Forests. Natal Blue Book.* W. Watson, Pietermaritzburg.

Hamilton, A. C. 1988. Guenon evolution and forest history. In: A. Gautier-Hion, F. Bourlière, J.-P. Gautier, and J. Kingdon (eds.), *A Primate Radiation: Evolutionary biology of the African Guenons,* pp. 13–34. Cambridge University Press, Cambridge.

Hanski, I. 1991. Single-species metapopulation dynamics: concepts, models and observations. *Biol. J. Linn. Soc.* **42**:17–38.

Hanski, I. 1992. Inferences from ecological incidence functions. *Am. Nat.* **139**:657–662.

Hanski, I. 1994a. Patch-occupancy dynamics in fragmented landscapes. *Trend. Ecol. Evolut.* **9**(4):131–135.

Hanski, I. 1994b. A practical model of metapopulation dynamics. *J. Anim. Ecol.* **63**:151–162.

Hanski, I., and Gilpin, M. 1991. Metapopulation dynamics: brief history and conceptual domain. *Biol. J. Linn. Soc.* **42**:3–16.

Hanski, I. A., and Gilpin, M. E. 1997. *Metapopulation Biology: Ecology, Genetics, and Evolution.* Academic Press, San Diego.

Hanski, I. A., and Simberloff, D. 1997. The metapopulation approach, its history, conceptual domain, and application to conservation. In: I. A. Hanski, and M. E. Gilpin (eds.), *Metapopulation Biology: Ecology, Genetics, and Evolution,* pp. 5–26. Academic Press, San Diego.

Harrison, S. 1991. Local extinction in a metapopulation context: an empirical evaluation. *Biol. J. Linn. Soc.* **42**:73–88.

Harrison, S., Murphy, D. D., and Ehrlich, P. R. 1988. Distribution of the bay checkerspot butterfly, *Euphydryas editha bayensis*: evidence for a metapopulation model. *Am. Nat.* **132**:360–382.

Harrison, S., and Taylor, A. D. 1997. Empirical evidance for metapopulation dynamics. In: I. A. Hanski, and M. E. Gilpin (eds.), *Metapopulation Biology: Ecology, Genetics, and Evolution,* pp. 27–42. Academic Press, San Diego.

Henien, K., and Merriam, G. 1990. The elements of connectivity where corridor quality is variable. *Landscape Ecol.* **4**:157–170.

Hess, G. R. 1994. Conservation corridors and contagious disease: A cautionary note. *Conserv. Biol.* **8**:256–262.

Laurance, W. F. 1990. Comparative responses of five arboreal marsupials to tropical forest fragmentation. *J. Mammal.* **71**:641–653.

Lawes, M. J. 1990. The distribution of the samango monkey (*Cercopithecus mitis erythrarchus* Peters, 1852 and *Cercopithecus mitis labiatus* I. Geoffroy, 1843) and forest history in southern Africa. *J. Biogeogr.* **17**:669–680.

Lawes, M. J. 1991. Diet of samango monkeys (*Cercopithecus mitis erythrarchus*) in the Cape Vidal dune forest, South Africa. *J. Zool., Lond.* **224**:149–173.

Lawes, M. J. 1992. Estimates of population density and correlates of the status of the samango monkey *(Cercopithecus mitis)* in Natal, South Africa. *Biol. Conserv.* **60**(3):197–210.

Lawes, M. J., Eeley, H. A. C., and Piper, S. E. 2000a. The relationship between local and regional diversity of indigenous forest fauna in KwaZulu-Natal Province, South Africa. *Biodiv. Conserv.* **9**:683–705.

Lawes, M. J., Mealin, P. E., and Piper, S. E. 2000b. Patch occupancy and potential metapopulation dynamics of three forest mammals in fragmented Afromontane forest in South Africa. *Conserv. Biol.* **14**:1088–1098.

Leakey, M. 1988. Fossil evidence for the evolution of the guenons. In: A. Gautier-Hion, F. Bourlière, J.-P. Gautier, and J. Kingdon (eds.), *A Primate Radiation: Evolutionary Biology of the African Guenons*, pp. 7–12. Cambridge University Press, Cambridge.

Macfarlane, D. M. 2000. *Historical Change in the Landscape Pattern of Indigenous Forest in the Karkloof/ Balgowan Region in the Midlands of KwaZulu–Natal.* M.Sc. dissertation, School of Botany and Zoology, University of Natal, Pietermaritzburg.

Moreau, R. E. 1966. *The Bird Faunas of Africa and Its Islands.* Academic Press, London.

Noss, R. F. 1991. Landscape connectivity: Different functions at different scales. In: W. E. Hudson (ed.), *Landscape Linkages and Biodiversity*, pp. 27–39. Island Press, Washington, DC.

Onderdonk, D. A., and Chapman, C. A. 2000. Coping with forest fragmentation: The primates of Kibale National Park, Uganda. *Int. J. Primatol.* **21**:587–611.

Pulliam, H. R. 1988a. Sources, sinks and population regulation. *Am. Nat.* **132**:652–661.

Pulliam, R. 1988b. Sources, sinks and population regulation. *Am. Nat.* **132**:652–661.

Recher, H. F., Shields, J., Kavanagh, R., and Webb, G. 1987. Retaining remnant mature forest for nature conservation at Eden, New South Wales: a review of theory and practice. In: D. A. Saunders, G. W. Arnold, A. A. Burbidge, and A. J. M. Hopkins (eds.), *Nature Conservation: The Role of Remnants of Native Vegetation*, pp. 177–194. Surrey Beatty, Sydney.

Rudran, R. 1978. Intergroup dietary comparisons and folivorous tendencies of two groups of blue monkeys (*Cercopithecus mitis stuhlmanni*). In: G. G. Montgomery (ed.), *The Ecology of Arboreal Folivores*, pp. 483–503. Smithsonian Institute Press, Washington, DC.

Ruvolo, M. 1988. Genetic evolution in the African guenons. In: A. Gautier-Hion, F. Bourlière, J.-P. Gautier, and J. Kingdon (eds.), *A Primate Radiation: Evolutionary Biology of the African Guenons*, pp. 127–139. Cambridge University Press, Cambridge.

Rycroft, H. B. 1944. The Karkloof Forest, Natal. *J. S. Afr. For. Assoc.* **11**:14–25.

Simberloff, D., and Cox, J. 1987. Consequences and costs of conservation corridors. *Conserv. Biol.* **1**:63–71.

Sjögren-Gulve, P., and Ray, C. 1996. Using logistic regression to model metapopulation dynamics: large-scale forestry extirpates the pool frog. In: D. R. McCullough (ed.), *Metapopulations and Wildlife Conservation*, pp. 111–138. Island Press, Washington, DC.

Skorupa, J. P. 1986. Responses of rainforest primates to selective logging in Kibale forest, Uganda: a summary report. In: K. Bernirschke (ed.), *Primates: The Road to Self-sustaining Populations*, pp. 57–70. Springer-Verlag, New York.

Stephens, P. A., and Sutherland, W. J. 1999. Consequences of the Allee effect for behaviour, ecology and conservation. *Trend. Ecol. Evolut.* **14**:401–405.

Struhsaker, T. T. 1978. Food habits of five monkey species in the Kibale forest, Uganda. In: D. J. Chivers, and J. Herbert (eds.), *Recent Advances in Primatology*, pp. 225–248. Academic Press, New York.

Swart, J., and Lawes, M. J. 1996. The effect of habitat patch connectivity on samango monkey (*Cercopithecus mitis*) metapopulation persistence. *Ecol. Model.* **93**:57–74.

Swart, J., Lawes, M. J., and Perrin, M. R. 1993. A mathematical model to investigate the demographic viability of low density samango monkey (*Cercopithecus mitis*) populations in Natal, South Africa. *Ecol. Model.* **70**(3–4):289–303.

Taylor, P. (1998). *The Smaller Mammals of KwaZulu-Natal.* University of Natal Press, Pietermaritzburg.

Taylor, P. D., Fahrig, L., Henien, K., and Merriam, G. 1993. Connectivity is a vital element of landscape structure. *Oikos* **68**:571–573.

Thomas, C. D. 1994a. Difficulties in deducing dynamics from static distributions. *Trend. Ecol. Evolut.* **9**:300.

Thomas, C. D. 1994b. Extinction, colonization, and metapopulations: Environmental tracking by rare species. *Conserv. Biol.* **8**:373–378.

Thomas, C. D., Thomas, J. A., and Warren, M. S. 1992. Distributions of occupied and vacant butterfly habitats in fragmented landscapes. *Oecologia* **92**:563–567.

White, F. 1981. The history of the Afromontane archipelago and the scientific need for its conservation. *Afr. J. Ecol.* **19**:33–54.

Wiens, J. A. 1997. Metapopulation dynamics and landscape ecology. In: I. A. Hanski, and M. E. Gilpin (eds.), *Metapopulation Biology: Ecology, Genetics, and Evolution*, pp. 43–62. Academic Press, SanDiego.

Assessing Extinction Risk in *Cercopithecus* Monkeys

25

THARCISSE UKIZINTAMBARA and CHRISTOPHE THÉBAUD

Introduction

The African continent is undergoing rapid habitat decline due to demographic, socio-economic and political problems, and climatic changes. The Food and Agriculture Organization of the United Nations (FAO, 1999) reported that sub-Saharan Africa lost *ca.* 0.7% of its forest cover annually between 1990 and 1995. The deforestation rate in Africa is directly linked to the annual population growth rate of 2.7% predicted by the World Resources Institute (1998) for the period between 2000 and 2005, the highest of all continents. At this level, the survival of species is seriously affected (Myers, 1986). As it is difficult to monitor and manage every aspect of biodiversity (Moritz and Faith, 1998), conservation priorities should focus on species hotspots, such as areas of high specific richness, rarity and endangerment (Rabinowitz, 1981; Mace and Lande, 1991; Gaston and Blackburn, 1996; Hacker *et al.*, 1998; Olson and Dinerstein, 1998). Reacting to the continued dwindling of forests, the International Union for Conservation of Nature and Natural Resources (IUCN, 1994) proposed the assessment

THARCISSE UKIZINTAMBARA • School of Biological Sciences, University of East Anglia, Norwich NR4 7TJ, UK and Station d'Etudes des Gorilles et des Chimpanzés, Lopé, BP 7847 Libreville, Gabon. CHRISTOPHE THÉBAUD • UMR 5552 CNRS/UPS, 13 Avenue du Colonel Roche, BP 4072, F-31029, Toulouse Cedex 4, France.
The Guenons: Diversity and Adaptation in African Monkeys, edited by Glenn and Cords. Kluwer Academic/Plenum Publishers, New York, 2002.

of changes in habitat size and quality as perhaps one criterion in determining extinction risks and setting conservation policies.

African primates account for 30% of extant primate species (Wolfheim, 1983; Nowak, 1999). *Cercopithecus* monkeys belong to one especially species-rich genus, and occur in most forests of sub-Saharan Africa, where anthropogenic activities are threatening them, sometimes to extinction. For example, Sclater's monkey (*Cercopithecus sclateri*), one of the most restricted and rare primates in Africa, is suffering important habitat disturbance and widespread hunting pressure in Nigeria (Oates *et al.*, 1992). The habitat of other species, such as *Cercopithecus mitis*, has been highly fragmented across their range (Eeley, 1994). In this chapter, we aim to (1) determine the intrinsic causes of *Cercopithecus* monkeys' vulnerability, (2) analyze effects of habitat decline on specific distribution and survival, and (3) evaluate the impact of socio-economic and political factors on *Cercopithecus* monkey conservation. We examine deforestation rates and habitat decline and conduct statistical analyses to assess relationships between published data on anthropogenic factors and specific variables and their potential effects on the survival of *Cercopithecus* species and their habitat.

Methods

Data Sources and Restrictions

To determine the intrinsic causes of *Cercopithecus* monkeys' vulnerability, we analyzed the relationship between published data on species characteristics and variables such as deforestation, hunting, civil wars, human population density, per capita gross national product, and protected areas. For species characteristics, we present the mean for variables whose values differ according to areas and authors. Our data set includes continuous data on group size, home range and bodily mass for each species (Table I). We considered only female bodily mass because it is more influenced by the ecological and physiological demands of lactation whereas male bodily mass can be strongly influenced by sexual selection (Ganzhorn, 1999). Discrete data include four major food types: fruit, seeds, leaves and insects, and five habitat types: primary, secondary and gallery forests, savannah woodland and bamboo forest (Eeley, 1994). We use logarithmic and square root transformations to normalize skewed variables such as bodily mass, areas of occupancy, human population size and density.

To evaluate the impacts of socio-economic and political factors on *Cercopithecus* monkeys, we used species distribution maps from Hill (1966), Kingdon (1997) and Boitani *et al.* (1998). Boitani *et al.* (1998) distinguish

Table I. Cercopithecus Monkeys' Characteristic Variables

Cercopithecus species	Extent of occurrence (km²)	Area of occupancy (km²)	Percentage deforestation rate/year[a]	Habitat persistence time (years)[a]	Major habitat types	Major food types	IUCN Categories	Group size	Home range (ha)	Female bodily mass (kg)
1. C. aethiops	13786087	10111221	0.69	2329	5	4	Low	40	42 (18-96)	5.6
2. C. ascanius	2903287	1507076	0.71	1996	3	3	Low	30	38 (22-55)	2.9 (1.8-4)
3. C. cephus	885572	622134	1.00	1327	2	3	Low	20	35 (18-45)	2.9 (2-4)
4. C. erythrotis	85124	45353	0.68	1571	2	2	At risk	12		2.9 (2.25-3.5)
5. C. erythrogaster	98379	46791	1.26	848	2	2	Vulnerable	5		2.4 (2-4)
6. C. petaurista	604750	422181	1.10	1171	3	2	Low	22	41	2.9 (2-3.5)
7. C. mitis	3272773	708864	0.71	1891	4	3	Low	41	65	4.2 (3-5.5)
8. C. nictitans	2148145	1272605	0.60	2336	3	3	Low	33	65 (37-75)	4.3 (2.7-5)
9. C. campbelli	534030	292167	1.12	1117	2	2	Low	19	40	3.0
10. C. mona	630540	220135	0.91	1346	3	2	Low	12	3	2.5 (2-3.6)
11. C. pogonias	2925818	1563636	0.72	1848	3	3	Low	19	80 (55-100)	3 (2.8-3.6)
12. C. neglectus	2895342	1004851	0.62	2222	3	4	Low	4	7 (6-13)	4.1 (4-5)
13. C. hamlyni	318877	173751	0.70	1718	2	2	At risk	8		5 (4.5-6)
14. C. diana	313968	70858	1.00	1111	2	2	Vulnerable	23	93 (37-175)	3.9 (3-4.5)
15. C. dryas	149511	21256	0.70	1418	2	2	At risk	10		2.26 (2-2.5)
16. C. lhoesti	438077	186086	0.68	1778	3	2	At risk	11	100	3.5 (3-4.5)
17. C. preussi	58528	1304	0.69	1036	2		Vulnerable	5		3.0
18. C. solatus	14515	11303	0.50	1862	1		Vulnerable	10		3.9
19. C. sclateri	29170	8308	0.90	998	1		Endangered	22		3 (2.5-3.5)

[a] Values calculated using our models.
Sources: Napier, 1981; Lee et al. 1988; Oates, 1988, 1996; Eeley, 1994; Baillie and Groombridge, 1996; Kingdon, 1997; Boitani et al., 1998; Kaplin, 1998; Eeley and Lawes, 1999.

the extent of occurrence for a species from the real area of occupancy, in order to provide an adequate picture of the actual distribution of species. The extent of occurrence is the area where the probability of finding the species is high based on habitat type characteristics, whereas the real area of occupancy refers to the area in which the species has actually been found. Using geographical information system (GIS) analysis, Boitani *et al.* (1998) overlay the extent of occurrence with maps that contain all areas of actual species presence. The robustness of estimates of the area of occupancy depends on many factors, including the biology of species and quality of available data. Boitani *et al.* (1998) studied all the medium and large-sized mammals of Africa; extracting a small data set on *Cercopithecus* monkeys may have an effect on our interpretation. We used data mainly from Boitani *et al.* (1998), but verified this dataset by comparing different information sources such as Kingdon (1997) as well as Hill (1966) for species with no recent distribution data available. We based our analysis on the real area of occupancy which represents areas actually occupied by a species within the extent of occurrence (Gaston, 1991).

Taxonomists disagree about how many species or subspecies of *Cercopithecus* monkeys should be recognized (Groves, 2001). Our data conform to the Orlando taxonomy used throughout this volume and include 19 species of *Cercopithecus* monkeys.

Habitat Characteristics and Modeling

To analyze the effects of habitat decline on the distribution and survival of species, we digitized the species' ranges of *Cercopithecus* monkeys on ArcInfo Version 7.2.1 Patch 2 (Environmental Systems Research Institute, Inc. ESRI) via a DEC 2000 workstation. The default scale is based at the Lambert–Azimuth Projection (ESRI, 1994) for equatorial zones. We generated summary statistics for national ranges for species occupying multiple countries using ArcInfo 7.2.1. To determine areas where most species occur, we overlaid specific coverages in ArcInfo 7.2.1. and edited them in Arcview 3.1 (Fig. 1). We report continuous values on country size and the rate of deforestation per year according to the FAO (1999) estimates for African counties that we group in four regions: western, central, eastern and southern (Table II). FAO data focus on food, agricultural and human population issues and therefore may underestimate the extent of deforestation. Continuous data for gross national product per capita in 1995 (in US$), population density and percentage of national territory in protected areas are taken from IUCN (1991), Siegfried *et al.* (1998), and WRI (1998). We categorize conservation awareness and peacefulness levels via national ratification of the African Convention, CITES, Ramsar, Biosphere Reserves and World Heritage, and extent of civil strife (IUCN, 1991).

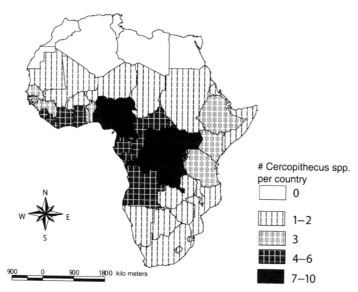

Fig. 1. The distribution pattern of *Cercopithecus* monkeys in Africa. Hotspots of *Cercopithecus* species diversity are those with more than four species (dark areas).

Finally, we created several models to analyze habitat decline which all assume that current rates of deforestation (r) remain constant. The first model predicts the size of suitable habitat (A_n) that will remain from the present area of occupancy (A_0) after an n-year period: $A_n = A_0(1-r)^n$. The second model predicts the habitat persistence time (t) i.e., the time during which all suitable habitat will be cleared in all countries for each *Cercopithecus* species: $t = -\log A_0/(\log(1-r))$. The third model, for species which occur in many countries, calculates the mean habitat deforestation rate r in every species' area of occupancy: $\bar{r} = \sum_{i=1}^{m} A_{0i} r_i / \sum_{i=1}^{m} A_{0i}$, in which A_{0i} represents the area of occupancy of a species in country i, r_i, the percentage deforestation in i, and m the total number of countries.

Barnes (1990) found that the area of forest felled per year in sub-Saharan Africa follows a predator–prey relationship in which a predator kills a constant amount of prey each year, and that the absolute area deforested in km^2 is a better predictor of habitat decline than the percentage deforestation. Therefore, in addition to our models explained above, we also used Barnes' model to calculate the absolute area deforested (ADF), which is the amount of forest felled per year for each country (Table II). Using Barnes' model, habitat persistence time t is calculated by the equation $t = A_0/ADF$. Using this equation, the habitat persistence time is the same for all species occurring in the same country.

Table II. Country Specific and Socio-Economic Variables

Region and country	Country size (km²)	Species present[a]	Mean AO (km²)	No. human pop. (× 1,000,000)	Pop. density	Pop. growth (%/year)	GNP (in US$)	%PA	%DR (/year)	War zone	No. CCR	BMT	HPT (year)
West Africa													
1. Mauritania	10,252,20	1	4771	1.8	1.82	2.9	410	0.5		No	3		142
2. Senegal	197,000	1,6,9	55480	8.4	44.30	2.9	600	11.30	0.70	Yes	5	Yes	111
3. Gambia	10,689	1,6,9	4430	1.1	114.10	2.60	320	2.20	0.90	No	2		90
4. Guinea	246,000	1,6,9	92710	6.7	30.60	3.12	550	0.70	1.10	No	3		250
5. Guinea Bissau	36,000	1,6,9	17875	1.1	38.80	2.20	250	2.00	0.40	No	2		33
6. Sierra Leone	72,000	1,6,9,14	37300	4.7	50.00	2.75	180	1.10	3.00	Yes	1	Yes	166
7. Liberia	111,000	1,6,9,14	36342	3.0	23.30	3.30	450	1.30	0.60	No	1	Yes	100
8. Mali	1,240,192	1	321500	9.0	9.10	3.00	250	3.70	1.00	No	5		142
9. Burkina Faso	274,200	1	209475	10.3	39.40	2.80	230	10.40	0.70	No	3		166
10. Ivory Coast	322,000	1,6,9,14	96020	14.5	44.10	3.83	660	6.20	0.60	No	4	Yes	76
11. Ghana	239,000	1,6,9,10,14	59040	17.6	78.40	3.10	390	4.80	1.30	No	5	Yes	71
12. Togo	75,000	1,6,10	21740	4.1	77.20	3.22	810	7.90	1.40	No	2		83
13. Benin	112,620	1,6,10	34850	4.7	50.30	3.00	370	7.00	1.20	No	4		111
14. Nigeria	394,000	1,4,5,8,10,11,17,19	67400	127.7	126.30	3.17	260	3.30	0.90	Yes	4	Yes	
15. Niger	1,267,000	1	715546	9.1	7.50	3.00	220	7.70	0.00	No	4	Yes	
Central Africa													
16. Cameroon	4,750,00	1,3,4,10,11,12,17	97434	14.0	29.10	3.48	650	4.50	0.60	No	4	Yes	166
17. Chad	1,284,000	1	367486	6.3	5.20	2.50	180	9.10	0.80	No	3		125
18. CAR	623,000	1,2,3,8,11,12	126464	3.5	5.40	2.98	340	8.20	0.70	Yes	4	Yes	142
19. Sudan	2,505,700	1,12	399600	29.1	11.50	2.87	340	3.60	0.80	Yes	4		125
20. Gabon	265,000	3,8,11,12,18	110827	1.4	4.30	3.09	3490	2.80	0.50	No	4	Yes	200
21. R. Congo	342,000	1,2,3,8,11,12	178865	2.7	20.60	3.35	680	4.50	0.20	Yes	4	Yes	500

(Cont.)

Region and country	Country size (km²)	Species present[a]	Mean AO (km²)	No. human pop. (× 1,000,000)	Pop. density (km²)	Pop. growth (%/year)	GNP (in US$)	%PA	%DR (/year)	War zone	No. CCR	BMT	HPT (year)
22. DRC	2,645,000	10 species[b]	4103558	41.8	5.80	3.25	120	4.50	0.70	Yes	4	Yes	142
23. Eq.Guinea	280,00	3,8,11,12,17	11216	0.4	14.60	2.60	380	0.50	0.50	No	1	Yes	200
24. Ethiopia	1,222,000	1,7,12	237530	57.1	58.20	2.99	100	5.50	0.50	Yes	2		200
25. Eritrea	1,174,000	1	45880	3.5	32.80		112	5.00	0.00	Yes	1		
26. Somalia	637,000	1,7	25310	6.7	45.70	2.40	170	0.30	0.20	Yes	1		500
27. Kenya	583,000	1,7,12	164440	29.0	48.80	3.81	280	6.20	0.30	No	5		333
28. Uganda	236,000	1,2,7,11,12	43710	22.7	101.50	3.47	240	9.60	0.90	Yes	5		111
29. Tanzania	945,000	1,2,7	269728	32.8	34.90	3.68	120	15.60	1.00	No	4		100
30. Rwanda	26,000	1,2,7,13,16	9260	8.6	248.80	3.41	180	14.70	0.20	Yes	3		500
31. Burundi	28,000	1,2,7,16	10460	6.4	242.30	2.91	160	5.60	0.40	Yes	3		250
Southern Africa													
32. Zambia	7,529,57	1,7	323857	10.2	11.10	3.65	400	8.60	0.80	No	4		125
33. Malawi	1,184,84	1,7	57195	10.0	104.60	3.60	170	11.30	1.60	No	3		62
34. Mozambique	802,000	1,7	347460	17.9	22.70	2.70	80	6.30	0.70	No	3	Yes	142
35. Botswana	581,730	1	72350	1.5	2.60	1.05	3020	18.50	0.50	No	2		200
36. Zimbabwe	390,759	1,7	142575	11.3	29.60	2.80	540	7.90	0.60	No	2		166
37. Swaziland	173,64	1,7	8182	0.9	51.20	3.60	1170	2.00	0.00	No	1		
38. Lesotho	30,355	1,7	6326	2.6	68.50	2.80	770	0.20	0.00	No	0		
39. South Africa	1,123,226	1,7	221970	41.7	34.70	2.70	3160	5.40	0.20	No	2	Yes	500
40. Namibia	8,242,92	1	64342	1.6	4.90	3.10	2000	12.90	0.30	No	1	Yes	100
41. Angola	1,24,700	1,2,3,7,11,12	187285	11.5	9.00	2.85	410	6.60	1.00	Yes	0		100

[a] The code numbers stand for species as in Table I.
[b] The Democratic Republic of Congo has 10 species: number 1,2,3,7,11,12,13,15, and 16.
AO stands for specific area of occupancy; pop, population; GNP, gross national product per capita; PA, protected areas; DR, deforestation rate; CCR, conservation conventions ratified per country; BMT, bushmeat trade (Oates, 1996, Kingdon, 1997) and HPT, habitat persistence time *sensu* Barnes (1990).
Sources: Barnes, 1990; IUCN, 1991; Boitani *et al.*, 1998, WRI, 1998; FAO, 1999.

Results

Large Scale Distribution Patterns

Cercopithecus monkeys occur in 41 African countries (Fig. 1). Current areas of occupancy range from 1304 km^2 for *Cercopitheuc preussi* to 10.1 million km^2 for *C. aethiops* (the median range size for all species is 0.53 million km^2, $n = 19$). Ranges may be restricted, as in *Cercopithecus solatus* which is found only in Gabon, *C. dryas* in the Democratic Republic of Congo and *C. sclateri* in Nigeria, or widespread, as in *C. aethiops* which has a broad distribution covering 39 countries (Table I). *Cercopithecus* species live from sea level up to 4500 m elevation, as in *Cercopithecus hamlyni* in the Ruwenzori Mountains, Uganda. However, species' ranges are often locally very fragmented. The Congo Basin, Mount Cameroon and Upper Guinea areas each have at least four species (21% of *Cercopithecus* species) and consequently represent hotspots of diversity for *Cercopithecus* monkeys (Fig. 1).

Area of Occupancy

We found that the present area of occupancy increases with the extent of occurrence across species ($R^2 = 0.864$, $\beta = 0.929$, $p \leqslant 0.0001$, $n = 19$). However, small-ranging species tended to show large residuals in the analysis. For example, *Cercopithecus preussi*, which is restricted to densely human populated areas of southern Nigeria and southwestern Cameroon where intense deforestation has occurred, occupies only 2% of its extent of occurrence. In contrast, *Cercopithecus solatus*, which is restricted to near pristine forest of Central Gabon, occupies 78% of its 14,515 km^2 extent of occurrence. On average across species, the area of occupancy represents 46% ($\pm 21.08\%$, $n = 19$) of the extent of occurrence.

Habitat Loss and Persistence

Over the 1990–95 period, our analyses show that *Cercopithecus* monkeys have lost on average 0.8% ($\pm 0.20\%$, $n = 19$) of their habitat per year. Only *Cercopithecus aethiops*, *C. erythrotis*, *C. nictitans* and *C. solatus* have experienced rates of habitat loss <0.7%, whereas, *C. erythrogaster*, *C. campbelli* and *C. diana* have lost >1% in West Africa (Table I).

Based on our models, trends of predicted habitat decline are the same for large and small areas of occupancy that are subject to similar rates of deforestation. However, when we increase deforestation rates four-fold, the habitat decreases dramatically (Fig. 2). According to our models, the minimum time in which suitable habitat for most restricted *Cercopithecus* species will disappear is 300 years in West Africa, 630 in Central Africa, 750 in

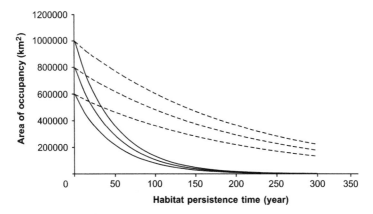

Fig. 2. The effect of two different deforestation rates on three different sizes of initial areas of occupancy. The broken lines represent the rate of habitat loss of 0.5% per year whereas continuous lines represent 2% annual deforestation rate effect on 1 million, 800,000 and 600,000 km² respectively.

East Africa and 625 in Southern Africa. However, considering the Barnes' model of absolute area deforested, we found that the minimum habitat persistence time for *Cercopithecus* species drops approximately ten-fold to 33, 125, 100 and 62 years for West, Central, East and South Africa, respectively, with a maximum duration of about 500 years.

Intrinsic factors of Cercopithecus *monkeys' vulnerability*

We plotted the specific area of occupancy against female bodily mass and found that *Cercopithecus* species aggregate into three distinct species clusters (Fig. 3). According to IUCN (1996), species with both large areas of occupancy and large bodily mass tend to be less threatened with extinction. Species with large areas of occupancy and small bodily mass are also likely to be less threatened. In contrast, species with small areas of occupancy also tend to have small bodily mass, and this group is generally more threatened. A similar triangular relationship (Fig. 3) was found for most African primate species by Eeley and Lawes (1999) although analyses differ in that these authors used somewhat different variables (i.e., the relationship between species range sizes, local species population density and food types).

Using a multiple linear regression analysis, we also found that more generalist *Cercopithecus* species in terms of food and habitat types occupy large areas of occupancy ($R^2 = 0.798$, $F = 25.67$, $p \leqslant 0.0001$, $n = 16$). Furthermore, the larger-bodied *Cercopithecus* species have longer habitat persistence time ($r = 0.629$, $p \leqslant 0.001$, $n = 19$), indicating that larger species may be losing less habitat than smaller ones which live in highly deforested areas. Finally, species with specific dietary preferences are most endangered according to IUCN

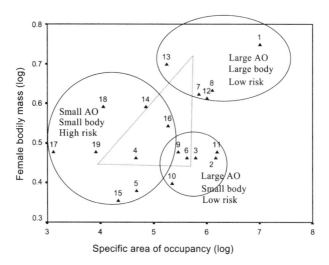

Fig. 3. Triangular relationship between specific area of occupancy, bodily mass and threat categories. The relationship between evolutionary, ecological and anthropogenic effects on species adaptation and survival is complex.

threat categories ($r = -0.771$, $p \leqslant 0.0001$, $n = 19$) whereas food generalists seem to be less threatened.

Extrinsic factors of species vulnerability

We conducted a multiple regression analysis in which deforestation rate is the dependent variable and area of occupancy, human population size and density, and per capita gross national product are the four independent variables. We found that only 24% of the variance in deforestation can be explained by these four factors together (Table III). Although this percentage is relatively low, it is evident that habitat loss together with socio-economic and demographic constraints directly or indirectly affect the survival of species and their habitat. In general, countries with no or very low defore-station rates in Southern Africa (Table II) have few species or little remnant forest or both. The bushmeat trade is well developed in more than 25% of countries where *Cercopithecus* monkeys are found, and civil strife occurs in 50% of them (Table II). The bushmeat trade and unrest added to the effects of deforestation may result in unprecedented decrease of *Cercopithecus* monkey populations in Africa. The rate of deforestation, however, appears to vary independently of bushmeat trade ($t = -0.422$ and $F = 0.339$, $t = -0.553$, $p > 0.1$, $n = 40$) or warfare ($t = -0.553$, $F = 0.024$, $p > 0.1$, $n = 40$). Nevertheless, war zones are especially prevalent in *Cercopithecus* monkey hotspot areas (Mann-Whitney $U = 93.5$, $p \leqslant 0.05$, $n = 40$) and can lead to uncontrolled intensive local hunting over short periods.

Table III.   Factors Influencing the Annual National Deforestation Rate

Parameter	F	t	p
Intercept	0.345	0.588	0.561
Area of occupancy ($\log_{10}$)	5.051	−2.247	0.032
Human population size ($\log_{10}$)	4.762	2.187	0.037
Human population density ($\log_{10}$)	5.438	−2.332	0.026
Per capita gross national product	4.429	−2.105	0.044

General Linear Model, A Univariate Regression Analysis, $R^2 = 0.238$.

By using a stepwise multiple linear regression analysis we found that the more restricted *Cercopithecus* species are more vulnerable because they occur in countries with high human population density ($t = -4.542$) and in countries with very few and small protected areas ($t = 3.489$) ($R^2 = 0.471$, $n = 39$ and $p \leqslant 0.0001$).

Setting Conservation Priorities

We qualitatively classify *Cercopithecus* monkeys into four broad categories of vulnerability by plotting the area of occupancy vs. deforestation rate: (1) species with a narrow distribution in highly deforested habitats, (2) species with a narrow distribution in less deforested habitats, (3) species with a wide distribution in less deforested habitats, and (4) species with a wide distribution in highly deforested habitat. These categories are based on the mean deforestation rate (0.8%) in all species' areas of occupancy and the median area of occupancy (534,030 km^2) respectively (Fig. 4).

Similarly, by considering national deforestation rate and the coverage of protected areas where species are believed to be sustained, we determine four categories: (1) countries with a low percentage of protected areas and a high rate of deforestation, (2) countries with a high percentage of protected areas and a high deforestation rate, (3) countries with a low percentage of protected areas and a low deforestation rate, and (4) countries with a high percentage of protected areas and a low deforestation rate (Fig. 4). Our results show that most range-restricted *Cercopithecus* species live in highly deforested regions (i.e., $>0.7\%$, the average rate of deforestation per annum in sub-Saharan Africa). They also occur in countries whose minimum size of national protected areas is below the 10% of national territory recommended by the FAO.

Discussion

Cercopithecus monkeys, including the savannah dweller *Cercopithecus aethiops*, are always associated with woodland and forested habitats for food,

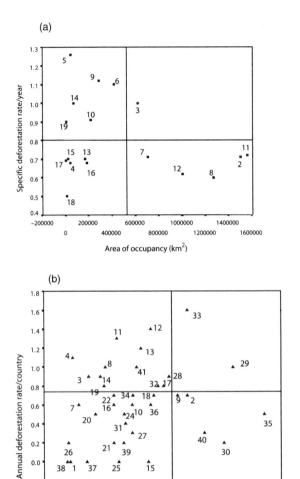

Fig. 4. Annual percentage of specific habitat loss vs. (a) size of specific area of occupancy and (b) national rate of habitat loss vs. percentage of protected areas. The species and country numbers in these figures refer to Table I for species [Fig. 4(a)] and Table II for country [Fig. 4(b)] names respectively.

cover and sleeping sites. Therefore, all of them are seriously affected by human induced deforestation. Our results show a great deal of variation in their potential for specific long-term survival, as summarized below.

• Small-bodied and range restricted *Cercopithecus* species are the most vulnerable whereas widespread species whether small or large in terms of bodily mass seem to be less vulnerable (Fig. 3).

- Food and habitat generalist species occupy larger areas than more specialist species and are therefore less vulnerable.
- Habitat persistence time is predicted to be higher for large-bodied species owing to their wider distribution across distinct regions and their broader diet.
- More species occur in countries where human population growth is highest.
- Most rare species live in countries that are being highly deforested and that have few protected areas.
- The intimate relationship of *Cercopithecus* monkeys' habitat with human population dynamics and the socio-economic issues prevailing in Africa could make their conservation difficult.

We will discuss these results in the context of *Cercopithecus* monkey ecology and conservation by (1) explaining variation in *Cercopithecus* monkeys' rarity, (2) proposing conservation priorities for African hotspots, and (3) providing an outlook for conservation in Africa.

Pattern of Cercopithecus *Monkeys' Rarity*

Area of occupancy is a useful concept in representing species' distributions and setting conservation priorities. According to Boitani *et al.* (1998), the estimates of *Cercopithecus* monkeys' areas of occupancy are generally very reliable except for species that occupy very small ranges (Boitani *et al.*, 1998), such as gallery forests that are too small for map resolution. Hence, interpretations and general conclusions drawn from national distribution data on *Cercopithecus* spp. are unlikely to be much affected by bias in range estimates from local populations, at least with regard to national conservation policy.

Instead of using historical deforestation rates (Cowlishaw, 1999) that are not available for most countries, we modeled the most recent estimates of annual deforestation rates to predict the *Cercopithecus* monkeys' extinction risk. Deforestation in Africa is one of the most important factors affecting the survival of species (Myers, 1986; Barnes, 1990). The habitat persistence time could be maximized if the rates of deforestation are significantly reduced. In reality, however, restricted species suffer more than widespread species from loss of their limited suitable habitat, as with *Cercopithecus preussi*. Obviously, a population cannot remain viable for long if its habitat has been cleared. Nevertheless, if the rate of deforestation is negligible, species such as *Cercopithecus solatus* with restricted ranges or narrowly endemic species may survive and remain locally abundant (Primack, 1998).

Widespread species occur in both high and low densities throughout their range (Eeley and Lawes, 1999), while species with narrow distribution, such as *Cercopithecus erythrogaster* (Fig. 3), are typically at low density (Gaston, 1994). This pattern conforms to the general distribution pattern of the African primates found by Eeley and Lawes (1999). Furthermore, with regard to habitat and diet, generalist species might be expected to occupy relatively large geographical areas and exploit a variety of resources and therefore may persist through time longer than more specialized species (Eeley, 1994).

In general, the results of our study on deforestation alone suggest that there is an alarming conservation problem regarding *Cercopithecus* monkeys. Using three methods to project impending specific extinction, Lawton and May (1995) also estimated the average lifespan of avian and mammalian species to be 200–400 years if current trends in environmental degradation continue. When considering the absolute area deforested (Barnes, 1990), our results indicate that some habitats may be cleared in *ca.* 30 years, particularly in West Africa, resulting in local extinction.

Setting Conservation Priorities for Cercopithecus Monkeys

After examining intrinsic causes of vulnerability, we suggest that conservation efforts should be concentrated, albeit not exclusively, on threatened *Cercopithecus erythrogaster, C. diana, C. campbelli, C. petaurista, C. mona* and *C. sclateri* in West Africa. Threats to *Cercopithecus* monkey survival are greatest in Sierra Leone, Ghana, Guinea, Angola, Benin, Nigeria and Democratic Republic of Congo. Governmental and non-governmental conservation bodies and the implementation of the 1997 Kyoto Protocol should encourage countries to reach the FAO goal of placing 10% of national land in protected areas, which will probably help in increasing *Cercopithecus* monkeys' habitat persistence time and thus species survival.

Outlook for Conservation in Africa

In addition to threats imposed by deforestation and other human activities, forest-based insurgences are currently annihilating many conservation efforts in sub-Saharan Africa (Kanyamibwa, 1998). Although there is no significant difference between deforestation rates in countries with and without wars, it is evident that species under conditions of human unrest are experiencing high hunting pressure owing to ever-growing trafficking of firearms in more remote areas. Socio-economic factors are imperfect indicators of extinction, but it is clear that the correlation found between population growth, population density, poverty and guenon diversity persistence is a complicated puzzle, which conservation scientists will find hard to solve if socio-economic considerations are not taken into account. The long-term survival of species requires national and international practical efforts to protect rare species and diverse areas. Moreover, for long-term success, sound conservation measures must occur hand-in-hand with alleviating poverty and advocating peace on the African continent

Summary

We conducted a comprehensive literature review of the biogeography of 19 *Cercopithecus* species. We investigated the socio-economic and political situation in countries where *Cercopithecus* monkeys occur and modeled national deforestation trends in specific areas of occupancy. We determined the intrinsic and extrinsic causes of *Cercopithecus* monkeys' vulnerability by analyzing specific characteristics and the effects of habitat loss on specific distribution and survival, and by evaluating the impact of socio-economic and political factors on species conservation in Africa. We found that specific area of occupancy is declining in most countries but substantial variation in habitat decline trends and specific long-term survival exists. Small-bodied *Cercopithecus* species with restricted distribution, and specialized habitat requirements and dietary preferences are the most threatened. Socio-economic constraints and lack of representative protected areas exacerbate the risk of extinction, as well as the additional occurrence of civil unrest. *Cercopithecus* monkeys are particularly at risk in the Upper Guinea refugium of West Africa, especially in Sierra Leone where species may be extinct in about 30 years if the present trends of deforestation and human disturbance continue. In addition, the bushmeat trade is widespread in specific hotspots, especially in areas controlled by rebels, such as in the Upper Guinea and Great Lakes regions.

ACKNOWLEDGMENTS

This study would not have been possible without the assistance and collaboration of several institutions and individuals. Special thanks go to the University of East Anglia, especially members of the Departments of Biological and Environmental Sciences, World Conservation Monitoring Centre, Cambridge University Library, Natural History Museum (London), Musée Royal d'Afrique Centrale, David Brugière, Paula Jenkins, Beth Kaplin, Michel Louette, John Oates, Liz Rogers and Rosie Trevelyan for assistance during data collection. We are grateful to Mary Glenn and Marina Cords for inviting Tharcisse Ukizintambara to attend the Guenon Update Symposium at the XVIIIth Congress of the International Primatological Society in Adelaide, Australia. His attendance was sponsored by the University of East Anglia, Primate Conservation Inc., Melbourne Zoo and the Gorilla Haven. We extend our gratitude to Kate Abernethy, Allard Blom, Patricia Boulhosa, Beth Kaplin,

Natasha Shah, Phil Stephens, Gilla Sunnenberg, Caroline Tutin and Lee White for detailed comments on the manuscript. We are also thankful to Diana Bell, Jane and Steuart Dewar, Sam Kanyamibwa, Beth Kaplin, Derek Pomeroy, Peter Smith-Temple and Kaoru Yokotani for their various and continuous support. This study was financially supported by the Wellcome Trust of the United Kingdom (Grant Reference Number: 057631/Z/99/Z).

References

Baillie, J., and Groombridge, B. 1996. *1996 IUCN Red List of Threatened Animals*. IUCN Species Survival Commission, Cambridge.

Barnes, R. F. W. 1990. Deforestation trends in tropical Africa. *Afr. J. Ecol.* **28**:161–173.

Boitani, L., Corsi, F., De Biase, A., Carranza, I. D., Ravagli, M., Reggiani, G., Sinibaldi, I., and Trapanese, P. 1998. *A Databank for the Conservation and Management of the African Mammals. A Guide to the Methods and the Data*. Institute of Applied Ecology, Rome.

Cowlishaw, G. 1999. Predicting the pattern of decline of African primate diversity: an extinction debt from historical deforestation. *Conserv. Biol.* **13**:1183–1193.

Eeley, H. A. 1994. *Ecological and Evolutionary Patterns of Primates Species Area*. Ph.D. Thesis, University of Cambridge, Cambridge.

Eeley, H. A., and Lawes, M. J. 1999. Large-scale patterns of species richness and species range size in anthropoid primates. In: J. G. Fleagle, C. Janson, and K. E. Reed (eds.), *Primate Communities*, pp. 191–219. Cambridge University Press, Cambridge.

ESRI. 1994. *GIS by ESRI. Map Projections. Georeferencing Spatial Data*. Environmental Systems Research Institute, Inc., USA.

FAO. 1999. *State of the World's Forests*. Rome, Italy.

Ganzhorn, J. U. 1999. Body mass, competition and the structure of primate communities. In: J. G. Fleagle, C. Janson, and K. E. Reed (eds.), *Primates Communities*, pp. 141–157. Cambridge University Press, Cambridge.

Gaston, K. J. 1991. How large is a species' geographic range? *Oikos* **61**:434–438.

Gaston, K. J. 1994. *Rarity*. Chapman & Hall, London.

Gaston, K. J., and Blackburn, T. M. 1996. The relationship between geographic area and the latitudinal gradient in species richness in the New World. *Ecology* **11**:195–204.

Groves, C. P. 2001. *Primate Taxonomy*. Smithsonian Institute Press, Washington DC.

Lawton H. J. and May R. M. (eds.) 1995. *Extinction Rates*. Oxford University Press, Oxford.

Hacker, J. E., Cowlishaw, G., and Williams, P. H. 1998. Patterns of African primate diversity and their evaluation for the selection of conservation areas. *Biol. Conserv.* **84**:251–262.

Hill, W. C. O. 1966. *Primate. Comparative Anatomy and Taxonomy. VI. Catarrhini, Cercopithecoidea, Cercopithecinae*. Edinburgh University Press, Edinburgh.

IUCN. 1991. *Protected Areas of the Qorld. A Review of National Systems. Vol. 3. Afrotropical*. IUCN, Gland.

IUCN. 1994. *IUCN Red List Categories*. IUCN, Gland.

IUCN. 1996. *1996 IUCN Red list of Threatened Animals*. IUCN, Gland.

Kanyamibwa, S. 1998. Impact of war on conservation: Rwandan environment in agony. *Biodivers. Conser.* **7**:1399–1406.

Kaplin, B. A. 1998. *Ecology of Two African Forest Monkeys: Temporal and Spatial Patterns of Habitat Use, Foraging Behaviour, and Seed Dispersal*. Ph.D. Thesis, University of Wisconsin.

Kaplin, B. A., and Moermond, T. C. 2000. Foraging ecology of the mountain monkey (*Cercopithecus l'hoesti*): implications for its evolutionary history and use of disturbed forest. *Am. J. Primatol.* **50**:227–246.

Kingdon, J. 1997. *The Kingdon Field Guide to African Mammals*. Academic Press, London.

Lee, P. C., Thornback, J., and Bennett, E. L. 1988. *Threatened Primates of Africa: The IUCN Red Data Book*. IUCN, Gland.

Mace, G. M., and Lande, R. 1991. Assessing extinction threats: toward a reevaluation of IUCN threatened species categories. *Conserv. Biol.* **5**:148–157.

Moritz, C., and Faith, D. P. 1998. Comparative phylogeography and the identification of genetically divergent areas for conservation. *Molec. Ecol.* **7**:419–429.

Myers, N. 1986. Tropical deforestation and a mega-extinction spasm. In: M. E. Soulé (ed.), *Conservation Biology: The Science of Scarcity and Diversity*, pp. 394–409. Sinauer Press, Massachusetts.

Nowak, R. 1999. *Walker's Primates of the World*. The Johns Hopkins University Press, Baltimore.

Oates, J. 1988. The distribution of Cercopithecus monkeys in West African forests. In: A. Gautier-Hion, F. Bourlière, J.-P. Gautier, and J. Kingdon (eds.), *A Primate Radiation: Evolutionary Biology of the African Guenons*, pp. 79–103. Cambridge University Press, Cambridge.

Oates, J. F., Anadu, P. A., Gadsby, E. L., and Were, J. L. 1992. Sclater's guenon. *Natn. Geogr. Res. Explr.* **8**:476–491.

Olson, D. M., and Dinerstein, E. 1998. The global 2000: A representation approach to conserving the earth's most biologically valuable ecoregions. *Conserv. Biol.* **12**:502–515.

Primack, R. B. 1998. *Essentials of Conservation Biology*, 2nd ed. Sinauer Associates, Inc., Massachusetts.

Rabinowitz, D. 1981. Seven forms of rarity. In: H. Synge (ed.), *The Biological Aspects of Rare Plant Conservation*, pp. 205–217. Wiley, Chichester.

Siegfried, R. W., Benn, G. A., and Gelderblow, C. M. 1998. Regional assessment and conservation implications of landscape characteristics of African national parks. *Biol. Conserv.* **84**:134–140.

Wolfheim, J. H. 1983. *Primates of the World: Distribution, Abundance, and Conservation*. University of Washington Press, Seattle.

World Resources Institute. 1998. *World Resources 1998–99*. Oxford University Press, New York.

Conservation of the Guenons: An Overview of Status, Threats, and Recommendations

26

THOMAS M. BUTYNSKI

Introduction

All but a few of the 23 species (Grubb *et al.*, 2002) of guenons live within the tropical forests of Africa (Kingdon, 1997; Groves, 2001; Butynski, 2002). It is in these tropical forests that all threatened species and most threatened subspecies of guenons live (Lee *et al.*, 1988; Oates, 1996a). In this chapter, I review the conservation status of the guenons and provide an overview of the human activities that threaten them. I also make recommendations for conservation action to help retain the full range of diversity currently found among the guenons.

Conservation Status

The degree of threat listings in the *2000 IUCN Red List of Threatened Species* (Hilton-Taylor, 2000) for the primates is based upon assessments provided

THOMAS M. BUTYNSKI • Zoo Atlanta's Africa Biodiversity Conservation Program, National Museums of Kenya, P.O. Box 24434, Nairobi, Kenya.
The Guenons: Diversity and Adaptation in African Monkeys, edited by Glenn and Cords. Kluwer Academic/Plenum Publishers, New York, 2002.

by members of the IUCN/SSC Primate Specialist Group using the 1994 *IUCN Red List Categories* and criteria (IUCN, 1994). Often there are few reliable data for a particular taxon on total numbers, geographic distribution, extent of habitat fragmentation, kinds and levels of threat, and rates of decline (Butynski, 2002). There are also important taxonomic questions that remain unresolved and which, in some cases, influence the degree of threat rating (e.g., roloway monkey, *Cercopithecus diana roloway*, Stampfli's putty-nosed monkey, *C. nictitans stampflii*, Bioko putty-nosed monkey, *C. nictitans martini*, Mt. Kahuzi owl-faced monkey, *C. hamlyni kahuziensis*) (Grubb *et al.*, 2002). In cases for which there is uncertainty as to which degree of threat rating is most appropriate, the Primate Specialist Group assigns the higher rating. Where there is inadequate information on which to base an assessment of a taxon's risk of extinction, the taxon is listed as "data deficient."

Table I is a list of critically endangered, endangered, vulnerable, and data deficient species and subspecies of guenons as presented in the *2000 IUCN Red List of Threatened Species*. In 2000, there were four endangered and two vulnerable species of guenons, and two critically endangered and 11 endangered subspecies of guenons. Thus, the total number of threatened taxa of guenons in 2000 was 19. A recently completed reassessment of the degree of threat rating for the guenons for the *2001 IUCN Red List of Threatened Species* indicates that this list will remain unchanged except for the removal of *Cercopithecus mitis maesi*, since it now appears to be a synonym of *C. m. stuhlmanni* (Grubb *et al.*, 2002).

The appendix summarizes the geographic ranges, habitats, threats and recommended conservation action for the six threatened species of guenons. Of the 23 species of guenons recognized by Grubb *et al.* (2002), 17% are endangered and 9% are vulnerable, whereas among the 55 sub-species of guenons, 4% are critically endangered and 18% are endangered. The taxonomy of Grubb *et al.* (2002) for the guenons is presented in Table I of Butynski (2002).

All 19 of the threatened taxa (Table I) are dependent on moist tropical forest (rainfall >100 cm/year). The geographic ranges of all six threatened species, and ten of the 13 threatened subspecies are in West Africa or western Central Africa. Figure 1 shows the rough geographic ranges for the six threatened species. These are regions where not only primate diversity and endemism are high, but also where levels of deforestation and hunting are extreme. Efforts to conserve the biodiversity of guenons need to focus on these species and regions.

Threats

The guenons listed in Table I face the same main threats that confront other primates that are forest-dependent, not only in tropical Africa (Lee *et al.*, 1988;

Table I. **Threatened Taxa of Guenons as Indicated in the *2000 IUCN Red List of Threatened Species* (Hilton–Taylor, 2000)**

Species

. **Endangered**

Cercopithecus diana	Diana monkey
Cercopithecus erythrogaster	White-throated monkey
Cercopithecus preussi	Preuss's monkey
Cercopithecus sclateri	Sclater's monkey

Vulnerable

Cercopithecus erythrotis	Red-eared monkey
Cercopithecus solatus	Sun-tailed monkey

Data deficient

Cercopithecus dryas	Dryad monkey

Subspecies

Critically endangered

Cercopithecus diana roloway	Roloway monkey
Cercopithecus nictitans stampflii	Stampfli's putty-nosed monkey

Endangered

Cercopithecus diana diana	Diana monkey
Cercopithecus erythrogaster erythrogaster	Red-bellied monkey
Cercopithecus erythrogaster pococki	Nigeria white-throated monkey
Cercopithecus erythrotis erythrotis	Bioko red-eared monkey
Cercopithecus preussi insularis	Bioko Preuss's monkey
Cercopithecus preussi preussi	Cameroon Preuss's monkey
Cercopithecus pogonia pogonias	Golden-bellied crowned monkey
Cercopithecus mitis kandti	Golden monkey
Cercopithecus mitis maesi	Lake Kivu blue monkey
Cercopithecus nictitans martini	Bioko putty-nosed monkey
Cercopithecus hamlyni kahuziensis	Mt. Kahuzi owl-faced monkey

Data deficient

Cercopithecus aethiops djamdjamensis	Bale Mountains grivet monkey
Cercopithecus mitis albotorquatus	Pousargues's Sykes's monkey
Cercopithecus cephus ngottoensis	White-nosed mustached monkey

Note: *Cercopithecus mitis maesi*, listed as "Endangered" in the 2000 Red List, is now treated as a synonym of *C. m. stuhlmanni* (Grubb *et al.*, 2002).

Ammann and Pearce, 1995; Oates, 1996a,b, 1999; Butynski, 1997, 2001; Bowen–Jones, 1998) but almost everywhere in the tropics (Wolfheim, 1983; Mittermeier and Cheney, 1987; Mittermeier and Konstant, 1997). These threats are forest loss and over-hunting, both of which are accelerating. The primary driving force and ultimate cause of this forest loss and over-hunting is a rapidly expanding human population. Since much has been written over the past two decades about these threats, I will only briefly review them here.

Human Population Pressures

Africa's human population increased several-fold during the 20th Century and the growth rate today stands at about 2.6% per year, by far

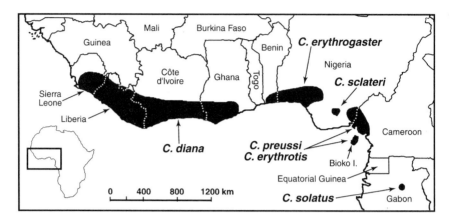

Fig. 1. The black areas indicate the geographic ranges of Africa's six most threatened species of guenon. Note that the area actually occupied by each species is, in all six cases, but a tiny fraction of the geographic range. Note also that all six species are found in the forests of West Africa and in western Central Africa. Adapted from Oates (1996). See Oates (1988) and Kingdon (1997) for more detailed maps of the geographic ranges for these six species. Map drawn by Stephen Nash.

the highest in the world (World Resources Institute, 1998). With this population increase has come a concomitant growth in demand for natural resources such as cropland, pasture land, lumber, and meat. The demand for Africa's natural resources by the industrial economies of Europe, Asia, and North America is also increasing rapidly. In addition, western technologies that facilitate the destruction of wildlife and forest are now being widely applied. Meeting the increasing internal and external demands for natural resources is the main cause of deforestation and local extinction of species in tropical Africa (Myers, 1993; Struhsaker, 1996, 1997; Butynski, 1997, 2001). This unsustainable exploitation of forests and wildlife has reduced options for development, destroyed the way of life for tens of thousands of indigenous people, left millions of Africans struggling to meet their short-term survival needs, and contributed to increased insecurity (World Resources Institute, 1998). In Africa, 22% of the people are poorer today than in 1997, food production per person is on the decline, one-third of the people are undernourished, and food production will need to increase 300% by 2050 to meet demand (Conly, 1998). It is not surprising that Africans have, by far, the lowest life expectancy in the world (World Resources Institute, 1998).

Loss and Degradation of Forest

More than two-thirds of the original forest cover in Africa has been removed by humans, mainly through their farming and logging activities

(World Resources Institute, 1998). Africans are losing their forests faster than any other peoples (Kemf and Wilson, 1997). Only six African countries have >20% of their original forest cover, and as many as 17 countries retain <10% of their forest cover (Sayer, 1992). Much of what remains is heavily degraded and fragmented (Bryant *et al.*, 1997). The World Resources Institute/ International Institute for Environment and Development (1988) stated that in Côte d'Ivoire and Nigeria, where tropical moist forest is being destroyed at an annual rate of roughly 5%, all moist forest could be lost by 2007. Côte d'Ivoire and Nigeria are two of the most important countries for the conservation of guenon diversity, yet the loss of forest alone could extirpate most of their species of guenons.

The main proximate causes of deforestation are industrial logging and clearing for agriculture. In Africa, about 61% of the remaining large, relatively undisturbed tracts of natural (frontier) forest is under threat from logging, while about 13% is under threat from agriculture (Bryant *et al.*, 1997). Thus, the stage is now set for most of the relatively intact forest ecosystems in tropical Africa to be degraded, fragmented, and lost over the next few decades. This will, of course, further disrupt essential ecological and evolutionary processes and patterns, and further imperil the indigenous peoples whose way of life depends upon these forests.

Logging roads and trucks give people access to vast areas of remote forest where they not only degrade and clear the forest, but also hunt guenons and other animals (Ammann and Pearce, 1995; Williams, 2000a; Anon., 2002). Foreign logging companies, working hand in hand with international aid agencies such as the European Union and World Bank, often construct these roads (Williams, 2000b). For example, the European Union has improved some 2000 km of roads in Cameroon for the benefit of European, mostly French, logging companies. These companies control more than 70% of Cameroon's logging industry (Anon., 2002).

Most of the timber taken from African forests is exported. The countries of the European Union represent the most important market for tropical timber from Africa, accounting for 87% of exports in 1991. The United Kingdom and France are the two largest importers of Africa's tropical timber (Ammann and Pearce, 1995). Illegal logging and corrupt practices are now rife in many areas (Williams, 2000a; Anon., 2002).

Hunting

Hunting of guenons and other primates is ongoing at an unsustainable level in many places. In Central Africa this is particularly so in logging concessions. Hunters work full-time to supply bushmeat to logging company employees, to expanding agricultural communities, and especially to people in distant towns and cities. Meat from guenons is considered to taste good, and guenons are relatively large-bodied and easy to hunt. As such, guenons are among the species most sought by hunters (Butynski and Koster, 1994; Usongo and Fimbel, 1995; Bowen–Jones, 1998). Logging roads, logging trucks, and guns enable the bushmeat trade (WSPA, 1996, 2000; Bryant *et al.*, 1997; Ape Alliance, 1998). Logging companies assist in every aspect of the bushmeat trade, often in direct contravention of the law (Ammann and Pearce, 1995; Bowen–Jones, 1998; Wilkie and Carpenter, 1999; Ammann, 2000; Butynski, 2001; Anon., 2002). As a result, now there are large areas of suitable habitat where, because of hunting, species of guenons that were once common are either at low densities or extirpated (Lee *et al.*, 1988; Butynski and Koster, 1994; Struhsaker and Oates, 1995; Usongo and Fimbel, 1995; McGraw *et al.*, 1998). The commercialization of bushmeat is probably a more significant and immediate threat than forest loss for the majority of guenons, as well as for many other primates and other mammals (Oates, 1996b; Butynski, 1997, 2001; Rose, 1997; Bowen-Jones, 1998; McGraw *et al.*, 1998; Wilkie and Carpenter, 1999).

Although all countries of West and Central Africa are signatories to the Convention on International Trade in Endangered Species of Wild Fauna and Flora (CITES) (Lee *et al.*, 1988; Butynski, 1996), these nations have not begun to meet their commitments to the Convention regarding the transborder trafficking in CITES-regulated species, particularly in the form of bushmeat. Few authorities in the countries of West and Central Africa have yet to demonstrate that they take conservation matters seriously. There is a general indifference and disregard for environmental and wildlife protection, or for the sustainable use of natural resources, at seemingly all levels of authority, from government ministers to judges, police officers, and administrators (Williams, 2000a; WSPA, 2000). As such, violations against CITES and those laws that should protect guenons and their habitats remain common in most, if not all, of the countries where guenons are found.

Since 1994, the World Society for the Protection of Animals (WSPA) has led a media campaign against unsustainable hunting of primates and other animals. The focus is to reduce hunting and to make governments and logging companies accountable for the consequences of their activities on wildlife and wildlife habitat (WSPA, 2000). Following the lead of the WSPA, many other organizations have adopted the bushmeat crisis as a major focus of their conservation efforts. For example, the Ape Alliance (1998), an international coalition of 34 organizations and consultants, is now working to halt illegal hunting of apes and other wildlife for meat. The Bushmeat Crisis Task Force is another international coalition with >100 member organizations and specialists dedicated to the conservation of wildlife populations threatened by commercial hunting of wildlife for sale as meat. Its primary goal is to "eliminate the illegal commercial bushmeat trade through the development of a global network that actively supports and informs nations, organizations, scientists and the general public" (www.bushmeat.com).

Despite much effort by the above-mentioned organizations and others over the past several years, there is no evidence that illegal hunting is abating in any country in tropical Africa. The illegal commercial exploitation of guenons and other wildlife is almost certainly greater today than a decade ago. A number of studies are reaching the conclusion that bushmeat hunting at present levels will continue to cause the extirpation of wildlife species from vast areas, and that some species of large mammals may become extinct within the next few decades as a result of over-hunting (Ammann and Pearce, 1995; Ape Alliance, 1998; Bowen–Jones, 1998; Bowen–Jones and Pendry, 1999; Butynski, 2001). If present trends continue, some species and subspecies of guenons will certainly be counted among the extinct taxa.

Recommendations for Conservation

The ultimate cause of the problems facing the survival of the guenons is Africa's rapidly expanding human population and the related poverty and insecurity. Curbing the growth of human populations and bringing them in line with the sustainable use of the natural resource base depends largely on effective action by governments. No African nation has a population policy. African governments need to develop and implement population policies (Myers, 1993; Struhsaker, 1997). Major international bodies such as the United Nations (UN), Organization of African Unity (OAU), European Union (EU), World Bank, African Development Bank (ADB), aid agencies, and non-government organizations (NGOs) all have an important role to play in this effort by providing technical and financial support for the implementation of national population policies, as well as for related initiatives concerned with socioeconomic advancement, education, family planning, child mortality, and women's rights.

There is an urgent need for the conservation community, national governments, donors, logging companies, trade organizations, and the public to give far more attention to the main proximate threats to populations of guenons and other African primates—forest degradation, forest loss, and hunting. Viewpoints and recommendations abound on the actions that might be taken to reduce these threats to acceptable levels (Oates, 1996a,b, 1999; Rose, 1996, 1997; Struhsaker, 1996, 1997; Butynski, 1997, 2001; Bowen–Jones, 1998; Bowen–Jones and Pendry, 1999; Wilkie and Carpenter, 1999; Ammann, 2000; WSPA, 2000). Here I summarize those recommendations that are likely to have the most impact on the conservation of the guenons and other African primates (Butynski, 2001). Even though many of the recommendations overlap and are interdependent, I have grouped them into four categories and ordered them according to priority for action.

Granting Agencies and Political Leaders

- International aid-granting agencies, particularly the United Nations (UN), European Union (EU), World Bank, United States Agency for International Development (USAID), German aid (GTZ), and Japan aid (JICA) should use their considerable political and financial resources to help ensure that no development project results in, or promotes, the unsustainable use of natural resources, particularly wildlife and natural forests.
- Major donors and conservation bodies should put into place trust funds to support the required law enforcement, research, monitoring, evaluation, information, and conservation education programs, especially in and around the protected areas.
- International aid-granting agencies should require that all development projects undergo a rigorous and independent environmental impact assessment (EIA) before approval to ensure that forests, wildlife, indigenous peoples, and local communities are not adversely affected.
- International aid-granting agencies should provide financial incentives in the form of conditional grants and loans for logging companies, hunters, and others who exploit tropical forests. In this case, those who support the protection of national parks and reserves, and who practice sustainable forest and wildlife use, become eligible for preferential financial assistance (Struhsaker, 1997).
- Global political, technical, and financial support for sustainable forest use and hunting practices should be promoted, particularly among political leaders, major decision-makers, financiers, and national governments.

Logging

- Logging companies should fully finance all programs necessary to ameliorate both the direct and indirect damage that their activities have on wildlife and on natural habitats.
- Logging companies should adhere to national wildlife and forestry laws and regulations, and implement corporate codes of conduct that help ensure sustainable forest use, including hunting, and the protection of biodiversity.
- A strong independent international certification program for logging companies should be established for sustainable tropical forest management. Bushmeat control and the maintenance of biodiversity, particularly for old growth forest communities, should be integral parts

of the accreditation process. The international conservation community should develop the criteria and performance indicators for this program.

- Import permits should be denied for uncertified timber.
- People in the more developed, industrialized countries should be made more aware of the negative impacts that their high levels of consumption of tropical timber and other natural resources from the tropical forests of Africa are having on the primates and biodiversity.

Protected Areas and Hunting

- National protected area systems should be improved, especially through the expansion of existing national parks and the establishment of additional national parks.
- Logging and hunting within protected areas should be prevented. Logging and hunting in buffer zones around protected areas should be limited.
- Law enforcement capabilities should be strengthened to reduce greatly illegal, unregulated, and unmanaged hunting.
- Domestic food alternatives to bushmeat should be developed and promoted, particularly around protected areas.

Research, Education and Lobbying

- Africa's populations of primates should be regularly surveyed and monitored so that numbers, distributions, trends, and threats are better known.
- People concerned with the disastrous consequences of unsustainable logging and hunting should do much more to bring these issues to public attention, and to lobby and pressure their own governments and the logging industry to take action to halt these practices.
- Support should increase for environmental education and public awareness programs, particularly around protected areas. The goal should be to change the attitudes of local peoples toward hunting and logging, foster interest in sustainable use, and promote local responsibility and accountability for the conservation of neighboring forests and wildlife.

Summary

Most of the 23 species and 55 subspecies of guenons inhabit the moist forests of West and Central Africa. As a result of high levels of hunting and forest loss and degradation in this region, six species (26%) and 12 subspecies

(22%) of guenons are threatened with extinction. There is a serious lack of reliable information on the distribution and numbers of African primates, including the guenons. More surveys are needed, particularly given the current rapid decline of guenon numbers and of guenon habitat due to hunting and logging. It is difficult to make effective conservation decisions without this information. Governments, donor agencies, conservation bodies, and the public must do much more to halt the great damage now being done to Africa's tropical forests and wildlife, both by hunters and loggers. In particular, government officers, conservationists, private individuals, hunters and loggers should work together to develop and implement the many activities and programs that will be required to safeguard Africa's forests and wildlife, and to make the extraction of wood and meat from these forests sustainable activities. If this can be accomplished, the guenons will be among the many beneficiaries.

ACKNOWLEDGMENTS

I thank John Oates, Mary Glenn, Marina Cords and two anonymous reviewers for their valuable comments on the draft manuscript, Stephen Nash for drawing the map in Figure 1, and Zoo Atlanta, Conservation International, The National Museums of Kenya, and the IUCN Eastern Africa Regional Office for support during the writing of this chapter.

Appendix

Summary of the geographic ranges, habitats, threats, and recommended conservation action for the six threatened and one "Data Deficient" species of guenon. This information compiled primarily from the following sources: Wolfheim, 1983; Colyn, 1988; Lee *et al.*, 1988; Oates, 1988, 1996a; Butynski and Koster, 1994; Butynski and Oates, pers. obs.

Diana monkey *Cercopithecus diana*: Endangered

- *Range/habitat*: SE Sierra Leone, Liberia, S Côte d'Ivoire, SW Ghana, and SE Guinea in canopy of primary and old secondary lowland moist forest, and riverine forest.
- *Threats*: Already rare in early 1960s and rapid decline has continued. Threatened by hunting, and habitat degradation and loss.
- *Conservation*: Protect: Gola Forest Reserve (FR), Tiwai Island Wildlife Sanctuary (Sierra Leone); Sapo National Park (NP), Mt. Nimba (Liberia); Bia NP, Ankasa Reserve, Nini-Suhien NP (Ghana); Mt Nimba

Strict Nature Reserve/Biosphere Reserve (Guinea); Taï Forest NP, Mt. Nimba Strict Nature Reserve/Biosphere Reserve (Côte d'Ivoire). Gazette: Mt Nimba NP, Liberia. Survey: SW Ghana; Lofa-Mano, Liberia; E and Central Côte d'Ivoire.

White-throated monkey *Cercopithecus erythrogaster:* Endangered

- *Range/habitat*: SW Nigeria and S Bénin in primary, secondary, and riverine lowland moist forest.
- *Threats*: Small, very scattered populations in decline due to hunting, and habitat loss and degradation.
- *Conservation*: Protect: Lama Forest (Bénin); Okomu NP and FR, Ifon FR, Omo FR (Nigeria), and all other *Cercopithecus erythrogaster* habitats in Bénin and Nigeria. Survey: Ondo FR, Ogun FR, and other forests in SW Nigeria.

Preuss's monkey *Cercopithecus preussi*: Endangered

- *Range/habitat*: SE Nigeria, W Cameroon (Mt. Cameroon, Bamenda Highlands, Cross River Highlands), and Bioko Island, Equatorial Guinea, in primary and old secondary medium-altitude and montane moist forest.
- *Threats*: Scattered, relict populations in decline due to hunting, and habitat loss and degradation.
- *Conservation*: Protect: Okwangwo Division of Cross River NP, Obudu Plateau (Nigeria); Takamanda FR, Mone River FR, Etinde forest, and forests of Bakossiland, Mt. Kupe and Mt. Oku (=Mt. Kilum) (Cameroon); Pico Basilé National Park, Gran Caldera and Southern Highlands Scientific Reserve (Bioko). Survey: Etinde forest, Cameroon; Pico Basilé, Bioko.

Sclater's monkey *Cercopithecus sclateri*: Endangered

- *Range/habitat*: Endemic to S Nigeria in swamp and riverine forest.
- *Threats*: In small, scattered, declining populations in highly threatened, fragmented forests. Threatened by hunting, and habitat loss and degradation.
- *Conservation*: Protect: Stubbs Creek Forest Reserve, Akpugoeze Forest. Survey: E Niger Delta.

Red-eared monkey *Cercopithecus erythrotis*: Vulnerable

- *Range/habitat*: SE Nigeria, SW Cameroon, and Bioko Island, Equatorial Guinea, in primary and secondary lowland moist forest.
- *Threats*: Widely scattered populations in decline due to hunting, and habitat loss and degradation.
- *Conservation*: Protect: Cross River NP, Mbe Mountains, Afi Mountain Wildlife Sanctuary (Nigeria); S Bakundu Reserve, Douala-Edea

Reserve, Korup NP (Cameroon); Pico Basilé National Park, Gran Caldera and Southern Highlands Scientific Reserve (Bioko).

Sun-tailed monkey *Cercopithecus solatus*: Vulnerable

- *Range/habitat*: Endemic to Gabon in lowland primary and secondary moist forest.
- *Threats*: Rare throughout very restricted geographic range. Threatened by hunting, and habitat loss and degradation.
- *Conservation*: Protect: Lopé Reserve, Lopé NP. Extend east boundary of Lopé NP to include part of Forêt des Abeilles.

Dryad monkey *Cercopithecus dryas*: Data Deficient

- *Range/habitat*: Endemic to Democratic Republic of Congo in primary and dense secondary lowland moist forest, and perhaps swamp forest.
- *Threats*: Poorest known guenon. Known only from Lomela and Wamba regions of Congo Basin. Likely very limited, perhaps fragmented, geographic range. Hunted.
- *Conservation*: Surveys needed to assess distribution and conservation status. Undertake socioecological and genetic research to determine habitat requirements and relationship with other guenons.

References

Ammann, K. 2000. Exploring the bushmeat trade. In: K. Ammann (ed.), *Bushmeat: Africa's Conservation Crisis*, pp. 16–27. World Society for the Protection of Animals, London.

Ammann, K., and Pearce, J. 1995. *Slaughter of the Apes: How the Tropical Timber Industry is Devouring Africa' Great Apes*. World Society for the Protection of Animals, London.

Anon. 2002. Africa's vanishing apes—A disastrous partnership of loggers and hunters in the Congo basin. *The Economist*, January 12–18, p. 59.

Ape Alliance. 1998. *The African Bushmeat Trade—A Recipe for Extinction*. Ape Alliance, London.

Bowen-Jones, E. 1998. A review of the commercial bushmeat trade with emphasis on Central/West Africa and the great apes. *African Primates* 3:S1–S42.

Bowen-Jones, E., and Pendry, S. 1999. The threat to primates and other mammals from the bushmeat trade in Africa, and how this threat could be diminished. *Oryx* 33:233–246.

Bryant, D., Nielsen, D., and Tangley, L. 1997. *The Last Frontier Forests: Ecosystems and Economies on the Edge*. World Resources Institute, Washington, DC.

Butynski, T. M. 1996. International trade in CITES Appendix II African primates. *African Primates* 2:5–9.

Butynski, T. M. 1997. African primate conservation—The species and the IUCN/SSC Primate Specialist Group network. *Primate Conservation* 17:87–100.

Butynski, T. M. 2001. Africa's great apes. In: B. Beck, T. S. Stoinski, M. Hutchins, T. L. Maple, B. Norton, A. Rowan, E. F. Stevens, and A. Arluke (eds.), *Great Apes and Humans: The Ethics of Coexistence*, pp. 3–56. Smithsonian Institution Press, Washington DC.

Butynski, T. M. 2002. The guenons: An overview of diversity and taxonomy. In: M. E. Glenn, and M. Cords (eds.), *The Guenons: Diversity and Adaptation in African Monkeys*, pp. 3–13. Kluwer Academic Publishers, New York.

Butynski, T. M., and Koster, S. H. 1994. Distribution and conservation status of primates in Bioko Island, Equatorial Guinea. *Biodivers. Conserv.* **3**:893–909.

Colyn, M. M. 1988. Distribution of guenons in the Zaire-Lualaba-Lomani river system. In: A. Gautier-Hion, F. Bourlière, J.-P. Gautier, and J. Kingdon (eds.), *A Primate Radiation: Evolutionary Biology of the African Guenons*, pp. 104–124. Cambridge University Press, Cambridge.

Conly, S. R. 1998. Sub-Saharan Africa at the turning point. *The Humanist* **58**:19–23.

Groves, C. P. 2001. *Primate Taxonomy*. Smithsonian Institution Press, Washington, DC.

Grubb, P., Butynski, T. M., Oates, J. F., Bearder, S. K., Disotell, T. R., Groves, C. P., and Struhsaker, T. T. 2002. An assessment of the diversity of African primates. Unpublished report. IUCN/SSC Primate Specialist Group, Washington, DC.

Hilton-Taylor, C. 2000. *2000 IUCN Red List of Threatened Animals*. IUCN, Gland, Switzerland.

IUCN. 1994. *The IUCN Red List Categories*. IUCN, Gland, Switzerland.

Kemf, E., and Wilson, A. 1997. *Great Apes in the Wild*. World Wide Fund for Nature, Gland, Switzerland.

Kingdon, J. 1997. *The Kingdon Field Guide to African Mammals*. Academic Press, London.

Lee, P. C., Thornback, J., and Bennett, E. L. 1988. *Threatened Primates of Africa: The IUCN Red Data Book*. IUCN, Gland, Switzerland.

McGraw, W. S., Monah, I. T., and Abedi-Lartey, M. 1998. Survey of endangered primates in the forests reserves of eastern Côte d'Ivoire. *African Primates* **3**:22–25.

Mittermeier, R. A., and Cheney, D. L. 1987. Conservation of primates and their habitats. In: B. B. Smuts, D. L. Cheney, R. M. Seyfarth, R. W. Wrangham, and T. T. Struhsaker (eds.), *Primate Societies*, pp. 477–490. The University of Chicago Press, Chicago.

Mittermeier, R. A., and Konstant, W. R. 1997. Primate conservation: A retrospective and a look into the 21st Century. *Primate Conservation* **17**:7–17.

Myers, N. 1993. Population, environment, and development. *Environ. Conserv.* **20**:205–216.

Oates, J. F. 1988. The distribution of *Cercopithecus* monkeys in West African forests. In: A. Gautier-Hion, F. Bourlière, J.-P. Gautier, and J. Kingdon (eds.), *A Primate Radiation: Evolutionary Biology of the African Guenons*, pp. 79–103. Cambridge University Press, Cambridge.

Oates, J. F. 1996a. *African Primates: Status Survey and Conservation Action Plan*, Revised Edition. IUCN/SSC Primate Specialist Group, IUCN, Gland, Switzerland.

Oates, J. F. 1996b. Habitat alteration, hunting and conservation of folivorous primates in African forests. *Aust. J. Ecol.* **21**:1–9.

Oates, J. F. 1999. *Myth and Reality in the Rain Forest: How Conservation Strategies are Failing in West Africa*. University of California Press, Berkeley.

Rose, A. L. 1996. The African great ape bushmeat crisis. *Pan Africa News* **3**:1–6.

Rose, A. L. 1997. Growing commerce in bushmeat destroys great apes and threatens humanity. *African Primates* **3**:6–12.

Sayer, J. 1992. A future for Africa's tropical forests. In: J. A. Sayer, C. S. Harcourt, and N. M. Collins (eds.), *Africa: The Conservation Atlas of Tropical Forests*, pp. 81–93. Macmillan, London.

Struhsaker, T. T. 1996. A biologists' perspective on the role of sustainable harvest in conservation. *African Primates* **2**:72–75.

Struhsaker, T. T. 1997. *Ecology of an African Rain Forest*. University Press of Florida, Gainesville.

Struhsaker, T. T., and Oates, J. F. 1995. The biodiversity crisis in south-western Ghana. *African Primates* **1**:5–6.

Usongo, L. and Fimbel, C. 1995. Preliminary survey of arboreal primates in Lobeke Forest Reserve, south-east Cameroon. *African Primates* **1**:46–48.

Wilkie, D. S., and Carpenter, J. F. 1999. Bushmeat hunting in the Congo Basin: An assessment of impacts and options for mitigation. *Biodivers. Conserv.* **8**:927–955.

Williams, J. 2000a. The lost continent: Africa's shrinking forests. In: K. Ammann (ed.), *Bushmeat: Africa's Conservation Crisis*, pp. 8–15. World Society for the Protection of Animals, London.

Williams, J. 2000b. Trouble in the pipeline. In: K. Ammann (ed.), *Bushmeat: Africa Conservation Crisis*, pp. 36–39. World Society for the Preservation of Animals, London.

Wolfheim, J. H. 1983. *Primates of the World: Distribution, Abundance, and Conservation*. University of Washington Press, Seattle.

World Resources Institute. 1998. *World Resources 1998–99*. Oxford University Press, New York.

World Resources Institute/International Institute for Environment and Development. 1988. *World Resources 1988–89*. Basic Books, New York.

World Society for the Protection of Animals (WSPA). 1996. *Wildlife and Timber Exploitation in Gabon: A Case Study of the Leroy Concession, Forest de Abeilles*. WSPA, London.

World Society for the Protection of Animals (WSPA). 2000. *Bushmeat: Africa's Conservation Crisis*. WSPA, London.

Epilogue

MARINA CORDS and MARY E. GLENN

Twenty years ago, when many of us began our studies of guenons, rather little was known about this group of monkeys, especially in comparison with several other primate taxa including other members of the Cercopithecinae. Molecular studies aimed at clarifying the evolutionary relationships among guenons were in their infancy, and only a handful of populations had been the subject of extended study, either in the field or in captivity. Accordingly, when researchers compared major primate groups, guenons were often exemplified by one or two relatively well-studied species, or simply overlooked. It is remarkable, for example, how many statements about cercopithecine monkeys ignore this large branch of the family.

The publication of *A Primate Radiation* and this volume clearly indicates that much has changed in the last two decades, and the authors whose work is represented here are largely responsible. Their reports are the fruit of long hours in the lab and in the field, where observation and collection conditions are far from ideal. We brought these researchers together to report on what they had found, and thus to update our knowledge about this group of monkeys. If we had to summarize what has been found in just one word, we would choose the word "variation." Components of this variation include whether various results or conclusions derive from different kinds of studies, whether given taxa show variation in their behavioral or ecological characteristics, or whether guenons deviate from patterns or expectations derived from theory or from other primate groups. This volume includes multiple examples of each component type.

In science, variation begs to be explained, and thus, while our authors were not asked to identify the agenda of the next generation of research on the African guenons, their results really cannot help but do so. We take this opportunity to summarize their recommendations, whether explicit or not, in hopes that these recommendations will in fact stimulate future work, and lead to a subsequent future volume in which the balance of questions answered and questions raised is, perhaps, more skewed toward the answers.

Evolutionary relationships are clearly central to all investigations in the fields of study represented in our volume. Recent studies have used new genetic loci, as well as more diverse sets of phenotypic traits, to evaluate phylogenetic relationships, but as the chapters by Tosi *et al.*, Disotell and Raaum, and Gautier *et al.* show, different kinds of data can lead to different results. The differences in results are not trivial: they influence the placement of species into genera, and relate to important questions like the evolution of arboreality, reproductive strategies, and vocalizations such as copulation calls within the guenon clade. Here is a case in which more data—from more taxa, and involving additional characters—are needed to resolve the different interpretations. A key species in these studies is *Cercopithecus lhoesti*, which has only recently become better known. Hybridization between species is widely acknowledged as a likely contributor to diversity among the guenons, but in-depth studies of hybridization, including genetic analyses, have yet to be undertaken. Detwiler's chapter on hybridization between *Cercopithecus ascanius* and *C. mitis* hybrids in East African forests reveals intriguing variation among communities in the frequency of hybridization, and raises many questions for further investigation.

Studies of biogeography are another important source of information related to understanding evolutionary history in any group of organisms. The biogeography of the guenons is complicated by the presence of intergradation and hybridization zones. Delineating biogeographical patterns remains a challenge as records on what forms are found in which places are still incomplete. Zones at the upstream ends of major rivers are especially problematical, and Colyn *et al.* show how their exclusion from the analysis clarifies the resolution of patterns downstream. In different ways, the chapters by Colyn *et al.* and by Kaplin show the value of incorporating multiple data sets in gaining understanding. An extension to a broader set of taxa (Colyn *et al.*) allows a more confident resolution of biogeographic zones. An understanding of ecological limitations (Kaplin) complements distributional data in developing explanatory frameworks. In the future, the addition of genetic data will surely add to our understanding of biogeographic history: Horsburgh *et al.* show this for one species, *Cercopithecus mona*, by using mitochondrial DNA to answer questions about the history of its distribution on islands.

Ecological studies can both inform and be informed by studies of biogeography. By now, there have been numerous field investigations of the feeding ecology of guenons, especially the forest species, which enables comparisons between populations, and within populations over time. As Chapman *et al.* demonstrate, the overall picture that emerges is one of considerable variation in diet, even at the level of the plant parts that are major dietary constituents. Lambert shows how this dietary flexibility can arise in the feeding strategies of individuals who are able to switch frequently between foods. It seems that flexibility is as much an attribute of the diet of forest guenons as frugivory, folivory or faunivory. The factors that drive and allow this flexibility are not yet very clear, however. The availability of an array of foods that differ in quality and distribution, and the digestive adaptations of our subjects, should be important variables, and yet we still know little about them. In animals with such catholic diets, merely defining "food" is problematical, and the web of relationships among foods that are differentially consumed may be exceedingly complex. Researchers will probably continue to look for the simpler patterns first, and because experimental study of food choice may be difficult, they will benefit from a comparative perspective. In that context, we are delighted to include in this book the first reports of Curtin's careful study of *Cercopithecus diana roloway*, and Gathua's comparative study of *C. ascanius*. There are still several guenon species whose feeding ecology has not been comprehensively described for even a single population, and we look forward to see that work completed. But to really understand both the general patterns and the variation that are likely to emerge, future investigations must include assessments of nutritional quality along with spatial distribution, as well as more comparative work on physiological and morphological digestive adaptations.

An understanding of functional anatomy is another window into the evolutionary process, important not only for interpreting the fossil record, but also in providing clues to relatively recent history in changing environments. McGraw's analyses suggest that correlations between morphology and behavior are not straightforward among the guenons, and do not always match patterns observed in other primate groups. This incongruity may be attributable to behavioral flexibility both between and within guenon species. With the analyses based on only six species, however, and in view of the fact that subsets of these six do not always reveal the same pattern as all six species combined, there is clearly a great need for more data from additional species and populations.

Since *A Primate Radiation* was published, there has been a small explosion of studies on the social behavior of guenons. Our understanding of guenon grouping has therefore become richer. As studies here attest, the notion that these monkeys live in rather small one-male/multi-female groups is an oversimplification that overlooks many of the interesting dynamics of their social systems. Glenn *et al.* describe well-developed all-male groups in mona monkeys, which appear to occur in only one other guenon species (*Erythrocebus patas*). The reasons for facultative male bonding are not yet well understood, and it would be interesting to find out more about the lives of solitary males. Cords documents multiple males joining heterosexual groups of blue monkeys during the breeding season, and the evidence is now quite strong that these influxes occur when defending multiple estrous females is difficult. Isbell *et al.* present a new hypothesis explaining the more consistent presence of multiple males in groups of vervet monkeys which stands in contrast to most other guenon species, and has so far eluded satisfactory explanation. To test their hypothesis, further data will be needed on kinship relations among males, as well as comparative data on vervet monkey populations that differ in terms of local group distribution and habitat. It is possible that comparisons to other riverine guenons, such as *Cercopithecus neglectus* and *Allenopithecus* would also be informative; these two species are little known at present. Windfelder *et al.* describe group fission in *Cercopithecus ascanius*; together with earlier reports, their results suggest that a group size of approximately 50 animals is an upper limit, at least for *Cercopithecus ascanius* and *C. mitis*, and that the animals will undertake the burdens of redefining social ties and territorial boundaries when their groups exceed this number. The immediate factors precipitating fission, and the immediate payoff to fissioning, remain obscure, partly because there is variation in the few observations available to date. Studies of guenon grouping focus on species inhabiting study sites where investigations have been ongoing for years, and habituation allows the researcher to monitor closely the behavior of individual monkeys and groups. The development of more long-term study sites including additional guenon species should be a high priority, although it is not a trivial undertaking.

Guenons are known for being less demonstrative and behaviorally conspicuous than papionin monkeys to which they are closely related. Understanding the social organization of guenons is therefore a challenge. Treves and Baguma advocate the use of simple measures, like spatial proximity between males and females and vigilance levels, to assess patterns of bonding and competition within species. Their suggestions beg corroboration with other taxa. On a finer scale, Chism and Rogers evaluate the importance of grooming as a behavioral mechanism fostering group cohesion in patas monkeys, and conclude that this social behavior, even though it is a common form of interaction, does not serve the hypothesized function. We think that grooming behavior, the most frequent form of clearly affiliative behavior in many guenon species, deserves further attention. It seems likely that individual grooming relationships, as well as other subtler signs of social bonding, will need to be considered to fully understand guenon social dynamics.

One type of social behavior in guenons is not at all inconspicuous, and this is the loud calling of adult males. While the anti-predator function of male loud calls has been well demonstrated by previous experimental and observational study, Zuberühler emphasizes possibly sexually selected functions of these calls as indicators of male quality. His hypotheses should stimulate further studies of inter-male variation in calls, fitness-related characters, and success in attracting mates and repelling rival males. Contextual studies of male loud calling are not easy in a forested environment, and yet they are important. Further complications arise from counter-calling between males of different species. We note also that it is not only adult males who vocalize in guenon groups, and very little is known about the function of quieter but still common calls given by females and juveniles, with the exception of alarm calls.

The reproductive biology of most guenon species remains poorly known, and available information comes mainly from studies of animals in captivity. The behavioral and hormonal study of wild *Cercopithecus mitis* by Pazol *et al.* is pioneering work, significantly revising prior estimates of gestation length for this species, and supporting at a physiological level the idea that sexual behavior of females serves purposes other than ensuring fertilization. It is time for others to adopt the research tools that allow 'field endocrinology', though again this will be possible only with well-habituated animals. A study like that of Macleod *et al.*, for example, would be much more conclusive if physiological data on female fertility had been available. Ultimately, an evaluation of the success of different male mating strategies will depend on measuring their success in terms of reproductive output. Simple models like those used by Macleod *et al.* may, however, guide the working hypotheses under consideration.

The fruit of mating is the production of the next generation and two papers in this book consider the social behavior of the young. Förster and Cords present the first systematic data on mother–infant relations from a forest species, finding many parallels with patterns described from other cercopithecines, and no support for the idea that arboreality retards the development of independence. Clearly, more data are needed before any meaningful comparisons with other primate groups can be made with confidence, and the challenge of finding a comparable measure of conspecific and predator threat remains. Worch's findings relate the frequency of play behavior among juveniles with dietary characteristics of different species. In light of the flexible dietary strategies that guenons appear to possess, a natural follow up to his study would include within-species comparisons of play under different ecological circumstances. Differences in the prevalence of social play, whether across or within species, also have unknown consequences; it is likely that longitudinal study will be required to unravel those consequences, or to determine that they are not important.

In general, there seem to be plenty of reasons to undertake further observational studies of guenons, but will there be guenon populations left to study? As Ukizintambara and Thébaud and also Butynski make clear, the prognosis for the continued survival of the full complement of currently existing guenon taxa is grim. Victims of human demands and human indifference, and, as pointed out by Lawes, perhaps also of their own ecological flexibility, guenons live in habitats that are being fragmented and destroyed, and where they are hunted. Their plight is not substantially different from that of many other African forest fauna, and the conservation problem is much larger than a guenon problem. It is not yet too late to save a great deal of the biodiversity in Africa's forests, but unless there are some significant changes in the way humans interact with forests and forest products, very soon it will be too late to ensure a different outcome. Only two years ago, Oates *et al.* announced the probable extinction of another African forest primate, Miss Waldron's red colobus (*Procolobus badius waldroni*).

Guenons have a role to play in motivating conservation action. Many of the researchers in this book devote a part of their professional life—or even most of it—to conservation efforts, whether directly or through the training of biological researchers, educators and conservationists in guenon-habitat countries and elsewhere. Several guenon biologists declined to contribute to this book, because their conservation-related work did not allow them the time they would have needed. We can only hope that as the world knows more about these fascinating creatures, the number of people acting to ensure their continued survival increases as well.

Index

431